Ion Channels

The Practical Approach Series

SERIES EDITORS

D. RICKWOOD
Department of Biology, University of Essex
Wivenhoe Park, Colchester, Essex CO4 3SQ, UK

B. D. HAMES
Department of Biochemistry and Molecular Biology
University of Leeds, Leeds LS2 9JT, UK

★ indicates new and forthcoming titles

Affinity Chromatography
Anaerobic Microbiology
Animal Cell Culture (2nd Edition)
Animal Virus Pathogenesis
Antibodies I and II
★ Basic Cell Culture
Behavioural Neuroscience
Biochemical Toxicology
★ Bioenergetics
Biological Data Analysis
Biological Membranes
Biomechanics — Materials
Biomechanics — Structures and Systems
Biosensors
★ Carbohydrate Analysis (2nd Edition)
Cell–Cell Interactions
Cell Culture Models of Epithelia
The Cell Cycle
Cell Growth and Apoptosis
Cellular Calcium
Cellular Interactions in Development
Cellular Neurobiology
Clinical Immunology
Crystallization of Nucleic Acids and Proteins
★ Cytokines (2nd Edition)
The Cytoskeleton
Diagnostic Molecular Pathology I and II
Directed Mutagenesis
★ DNA Cloning 1: Core Techniques (2nd Edition)
★ DNA Cloning 2: Expression Systems (2nd Edition)
★ DNA Cloning 3: Complex Genomes (2nd Edition)
★ DNA Cloning 4: Mammalian Systems (2nd Edition)
Electron Microscopy in Biology
Electron Microscopy in Molecular Biology
Electrophysiology
Enzyme Assays

Essential Developmental Biology

Essential Molecular Biology I and II

Experimental Neuroanatomy

★ Extracellular Matrix

★ Flow Cytometry (2nd Edition)

Gas Chromatography

Gel Electrophoresis of Nucleic Acids (2nd Edition)

Gel Electrophoresis of Proteins (2nd Edition)

★ Gene Probes 1 and 2

Gene Targeting

Gene Transcription

Glycobiology

Growth Factors

Haemopoiesis

Histocompatibility Testing

★ HIV Volumes 1 and 2

Human Cytogenetics I and II (2nd Edition)

Human Genetic Disease Analysis

Immunocytochemistry

In Situ Hybridization

Iodinated Density Gradient Media

★ Ion Channels

Lipid Analysis

Lipid Modification of Proteins

Lipoprotein Analysis

Liposomes

Mammalian Cell Biotechnology

Medical Bacteriology

Medical Mycology

★ Medical Parasitology

★ Medical Virology

Microcomputers in Biology

Molecular Genetic Analysis of Populations

★ Molecular Genetics of Yeast

Molecular Imaging in Neuroscience

Molecular Neurobiology

Molecular Plant Pathology I and II

Molecular Virology

Monitoring Neuronal Activity

Mutagenicity Testing

★ Neural Cell Culture

Neural Transplantation

Neurochemistry

Neuronal Cell Lines

NMR of Biological Macromolecules

★ Non-isotopic Methods in Molecular Biology

Nucleic Acid Hybridization

Nucleic Acid and Protein Sequence Analysis

Oligonucleotides and Analogues

Oligonucleotide Synthesis

PCR 1

★ PCR 2

★ Peptide Antigens

Photosynthesis: Energy Transduction

★ Plant Cell Biology

★ Plant Cell Culture (2nd Edition)

Plant Molecular Biology

Plasmids (2nd Edition)

- Pollination Ecology
- Postimplantation Mammalian Embryos
- Preparative Centrifugation
- Prostaglandins and Related Substances
- ★ Protein Blotting
- Protein Engineering
- Protein Function
- Protein Phosphorylation
- Protein Purification Applications
- Protein Purification Methods
- Protein Sequencing
- Protein Structure
- Protein Targeting
- Proteolytic Enzymes
- ★ Pulsed Field Gel Electrophoresis
- Radioisotopes in Biology
- Receptor Biochemistry
- Receptor–Ligand Interactions
- RNA Processing I and II
- Signal Transduction
- Solid Phase Peptide Synthesis
- Transcription Factors
- Transcription and Translation
- Tumour Immunobiology
- Virology
- Yeast

Ion Channels

A Practical Approach

Edited by

R. H. ASHLEY

Department of Biochemistry, University of Edinburgh, George Square, Edinburgh EH8 9XD, UK

IRL PRESS
at
OXFORD UNIVERSITY PRESS
Oxford New York Tokyo

Oxford University Press, Walton Street, Oxford OX2 6DP

Oxford New York
Athens Auckland Bangkok Bombay
Calcutta Cape Town Dar es Salaam Delhi
Florence Hong Kong Istanbul Karachi
Kuala Lumpur Madras Madrid Melbourne
Mexico City Nairobi Paris Singapore
Taipei Tokyo Toronto

and associated companies in
Berlin Ibadan

Oxford is a trade mark of Oxford University Press

Published in the United States
by Oxford University Press Inc., New York

A catalogue record for this book is available from the British Library

Library of Congress Cataloging in Publication Data
(Data available)

ISBN 0 19 963475 0 (Hbk)
ISBN 0 19 963474 2 (Pbk)

Typeset by Footnote Graphics, Warminster, Wilts
Printed in Great Britain by Information Press Ltd, Eynsham, Oxon.

Preface

Although at first glance they may appear to have relatively simple and restricted roles, ion channels are usually very clever molecules of immense intrinsic interest. And they will certainly never be short of attention from cell biologists, being central to so many fundamental biological processes. The investigator with an ion channel-specific problem has a wealth of techniques at his or her disposal, and the contributors to this book have provided detailed, thoughtful, and highly practical accounts of some of the most important approaches currently available. The reader will also note a refreshing lack of respect for traditional subject boundaries. We hope the result will be to encourage yet more bold enterprise and insight in a field already well-known for both.

Edinburgh
April 1995

R.A.

Contents

List of contributors xvii

Abbreviations xix

1. Patch-clamp recording 1

Alasdair J. Gibb

1. Introduction 1

2. Principles 1
Electrical considerations 1
The nature of the glass–membrane seal 2
Connecting the membrane patch to the electronics 4
Solutions for patch clamping 5

3. Patch-clamp recording configurations 5
Basic sealing procedure 5
Notes on the sealing procedure 6
Cell-attached patch configuration 7
Inside-out patch configuration 7
Whole-cell recording configuration 8
Perforated patch whole-cell recording 12
Outside-out patch configuration 13
Patch-clamp recording conventions 14

4. Making patch pipettes 15
Patch pipette glass 15
Patch pipette pulling procedure 17
Coating with Sylgard 18
Pipette fire-polishing 19

5. Building a patch-clamp set-up 20
Patch-clamp amplifiers 20
Patch-clamp amplifier headstages 21
Oscilloscopes 22
Amplifiers and filters 23
Tape recorders and data storage 24
Computers and on-line experimental control, data storage, and analysis 25
Vibration isolation system 25
Microscopes 26
Micromanipulators 27
Grounding and shielding the patch-clamp set-up 27

6. Preparation of cells and tissues for patch-clamp recording 28
Dissociation of peripheral tissues 29
Dissociation of brain tissues 32
Plating out dissociated cells 33
Immobilizing cells prior to recording 33
Preparation of brain slices for patch-clamp recording 34
Patch clamping *Xenopus* oocytes 39

7. Conclusion 40

References 41

2. The measurement of the function of ion channels reconstituted into artificial membranes 43

Alan J. Williams

1. Introduction 43

2. The production of artificial phospholipid bilayers 44
Bilayers formed by spreading phospholipid dispersions (painted bilayers) 44
Bilayers formed from monolayers (folded bilayers and bilayers on patch pipettes) 48

3. Ion channel insertion into bilayers 54
Channel insertion by vesicle fusion 55
Channel orientation in the bilayer 58

4. Patch clamping channels in liposomes 59
Freeze–thaw 59
Dehydration–rehydration 60
Patch clamping large liposomes 61

5. Monitoring current flow through ion channels 62
Electronics 62
Noise 64

6. General hints 64
Cleaning the apparatus 64
Solutions 65
Changing solutions 65
Agar bridges and silver–silver chloride electrodes 65

Acknowledgements 66

References 66

3. Concepts of single-channel analysis: inferring function from fluctuations 69

Guy W. J. Moss and Edward Moczydlowski

1. Introduction 69
   Single channels in practice and theory 69
   Basic considerations for recording and analysing data 71
2. Measurement of ionic selectivity 72
   What is selectivity? 72
   Continuum versus discrete-state models 74
   Classical definition of a permeability ratio 77
3. The rate-theory perspective on ion conduction 80
   A one-site channel compared to a one-site enzyme 80
   Multi-ion channels and ion–ion interactions 86
4. Probability distributions used to describe single-channel activity 88
5. Analysis of channel gating according to the Q-matrix formulation 94
   Channel gating can be described by a Markov process 94
   The probabilistic nature of ion channel gating 95
   Conditional probabilities and transition rates 96
   Derivation of the Q-matrix 97
   Relationship between conditional and macroscopic probability equations 99
   Eigenvalues and probability density functions of the Q-matrix 100
   Q-matrix analysis of a three-state scheme 104

Appendix 1: Alternative calculation of A_i 107

Appendix 2: A basic summary of matrix algebra 109

References 110

4. Assaying ion channels: design and practice 113

Michiki Kasai, Toru Ide, and William S. Agnew

1. Introduction 113
2. Ligand binding assays 113
   Introduction 113
   Theoretical considerations 114
   Practical guidelines for receptor assay protocols 118
3. Receptor assays 119
   Assay for inositol 1,4,5-trisphosphate receptors 119
   Assay for voltage-sensitive Na^+ channels using labelled tetrodotoxin 121
   Assay for voltage-sensitive Na^+ channels using radiolabelled STX 125

4. Functional assay by bilayer reconstitution 126

References 130

5. Channel cloning, mutagenesis, and expression 133

Paul J. Whiting, Keith A. Wafford, Iris Pribilla, and Thomas Petri

1. Introduction 133

2. Isolation of cDNAs by hybridization screening of cDNA libraries 134
- General principles 134
- DNA probes for screening cDNA libraries 134
- Screening of cDNA libraries 137
- Obtaining full length cDNAs 138

3. Site-directed mutagenesis 139
- General principles 139
- Oligonucleotide-directed mutagenesis 140
- PCR mutagenesis 141

4. Choice of expression system 144
- Overview 144
- *Xenopus* oocyte expression 144
- Transiently transfected cells 145
- Stably transfected cells 145
- Baculovirus expression 145

5. *Xenopus* oocyte expression system 146
- Overview 146
- Preparation and injection of oocytes 146
- Endogenous receptors and ion channels 152
- Electrophysiological recording from *Xenopus* oocytes 153
- Expression cloning 154

6. Transient and stable expression in mammalian cells 156
- Transient expression 156
- Stable expression 158

7. Baculovirus expression 161
- Introduction 161
- Construction of recombinant baculoviruses 161
- Protein production 167

Acknowledgements 168

References 168

6. Antibodies as tools for the study of the structure and function of channel proteins 171

F. Anne Stephenson

1. Introduction 171

2. The choice of the antigen 172
Bacterially derived fusion proteins 172
Synthetic peptides 174

3. Antibody characterization 176
Antibody production 176

4. Uses of antibodies 180
Identification of the subunit complements of native channels 180
Use of antibodies to map ligand binding sites and functional domains of proteins 185
Immunocytochemical studies 188

5. Summary 188

Acknowledgements 189

References 189

7. Assembly, post-translational processing, and subcellular localization of ion channels 191

Neil S. Millar, Stephen J. Moss, and William N. Green

1. Introduction 191
Scope of the chapter 191

2. Structural organization of ion channels 192
Neurotransmitter-gated ion channels 192
Voltage-gated calcium and sodium channels 193
Voltage-gated potassium channels 193
Other ion channels 193

3. Post-translational processing and folding 193
Introduction 193
Phosphorylation 194
Glycosylation and fatty acylation 200
Disulphide bond formation 202
Protein folding 203

4. Subunit assembly 203
Metabolic labelling of proteins expressed in mammalian cells 204
Cell lysis and solubilization of ion channel proteins 205
Immunoprecipitation 206
Sucrose gradient centrifugation 208
Overview of approaches to the study of subunit assembly 211

5. Subcellular localization of ion channels 212
Surface expression of subunit combinations 213
Immunofluorescence microscopy 213

6. Concluding remarks 215

Acknowledgements 215

References 216

8. Ion channels and signal transduction systems in plant and animal cells 219

Ian McFadzean and Colin Brownlee

1. Introduction 219

2. Recording from plant cells 219
Special considerations for recording from plant cells 219
Voltage clamping plant cells 220
Patch clamping plant cells 221

3. Measurement of cytosolic Ca^{2+} 224
Methods for measuring cytosolic Ca^{2+} 224
Use of fluorescent dyes 225

4. Animal cells 230
The use of patch clamping to determine the involvement of a diffusible second messenger 230
Ligand-gated ion channels 231
G proteins: membrane-delimited pathways 231
Pathways involving soluble second messengers 234

5. Conclusions 239

References 239

9. Channel structure 241

John B. C. Findlay and Derek Marsh

1. Introduction 241

2. General modelling principles 242
Topography 242
Conformation 244
'Sidedness'/orientation 245
Folding 246

3. Spectroscopic characterization of channel structure and assembly 246
Fourier-transform infrared (FTIR) spectroscopy 247
Spin-label electron spin resonance (ESR) spectroscopy 250

Nuclear magnetic resonance (NMR) spectroscopy 259
Circular dichroism (CD) spectroscopy 262
Fluorescence spectroscopy 263

References 264

10. Electron microscopical analysis of ion channels 269

Andreas Holzenburg

1. Introduction 269

2. Reconstitution of ion channels into 2D crystals 270
Reconstitution parameters 271
Monitoring progress 276
Trouble-shooting 279

3. Conventional electron microscopy 280
Microscope requirements 280
Specimen preparation 280

4. Image analysis and processing 283

5. High-resolution electron microscopy 285

Acknowledgements 288

References 288

Appendix 1 Addresses of suppliers 291

Index 295

Contributors

WILLIAM S. AGNEW
Department of Physiology and Neuroscience, Johns Hopkins University School of Medicine, 725 North Wolfe Street, Baltimore, MD 21205–2185, USA.

COLIN BROWNLEE
The Marine Biological Association, The Laboratory, Citadel Hill, Plymouth PL1 2PB, UK.

JOHN B. C. FINDLAY
The Research School of Biological Sciences, Department of Biochemistry and Molecular Biology, University of Leeds, Leeds LS2 9JT, UK.

ALASDAIR J. GIBB
Department of Pharmacology, University College London, Gower St., London WC1E 6BT, UK.

WILLIAM N. GREEN
Department of Pharmacological and Physiological Sciences, University of Chicago, Chicago, IL 60637, USA.

ANDREAS HOLZENBURG
Department of Biochemistry and Molecular Biology and Department of Genetics, University of Leeds, Leeds LS2 9JT, UK.

TORU IDE
Department of Biophysical Engineering, Faculty of Engineering Science, Osaka University, Toyonaka, Osaka 560, Japan.

MICHIKI KASAI
Department of Biophysical Engineering, Faculty of Engineering Science, Osaka University, Toyonaka, Osaka 560, Japan.

IAN McFADZEAN
Pharmacology Group, Division of Biomedical Sciences, King's College London, Manresa Road, London SW3 6LX, UK.

DEREK MARSH
Max-Planck-Institut für Biophysikalische Chemie, Abteilung Spektroskopie, D-37077 Göttingen, Germany.

NEIL S. MILLAR
Wellcome Laboratory for Molecular Pharmacology, Department of Pharmacology, University College London, London WC1E 6BT, UK.

EDWARD MOCZYDLOWSKI
Department of Pharmacology and Department of Cellular and Molecular Physiology, Yale University School of Medicine, PO Box 3333, New Haven, CN 06520–8066, USA.

GUY W. J. MOSS
Department of Pharmacology, Yale University School of Medicine, PO Box 3333, New Haven, CN 06520–8066, USA.

STEPHEN J. MOSS
MRC Laboratory for Molecular Cell Biology and Department of Pharmacology, University College London, London WC1E 6BT, UK.

THOMAS PETRI
Schering AG, Müllerstrasse 170, 1000 Berlin 65, Germany.

IRIS PRIBILLA
Schering AG, Müllerstrasse 170, 1000 Berlin 65, Germany.

F. ANNE STEPHENSON
Department of Pharmaceutical Chemistry, School of Pharmacy, 29/39 Brunswick Square, London WC1N 1AX, UK.

KEITH A. WAFFORD
Merck, Sharp & Dohme Research Laboratories, Neuroscience Research Centre, Terlings Park, Eastwick Road, Harlow, Essex CM20 2QR, UK.

PAUL J. WHITING
Merck, Sharp & Dohme Research Laboratories, Neuroscience Research Centre, Terlings Park, Eastwick Road, Harlow, Essex CM20 2QR, UK.

ALAN J. WILLIAMS
Department of Cardiac Medicine, National Heart and Lung Institute, Dovehouse Street, London SW3 6LY, UK.

Abbreviations

AEBSF	4-(2-aminoethyl)-benzenesulphonyl fluoride, Pefabloc©
AMPA	(±)-α-amino-3-hydroxy-5-methylisoxazole-4-propionic acid
BAPTA	1,2-bis(2-aminophenoxy)ethane tetraacetic acid
cAMP	cyclic adenosine 5′-phosphate
C_8E_5	octyl-pentaoxyethylene
$C_{12}E_8$	dodecyl-octaoxyethylene
CCD	charge-coupled device
CD	circular dichroism
cGMP	cyclic guanosine 5′-phosphate
CHAPS	3-[(3-cholamidopropyl)-dimethylammonio]-1-propane sulphonate
CLSM	confocal laser-scanning microscopy
c.m.c.	critical micelle concentration
CMV	cytomegalovirus
COSY	correlation spectroscopy
c.p.m.	counts per minute
CPMAS	cross-polarization magic-angle spinning
CTEM	conventional transmission electron microscope
DAG	diacylglycerol
DAT	digital audio tape
DIC	differential interference contrast
DMEM	Dulbecco's modified Eagle's medium
DMPC	dimyristoyl phosphatidylcholine
DMSO	dimethyl sulphoxide
DOXYL	4,4-dimethyloxazolidine-*N*-oxyl
DTT	dithiothreitol
EDAC	1-ethyl-3-(3-dimethylaminopropyl)-carbodiimide
EDTA	ethylene diamine tetraacetic acid
EGTA	ethylene glycol-bis(β-aminoethyl ether) tetraacetic acid
ELISA	enzyme-linked immunosorbent assay
EPPS	4-(2-hydroxyethyl)-1-piperazine propane sulphonic acid
ESR	electron-spin resonance
FCS	fetal calf serum
FDG	fluorescein di-β-D-galactopyranoside
FM	frequency modulated
FTIR	Fourier-transform infrared
GABA	γ-aminobutyric acid
8-Gluc	octyl-β-D-glucopyranoside
GST	glutathione *S*-transferase

Hepes	4-(2-hydroxyethyl)-1-piperazine ethane sulphonic acid
HETE	hydroxyeicosatetraenoates
HWHH	half-width at half-height
IBMX	isobutylmethyl xanthine
IP_3	inositol trisphosphate
KLH	keyhole limpet haemocyanin
LDR	lipid-to-detergent ratio
LPC	Lubrol–phosphatidylcholine
LPR	lipid-to-protein ratio
LT	leukotriene
12-malt	dodecyl-β-D-maltopyranoside
MAPS	multiple antigen presentation system
MBS	modified Barth's solution
MMTV	mouse mammary tumour virus
m.o.i.	multiplicity of infection
MOPS	4-morpholinepropane sulphonic acid
MTT	3-[4,5-dimethylthiazol-2-yl]-2,5-diphenyltetrazolium bromide
nAChR	nicotinic acetylcholine receptor
NDGA	nordihydroguaiaretic acid
NMDA	*N*-methyl-D-aspartate
NMR	nuclear magnetic resonance
NOESY	nuclear Overhauser enhancement spectroscopy
NSAI	non-steroidal anti-inflammatory
OAG	1,2-oleoylacetyl glycerol
PBS	phosphate-buffered saline
PCR	polymerase chain reaction
PDAc	phorbol 12,13-diacetate
PDBu	phorbol 12,13-dibutyrate
4α-PDD	4-α-phorbol 12,13-didecanoate
pdf	probability density function
p.f.u.	plaque-forming units
PG	prostaglandin
p.i.	post infection
PIP_2	phosphatidyl inositol 4,5-bisphosphate
PKC	protein kinase C
PMA	phorbol 12-myristate-13-acetate
PMSF	phenylmethylsulphonyl fluoride
8-POE	octyl-(polydisperse) oligooxyethylene
PVP	poly(vinyl pyrrolidone)
PVSG	post-vesiculation size growth
RACE	rapid amplification of cDNA ends
rms	root mean square
SDM	site-directed mutagenesis

SDS	sodium dodecyl sulphate
SDS-PAGE	sodium dodecyl sulphate polyacrylamide gel electrophoresis
SSPE	saline/sodium phosphate/EDTA
STEM	scanning transmission electron microscope
STX	saxitoxin
TE buffer	Tris-EDTA buffer
TEM	transmission electron microscope
TLC	thin-layer chromatography
TNM-FH	*Tric.ni* medium (modified Grace's medium)
TPA	12-*o*-tetradecanoylphorbol-13-acetate
Triton X-100	octylphenyl-nonaoxyethylene
TTX	tetrodotoxin
TX	thromboxane

1

Patch-clamp recording

ALASDAIR J. GIBB

1. Introduction

Patch-clamp recording is now widely used in the biological sciences. The technique was originally developed with the very specific intention of observing ionic currents flowing through single ion channels in cell membranes (1). From this origin the technique has developed into one of the most versatile and precise methods of measurement in biology. For example, no other technique allows the observation of the functioning of a single protein molecule in a lipid bilayer, yet can also allow monitoring of synaptic transmission between neurones in the brain, or permit monitoring of the change in cell membrane area during secretion. Underlying all the applications of patch clamping is the single principle that it provides an extremely sensitive means to measure ionic currents flowing through cell membranes. For their discovery and development of this technique and its application to a range of critical questions in biology, Neher and Sakmann were awarded the 1991 Nobel prize in Physiology or Medicine (1, 2, 3). Two excellent sources of detailed information on the patch-clamp technique are available: the 'blue book' guide to patch-clamp recording (4, a new edition of this book is due in 1995) and the recent 'Ion channels' volume of *Methods in enzymology* (5). In recent years various aspects of the patch-clamp technique have been reviewed in several different articles and books (6, 7, 8, 9). This chapter is intended to provide an overview, concentrating on the more practical aspects of the patch-clamp technique.

2. Principles

Before discussing the practical aspects of patch clamping in detail, it is useful to describe the electrical and physical principles underlying patch-clamp recording as it is only through a clear understanding of these that the experimenter can make best use of the technique.

2.1 Electrical considerations

One reason why patch clamping is so popular is because of the exquisite resolution it provides in measuring membrane currents—in fact, ionic currents of less than 1 pA (10^{-12} A) can be measured (2). This is achieved because it is possible to form a seal between the end of a glass recording pipette and the cell membrane which has an extremely high electrical resistance (a gigaohm seal $> 10^9\ \Omega$). This high-resistance seal is the main reason why the background current noise in patch-clamp recordings is so small.

The important point to appreciate about current noise in any electrical recording is that the thermal noise in a current *source* is determined by the source resistance (R) according to Equation 1

$$\sigma_n = \sqrt{\frac{4kT\Delta f}{R}} \qquad [1]$$

where σ_n is the r.m.s. (root mean square) of the current flowing through the resistance, R; k is Boltzmann's constant; T is the absolute temperature; and Δf is the bandwidth. Thus, the current noise in the recording decreases as the source resistance increases. In order to measure currents (at room temperature) of the order of a pA at a bandwidth of 1 kHz, the source resistance needs to be about 2 GΩ or higher (r.m.s. noise approximately 0.1 pA, about 0.6 pA peak-to-peak). In the case of a patch-clamp recording, the 'source resistance' is mainly the combination of the feedback resistor in the amplifier, the seal resistance, and the resistance of the preparation itself (cell or isolated membrane patch). The noise contributions from each component add together according to the relation

$$rms_T = \sqrt{rms_1^2 + rms_2^2 + \ldots rms_n^2} \qquad [2]$$

where rms_T is the total noise variance observed from n independent noise sources. This type of relationship means that the largest noise source tends to dominate and so for low-noise recordings it is essential to try to minimize *all* noise sources (see refs 8 or 10 for a detailed discussion of noise sources in patch-clamp recordings).

Thus, the essence of any patch-clamp recording is a high-resistance seal between a cell membrane and the glass of the recording pipette (the 'patch electrode') combined with a low-noise current-to-voltage converter amplifier. The electrical aspects of this system have been analysed in detail and are now well understood (7, 11, 12). However, the physical nature of the seal between glass and membrane is not clear.

2.2 The nature of the glass–membrane seal

One attractive idea is that the seal occurs when a lipid 'bleb' of cell membrane forms and flows into the tip of the patch pipette (*Figure 1*) in response to suction in the pipette or osmotic pressure (13). Perhaps the lipid bleb squeezes the water molecules and hydrated ions out from between itself and the glass surface forming the gigaohm seal. Once formed, the seal is mechanically

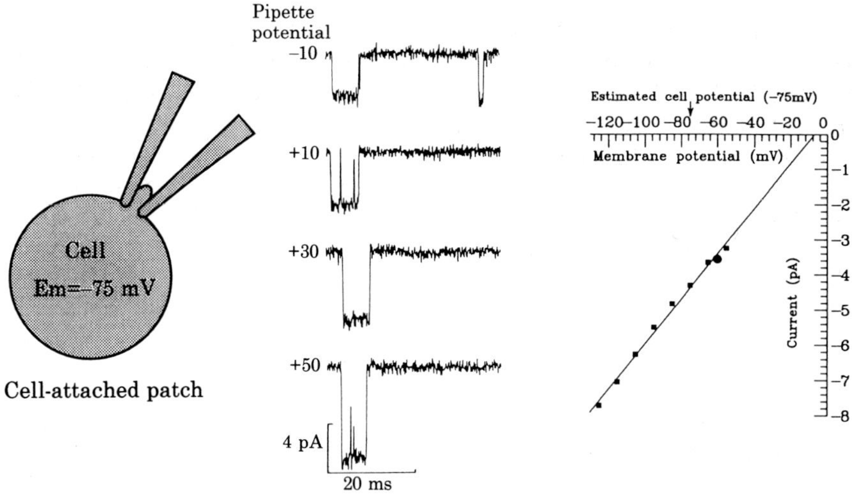

Figure 1. Cell-attached patch clamp configuration. A patch pipette is attached to the surface of a cell with a small piece of cell membrane drawn up into the tip of the patch electrode to make a gigaohm seal. The traces illustrated in the central panel show examples of single channel currents through NMDA receptor channels activated with 30 nM L-glutamate and 1 μM glycine in the pipette solution, recorded at different pipette potentials (V_p) as indicated. The traces have been inverted to comply with the standard patch-clamp recording conventions (Section 3.8). In the cell-attached configuration the cell membrane potential (E_m) is unknown. However, knowing that the reversal potential of the NMDA receptor channel is close to zero mV, it is possible to use the current–voltage relation for the single-channel amplitudes (which will be zero when $V_p = E_m$) to determine the cell resting potential. In this recording the cell membrane potential was predicted to be −75 mV as shown on the graph. The filled circle on the graph shows the single channel current recorded at a membrane potential of −60 mV after patch excision to form an inside-out patch. The fact that this point lies on the straight line fitted to the channel amplitudes measured in cell-attached mode confirms the estimate of −75 mV for the cell membrane potential. The solid line is a linear regression through the data points and has a slope of 64 pS (the slope conductance). This cell-attached recording was from an acutely dissociated hippocampal neurone (methods were as described in ref. 37).

stable and provides a molecular barrier such that ions or small charged molecules cannot pass from one side of the seal to the other (and so cannot pass from outside to inside the pipette or vice versa). It has also been shown that if nystatin (see Section 3.6) is applied to the cell outside the membrane patch, no nystatin pores appear within the patch, suggesting that the seal forms a barrier to diffusion even within the bilayer (14). Presumably, ion channels or other integral membrane proteins flow into the bleb of membrane along with the lipid, before formation of the seal, resulting in the isolation of a patch of membrane which is probably of very similar composition to the native cell membrane (although it has been suggested that the protein content of the membrane patch is lower than the surrounding cell membrane).

2.3 Connecting the membrane patch to the electronics

The patch pipette is filled with a solution which is wet and salty—it is well known that electronic circuits do not like salt water! Therefore, a 'go between' is required to carry the ionic currents to the headstage electronics. A chlorided silver wire inside the patch pipette forms a Ag/AgCl electrode allowing ionic currents (carried by Na^+ or Cl^- ions for example) flowing in the patch pipette solution to be seen as electrical currents by the headstage. A second Ag/AgCl wire or pellet in the recording bath is connected to the headstage ground point and completes the circuit from bath solution to pipette solution to headstage and back to the bath. The *pipette holder* has the combined function of providing a mechanically stable support for the pipette and allowing pressure or suction to be applied to the back of the pipette. Electrical noise is kept low by minimizing the capacitance between the silver wire and the pipette holder by

(a) using Teflon-coated silver wire (available for example from Clark Electromedical) with just a few millimetres of bare wire exposed at each end (the Teflon should be sealed to the wire with a little Sylgard (see Section 4.3) before chloriding);

(b) keeping the solution level low in the patch pipette (5–10 mm);

(c) manufacturing the pipette holder from low dielectric materials: typically a combination of Teflon and polycarbonate. In addition, one should avoid putting greasy finger prints on the pipette holder and care should be taken to ensure the holder is kept dry (e.g. by regularly blowing dry nitrogen or air through the holder).

2.3.1 Chloriding the silver wire

A variety of methods are available for making the silver chloride coating on a piece of silver wire, thus forming a Ag/AgCl electrode. A simple chloriding procedure is first to clean the silver wire by dipping it in HCl for a few minutes and then in a strong bleach solution for a few minutes (NB it is safest to use a cheap, plain bleach and not one of the more expensive varieties which contain a mixture of detergents that could possibly affect cell membranes). This will form a thin and quite robust coating of AgCl on the wire surface, visible by its grey/black colour. Only the end of the wire which is to be in contact with the pipette filling solution should be chlorided. The other end of the wire should be kept clean in order to ensure a good contact with the pin at the back of the holder. In some cases this is soldered to the pin at the back of the holder (using silver solder).

The chloride coating will last for several weeks if care is taken not to scratch it when fitting each new pipette into the holder. The back end of each pipette can be fire-polished in a Bunsen flame to smooth off any sharp edges before pulling. This not only protects the chlorided silver wire, but also prevents damage to both the Teflon coat on the silver wire and the rubber seal of the pipette holder. Fire-polishing also prevents scratching the inside of the holder. For some people these efforts are too much trouble! If so, it is best to dispense with the Teflon coat on the silver wire and to chloride the wire sufficiently far along its length to ensure that the pipette-filling solution does not contact bare silver wire. If this happens an offset potential will develop between the salty pipette-filling solution and the silver wire, making the membrane potential set on the patch-clamp amplifier meaningless.

2.4 Solutions for patch clamping

A very simple extracellular solution would be NaCl, 150 mM; KCl, 3.0 mM; $CaCl_2$, 1.0 mM; Hepes, 10 mM; pH 7.4 with NaOH. This solution may contain very little oxygen (e.g. if made up with freshly distilled water) and obviously it contains no glucose. Particularly for experiments with neurones or other cells which may have very little energy stores, oxygenation and the addition of glucose may be important. A more physiological solution might contain NaCl, 124 mM; $NaHCO_3$, 26 mM; KCl, 3 mM; $CaCl_2$, 1 mM; $MgCl_2$, 1 mM; and NaH_2PO_4, 1 mM bubbled with 95% O_2/5% CO_2 to maintain the pH at 7.4. With this solution, the recording chamber should be continuously perfused at a rate of 3–10 times the chamber volume per minute in order to maintain adequate pH control. An alternative to this is to blow humidified 95% O_2/5% CO_2 across the surface of a recording chamber which is not being perfused and, by the use of a pH-sensitive dye, ensure that the pH in the recording chamber is maintained at 7.4. Intracellular (pipette) solutions for patch clamping are described below in relation to whole-cell recording (Section 3.5) and outside-out patch recording.

3. Patch-clamp recording configurations

There are four basic patch-clamp recording configurations (15). Three of these are used to record single-channel currents in membrane patches and the fourth is used to record currents through the whole cell membrane:

- cell-attached patch
- inside-out patch
- outside-out patch
- whole-cell recording (records currents through the whole cell membrane).

3.1 Basic sealing procedure

As emphasized above, the formation of a high-resistance seal between the patch pipette and the cell membrane is fundamental to all patch-clamp recording. During seal formation progress should be monitored both visually, by looking down the microscope, and electrically by observing the pipette current in response to a repeated test voltage pulse (t-pulse). The basic procedure is outlined in *Protocol 1*.

Protocol 1. Basic patch-clamping procedure

1. Approach cell with positive pressure in pipette.
2. Observe dimple forming on cell surface as pipette tip presses on to cell (t-pulses become 10–20% smaller).
3. Remove positive pressure and apply gentle suction (e.g. by mouth); seal should begin to form (t-pulses become smaller).
4. Apply negative voltage in pipette (e.g. −55 mV).
5. Wait—remove suction when gigaohm seal forms (t-pulse becomes very small).

3.2 Notes on the sealing procedure

A glass patch pipette with its tip filled with extracellular solution is mounted in the pipette holder and positive pressure (between 20 and 100 millibar) applied to the interior of the pipette by blowing into a long piece of tubing connected to the pipette holder and then clamping the tube closed. The pipette tip is then inserted into the recording bath and manoeuvred carefully to within a few μm of the cell of interest while viewed with a high magnification microscope. Then using the fine control of the micromanipulator, the pipette tip is gently pressed up against the cell membrane. At this point it is possible to see a dimple form on the cell surface which is caused by the stream of solution flowing out of the end of the pipette.

3.2.1 Test pulses (t-pulses)

The electrical resistance of the pipette is monitored by observing the size of the current pulse produced by a repeated test voltage pulse or 't-pulse' (e.g. 1 mV of 5 ms duration every 20 ms). Many patch-clamp amplifiers have the facility to produce suitable test pulses or they can be produced by a suitable pulse generator. When the pipette tip presses against the cell surface the size of the t-pulse will fall by about 10%. The positive pressure is then released and the dimple should disappear as the cell recoils against the pipette tip. At the same time the t-pulse should become 10–20% smaller. Gentle suction is now applied to the interior of the pipette and a seal should begin to form. When the seal resistance is greater than about 100 MΩ (i.e. the t-pulse has reduced to less than about 10% of its original size) the formation of a gigaohm seal can be speeded up by the application of some negative voltage in the pipette (about −55 mV works well).

3.3 Cell-attached patch configuration

Cell-attached patch (*Figure 1*) is the simplest configuration and the one which probably causes least disturbance to the environment of the ion channel protein (2). This configuration is formed using the basic technique for obtaining seals in all types of patch clamping (*Protocol 1*). As soon as a seal resistance of greater than a few gigaohms is obtained the suction should be stopped and the pressure in the pipette returned to atmospheric. When a seal forms the current responses to the t-pulses almost disappear and any offset current (due to the −55 mV applied to the pipette) will disappear leaving only small blips produced by the pipette capacitance charging and discharging in response to each t-pulse. These transients should then be cancelled using the pipette capacitance cancellation controls on the patch-clamp amplifier. The pipette is now in the *cell-attached* recording configuration.

3.4 Inside-out patch configuration

The inside-out patch configuration (*Figure 2*) is formed by quickly pulling the patch pipette away from the cell when in the cell-attached recording configuration. This is most easily done when using a hydraulic micromanipulator but can also be achieved with care using any reasonable mechanical manipulator. In *Figure 2* single channel currents recorded from an inside-out patch are shown after inversion (see Section 3.8) to give the correct convention of inward currents as downward deflections of the current trace. Note also that the true membrane potential is given (by convention with respect to a

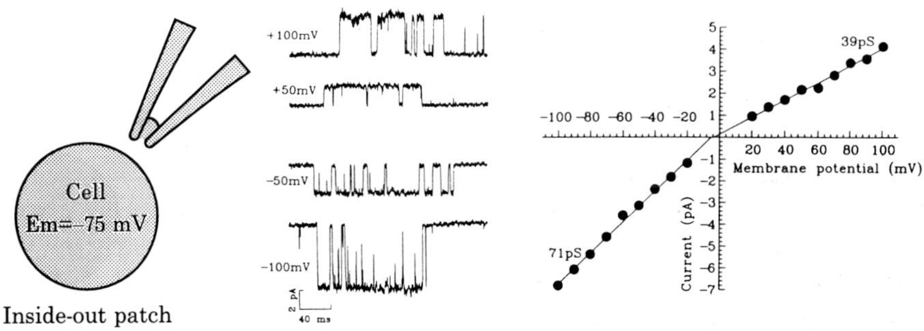

Figure 2. Inside-out patch configuration. The patch pipette has been pulled back from the cell to form an inside-out patch. Examples of single NMDA receptor currents activated by 30 nM L-glutamate and 1 μM glycine are shown, recorded in symmetrical salt solutions containing 118 mM NaCl, 30 mM $NaHCO_3$, 20 mM glucose, and 10 mM EDTA, bubbled with 95% O_2 and 5% CO_2. With identical solutions in the patch pipette and in the bath, ion channel currents are expected to reverse at zero mV. Traces at different membrane potentials are shown after inversion to illustrate the change in channel amplitude with membrane potential (shown as $E_m = -V_p$). The channel current/voltage relationship is shown plotted on the right. Note that for the NMDA channel inward and outward currents have a different single-channel conductance. Adapted from ref. 45.

ground on the outside of the membrane) although the pipette potential shown on the front panel of the patch-clamp amplifier in this recording configuration was the negative of the membrane potential.

3.5 Whole-cell recording configuration

The whole-cell recording configuration is obtained via the cell-attached configuration by applying short pulses of suction (by mouth or with a syringe) to the interior of the patch pipette (16). These pulses of suction break the patch of membrane inside the pipette tip so that the pipette interior is then in electrical continuity with the interior of the cell (*Figure 3*). The moment of breakthrough is detected using a large t-pulse (e.g. −10 mV, after cancellation of the pipette capacitance transients). There will be a *sudden* appearance of a large capacity transient. An example of a whole-cell current recording is shown in *Figure 3*. In this experiment, a whole-cell recording from a neurone in a slice of the rat medial habenula was made and the bath solution changed for the period indicated by the solid bar to a solution containing 20 μM of the nicotinic agonist cytisine. At the membrane potential of this experiment, opening of the non-selective cation channels of the neuronal nicotinic acetylcholine receptors caused an inward current which lasted for the duration of the cytisine application.

3.5.1 Whole-cell capacitance transient cancellation

On breakthrough, large and slower capacitance transients (due to the charging and discharging of the cell capacitance) suddenly appear. These transients

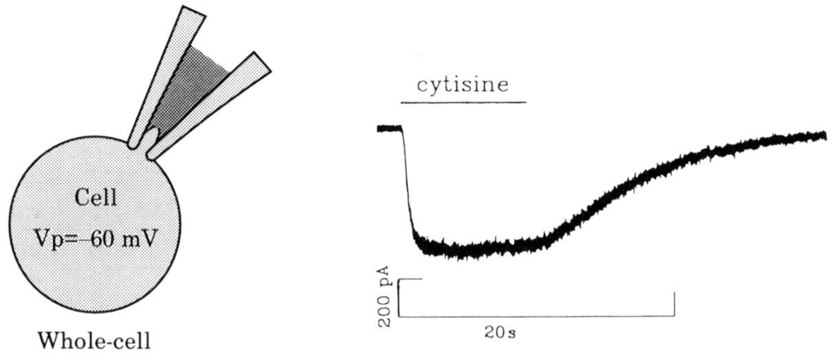

Figure 3. Whole-cell patch-clamp configuration. The illustration shows a patch pipette attached to a cell after rupture of the patch of membrane inside the pipette tip to form the whole-cell configuration. The cell membrane potential is now clamped at the pipette potential $V_p = -60$ mV. Notice that in the whole-cell configuration the pipette solution slowly dilutes and exchanges with the intracellular solution. The trace on the right illustrates the response obtained from a neurone in a brain slice of the rat medial habenula (46) to bath application of the nicotinic acetylcholine receptor agonist cytisine (20 μM). Cytisine is applied in the bath solution for the duration of the solid bar and generates a sustained inward current at this membrane potential (methods as described in ref. 46).

can then be cancelled using the patch-clamp amplifier 'cell' or 'slow' capacitance cancellation circuitry. It is important to use these controls as they allow estimation of the size of the cell membrane capacitance (C_m) and the resistance between the pipette interior and the interior of the cell: the *series resistance* (R_s). Also, if the cell capacitance transients are cancelled at the beginning of a recording then it is easy to check during a recording for changes in the series resistance which indicate if the access to the cell interior is blocking up, or the membrane sealing over. A change in R_s will be obvious because the shape of the residual transient left after capacitance cancellation will change. In a good recording, R_s is typically around twice the value of the pipette resistance measured before seal formation but if the situation deteriorates, R_s can often be considerably higher.

C_m and R_s are two fundamental factors in determining the fidelity of recordings made using the patch-clamp whole-cell recording configuration. The speed of the clamp response to any currents flowing in the cell is determined by C_m and R_s. The time constant of the clamp response is given by $\tau = C_m R_s$. Alternatively, this can be expressed as the bandwidth of the clamp $f_c = 1/2\pi C_m R_s$. Thus, for a 20 pF cell and a 10 MΩ series resistance, $\tau = 0.2$ ms (indicating a voltage step will require 1 ms or longer to reach the new value) and $f_c = 796$ Hz. In addition, the series resistance creates an error in the cell membrane potential whenever current is flowing in the pipette. This error, which can be substantial, can be partly corrected by the use of *series resistance compensation*. For a pipette potential of −60 mV, a series resistance of 10 MΩ and a pipette current of −1 nA, the error voltage will be − 10 mV and so the actual membrane potential will be −50 mV and not −60 mV as hoped. Clearly, large currents are not, in general, well clamped using the patch-clamp technique. For currents larger than a few nA it is preferable to use a switched, single electrode voltage clamp or a two-microelectrode voltage clamp (see, e.g. ref. 8).

3.5.2 Series resistance compensation

It is very important to realize that in the whole-cell recording configuration, the membrane potential (the command potential) set on the patch-clamp amplifier will not be the same as the potential across the cell membrane. This is because the current–voltage converter amplifier configuration clamps the potential in the pipette. The series resistance is so called because it is in *series* with the resistance across the cell membrane. Any current (I_p) flowing between the pipette and the cell interior will generate a voltage drop (V_{RS}) across R_s. By Ohm's law, $V_{RS} = I_p R_s$. This must be subtracted from the command potential to determine the potential across the cell membrane.

Series resistance compensation is easy to understand in principle. The series resistance compensation circuit uses the value of R_s dialled up on the front of the patch-clamp amplifier during cancellation of the cell capacitance transients. The pipette current is applied to this resistance and the voltage generated is summed with the command voltage. This means that the command voltage is increased by a factor equivalent to the expected V_{RS} so that the cell potential will equal the intended membrane potential.

As mentioned in connection with capacitance cancellation, the series resistance forms a low-pass filter of time constant $\tau = C_m R_s$. Thus, the clamp response becomes progressively slower as R_s increases. Patch-clamp amplifiers allow variable series resistance compensation from 0 to 100% (0–100% of the voltage V_{RS} is summed with the command). It is important to note (and easy to verify in practice!) that it is not possible to achieve 100% compensation. At 100% compensation the effective R_s would be zero ($\tau = 0$) and the bandwidth would be infinite. In this situation the compensation circuit will oscillate (in fact due to the non-ideal nature of R_s and the presence of capacitance associated with the pipette, the system generally oscillates at less than 100% compensation). To improve stability, the V_{RS} signal can be filtered before summing with the command voltage (this is referred to as a *lag* circuit on some patch-clamp amplifiers). This is extremely useful: until the introduction of the lag circuit (e.g. on the Axopatch 1C from Axon Instruments), most series resistance compensation circuits were difficult to use in practice due to the tendency of the clamp to go into oscillation. Finkel (7) has provided a good discussion of series resistance compensation.

3.5.3 Measurement of cell capacitance to monitor secretion

The ability to estimate cell capacitance with high resolution using the patch-clamp technique was first recognized by Neher and his co-workers (reviewed in ref. 17). The cell capacitance is a useful measure of the cell surface area since it has been found in a variety of different membranes that membrane capacitance is fairly constant around 1 μFcm^{-2} (0.01 pFμm^{-2}). Thus it has been observed that cell capacitance increases during exocytosis of secretory vesicles, presumably due to the fusion of secretory vesicles with the cell membrane. The high resolution of patch-clamp recordings has made it possible to resolve the capacitance change resulting from the fusion of single secretory vesicles with the plasma membrane of bovine chromaffin cells and peritoneal mast cells. In mast cells the size of the unitary event is about 20 fF.

Cell capacitance can be measured in several different ways (see, e.g. ref. 17). The simplest (but lowest resolution) method is to compensate the capacity transient in response to a voltage step as described above (Section 3.5.1). Fitting the time course of the average current relaxation ($\tau = C_m R_s$) in response to a voltage step should be more accurate and can be done on-line by computer. However, this is still of relatively low resolution. Higher resolution methods all utilize the fact that cell capacitance can be calculated from the current response of a cell to a sinusoidal voltage command. This technique is more complicated, requiring a phase-sensitive or 'lock-in' amplifier and has been described in detail by Lindau and Neher (18). This gives two outputs which are proportional to current (in response to a small sinusoidal command) at two orthogonal frequencies and from these, plus a measure of the DC current in the cell, the series resistance, cell membrane resistance, and cell capacitance are calculated. The technique has also been implemented in software and is available commercially. It allows the cell membrane capacitance to be calculated at high resolution many times per second and so is the technique of choice in secretion experiments.

3.5.4 Pipette filling solution

The composition of the pipette filling solution deserves careful thought for whole-cell or outside-out patch recordings (2), because the pipette filling solution diffuses into the cell during whole-cell recording and is therefore the intracellular solution in these recording configurations. A typical pipette filling solution for both whole-cell recordings and outside-out patch recordings will contain KCl, 150 mM; NaCl, 5 mM; Hepes, 10 mM; EGTA, 10 mM; $CaCl_2$, 1.0 mM; $MgCl_2$, 1.0 mM (pH 7.2). The EGTA/Ca^{2+} combination gives a free calcium concentration of less than 100 nM. Note that a free intracellular calcium concentration of less than 100 nM may actually prevent some cellular responses being observed.

In principle, the pipette solution should be iso-osmotic with the cell cytoplasm and the bath solution (around 300 mosmol). If the pipette solution is hyper-osmotic, the cell will swell and burst. Many people suggest that making the pipette solution hypo-osmotic (usually by around 10%) can improve the stability of whole-cell recordings. A rapidly equilibrating Ca^{2+} buffer such as BAPTA may sometimes be preferable to EGTA, which has relatively slow kinetics. In addition, Mg-ATP may be added to provide an energy source, Cs^+ may be substituted for K^+ to block potassium currents, gluconate may be substituted for chloride to move the Cl^- reversal potential, and so on. One of the great advantages of the patch-clamp method is that it allows control over the intracellular environment.

3.5.5 Liquid junction potentials

Since the solution in the pipette is of different ionic composition from that in the bath, a liquid junction potential will form between pipette and bath (19) because of the differing mobilities of different ions in solution. This can be offset with the 'junction null' control on the patch-clamp amplifier while the pipette is in the bath but, of course, once the pipette is in the whole-cell configuration or outside-out configuration, the junction potential will quickly disappear as the solution in the pipette equilibrates with the inside of the cell. Therefore junction potentials should be measured by comparing the difference in pipette null potentials when the pipette is immersed in pipette solution compared to when in the normal extracellular solution. The junction potential can then be subtracted from the membrane potential when making experiments which are critically dependent on membrane potential. However, most pipette solutions produce a junction potential of less than 3 mV.

Depending on the object of the experiment, this is often small enough to be ignored.

3.5.6 Washout of the cell contents

Unfortunately, many responses which involve a diffusible intracellular second messenger may be blocked or disrupted because the interior of the cell has been diluted by the pipette filling solution (2, 20). The time required to dialyse the cell interior depends on the series resistance (see above) which is related to the pipette resistance. Low resistance pipettes (<5 MΩ) often give series resistances of < 10 MΩ and will allow equilibration of ions and small molecules between the pipette and the inside of the cell within 30 sec to 1 min (20). However, for larger molecules such as enzymes several minutes may be necessary. If one suspects washout is preventing a desired response, the use of higher resistance pipettes (to slow down washout of the cell contents) may allow a response to be observed. A more sophisticated approach to this problem is to try the perforated patch whole-cell recording technique (see below).

3.6 Perforated patch whole-cell recording

The perforated patch technique is an ingenious method of avoiding the effects of washout of the cell contents by the pipette filling solution. It was devised by Horn and Marty (21) and is based on the fact that nystatin or amphotericin B act by inserting themselves in cell membranes to generate ion channels which are selective for monovalent ions. They are useful because they do not pass through the membrane into the cell, but stay in the membrane patch under the patch pipette. Nor do they diffuse *within* the membrane, beyond the bounds of the membrane patch (14). When many ion channels have been produced in the patch the resistance through the patch becomes low enough to allow recordings of ionic currents in the whole cell membrane. These channels have only a weak dependence on membrane potential and so can be used even in studies of voltage-dependent conductances. Details of the method have been published by Horn and Korn (22). The basic procedure is described in *Protocol 2*.

Protocol 2. Perforated patch recording

1. Make a stock solution of nystatin in DMSO (50 mg/ml) and sonicate before use. This can be kept in the freezer, but needs to be sonicated each day before use.
2. Dilute 20 μl of stock solution in 5 ml of pipette filling solution just before use, and sonicate (do not filter the pipette solution after this, as filtering will remove the nystatin). This solution can be used for about two hours after which new solution should be made up.

3. Dip the pipette tip (use filament glass) in nystatin-free pipette filling solution for a few seconds (the exact time needed depends on the shape of the pipette tip and the size of the tip opening). Fill the rest of the pipette as normal with nystatin-containing solution. The idea is to have nystatin-free solution in the pipette tip during seal formation. If nystatin is in the tip while attempting to make a seal, then a gigaohm seal will not form. If seals are a problem, it is worth increasing the amount of tip filling so that it takes longer for the nystatin in the pipette to diffuse down to the pipette tip.
4. Quickly insert the pipette into the bath and move the pipette tip to the cell. Do not apply positive pressure which will tend to force nystatin down to the pipette tip (and possibly spray the cell with nystatin!). The pipette should then be sealed on to the cell as quickly as possible.
5. Once in the cell-attached configuration, cancel the pipette capacity transient and set the pipette potential to around the expected cell resting potential.
6. Monitor the response to regular −10 mV hyperpolarizing t-pulses. There should be a gradual appearance of a slow capacity transient over the course of 5–30 min. This capacity transient should *gradually* become bigger and faster as the series resistance decreases due to more and more nystatin channels appearing in the patch. Note that if a sudden jump in the capacity transient occurs then this probably indicates that the patch has broken, producing a standard whole-cell recording configuration.
7. It is difficult to be sure that a perforated patch has not broken at some point during a recording unless lucifer yellow or some other dye is included in the pipette solution. With a fluorescent microscope it is then possible to observe if the dye has entered the cell, indicating that the perforated patch has ruptured.

3.7 Outside-out patch configuration

The outside-out patch configuration (*Figure 4*) is formed from the whole-cell configuration by simply pulling the pipette *very slowly* away from the cell. A hydraulic manipulator or piezoelectric manipulator is ideal for this but a steady hand and a good solid mechanical manipulator will also work. If this is done with the 'cell' capacitance cancellation circuitry switched off, then the formation of an outside-out patch can be monitored by the disappearance of the cell capacitance transients and a large reduction in the noise level of the recording. It is often a surprise to discover how far the pipette tip must be pulled away from the cell (50–200 μm) before complete loss of contact with the cell occurs. The noise level can be further reduced by raising the recording pipette until the tip is just below the surface of the bath solution. This some-

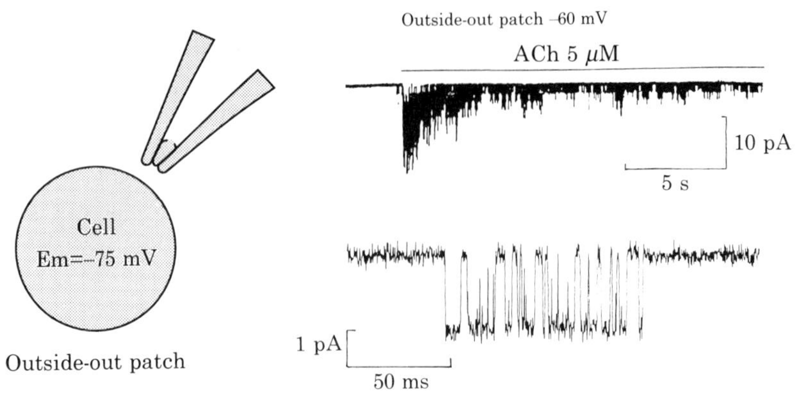

Figure 4. Outside-out patch clamp configuration. The illustration shows an outside-out patch formed by slowly withdrawing the pipette from the cell while in the whole-cell configuration. The cell membrane potential is no longer controlled. The cell contents will now have been at least partially replaced by the pipette solution. The membrane potential is determined by V_p and the pipette solution will quickly dilute any cell contents which had diffused into the pipette. The traces on the right show examples of the patch response to bath application of (5 μM) acetylcholine applied for the period indicated by the horizontal line. The single channel current levels are easily recognized and suggest that at least ten acetylcholine receptors are present in this patch of membrane. The lower trace shows single channel openings at higher gain and time resolution. The patch was isolated from a *Xenopus* oocyte previously injected with mRNA for the mouse muscle acetylcholine receptor subunits kindly supplied by Professor S. F. Heinemann, Salk Institute. Adapted from ref. 47.

times results in the pipette picking up mechanical noise from the bath perfusion system but the electrical noise is reduced because a low solution level minimizes the pipette capacitance and so reduces the background noise level. The pipette capacitance cancellation circuitry on the patch-clamp amplifier should then be adjusted.

Figure 4 shows traces of acetylcholine receptor channels recorded from an outside-out membrane patch. This patch of membrane was isolated from a *Xenopus* oocyte which had been injected five days previously with mRNA (kindly supplied by Professor S. F. Heinemann, Salk Institute) coding for the mouse embryonic acetylcholine receptor subunits. The upper trace shows channels activated by bath application of a relatively high concentration of acetylcholine (5 μM) on a slow time scale which illustrates that this patch probably contained at least 10 receptors whose activity rapidly desensitizes in response to this concentration of acetylcholine. The lower trace shows an expanded record of a single cluster of channel openings.

3.8 Patch-clamp recording conventions

Firstly, the cell membrane potential is always expressed as the potential inside the cell with respect to a ground or zero level outside the cell. When using the cell-attached configuration (*Figure 1*), if the actual cell membrane potential is −75 mV and a potential of +20 mV is applied to the interior of the patch pipette, then the potential difference across the membrane patch will be −95 mV (of course the potential across the rest of the cell membrane, outside the patch pipette, will remain at −75 mV regardless of changes in the pipette potential.

Secondly, by convention, positive ions flowing out of the patch pipette are measured as positive currents. In the cell-attached configuration, positive ions flowing out of the patch pipette and into the cell each time a cation channel opens, will be observed to give positive-going rectangular current pulses in the current trace (e.g. inward movement of Na^+ ions through a Na^+ channel or the outward movement of Cl^- ions through a Cl^- channel would give such a trace). The same current trace is observed with an inside-out patch. Thus, in *Figures 1* and *2*, the single channel currents have been inverted after recording so that the display shows these as inward currents (positive ions flowing from outside to inside of the cell membrane). However, in the outside-out configuration (*Figure 4*), positive ions flowing from the outside to the inside of the patch pipette will give negative current pulses on the current trace and so are displayed in *Figure 4* simply as recorded, without inversion.

4. Making patch pipettes

4.1 Patch pipette glass

Patch pipettes are generally referred to as electrodes although, strictly speaking, it is the chlorided silver wire inside the pipette which is the electrode. Pipettes can be made from a variety of different types of glass. The main variables are the geometry of the glass and the glass composition (23, 10).

4.1.1 Glass geometry

Patch pipettes are made from glass tubes with outer diameters of 1.5–2.0 mm and inner diameters of 0.8–1.5 mm. Most people use glass containing a small glass fibre fused to the inside of the tube as this makes it much easier to fill the pipette with solution without trapping bubbles near the tip. The fibre does not seem to affect the success rate of obtaining gigaohm seals but it is worth noting that it may carry fluid to the back of the pipette and result in moisture inside the holder and so increased background noise. Thick-walled tubing is preferable for single channel recordings because it has a lower noise level while thin-walled tubing gives a lower access resistance which is important for whole-cell recordings. If the glass tubes are bought as pieces 7.5 cm long (e.g. Clark Electromedical GC150F-7.5), then on pulling, each piece produces two pipettes of about the right length.

4.1.2 Pipette tip shape

To minimize access resistance the pipette should be as stubby as possible. However, this has to be compatible with obtaining a tip of around 1 μm in diameter after fire-polishing. In some situations (such as when working under a water-immersion objective) it is useful for the pipette to have a relatively long taper before the final constriction at the tip as this causes less optical distortion in the field of view. A steep taper angle is likely to be of advantage when making perforated patch recordings as the size of the patch of membrane accessible to the nystatin is likely to be bigger and the rate of diffusion of nystatin down to the pipette tip is likely to be faster.

The most relevant point to make about the pipette tip shape is that it should suit the type of cell under investigation. Generally, small cells are easier to record from with slightly smaller pipettes (say 5–10 MΩ rather than 2–5 MΩ) because there is a tendency for the cell to disappear up the pipette if the cell is small relative to the pipette tip size. To record from very small structures such as neuronal dendrites (24) may require even smaller pipettes.

4.1.3 Glass composition

Rae and Levis (10) have made detailed studies of the effect of glass composition on the patch pipette characteristics, particularly noise level and ease of obtaining seals. However, caution is required when attempting to relate glass composition to success rate in obtaining seals since, as Rae and Levis point out, these comparisons require that pipettes made from different glasses be manufactured to have identical tip shapes in order to rule out effects of tip geometry on seal success rate. This is a tall order since the tip geometry is very dependent on factors like the softening temperature of the glass which varies from around 600°C for lead glasses to over 900°C for aluminosilicate glasses.

The two physical parameters of glass that are most relevant to patch clamping are the loss factor of the glass and its softening temperature. These two parameters are to some extent inversely related, with some exceptions (10). High softening temperature glasses such as the aluminosilicate glasses have a low loss factor and consequently will exhibit lower noise. However, glasses with a high softening temperature require high heat settings on the pipette puller which will burn out the filament on the puller more readily and place extra strain on the coils and circuits in the puller. Pure quartz glass has the highest melting point of all (1580°C) and cannot be softened with a standard wire-filament puller. Instead, Sutter Instruments has developed a laser-scanning puller which uses the specific wavelength of a CO_2 laser to excite the silicon–oxygen bonds in the quartz to produce heat in the glass. Quartz glass pipettes can, in principle, give extremely low noise recordings but the specialist nature of their manufacture means they may not become generally used.

Glasses with lower melting points than quartz are made by mixing pure quartz with significant amounts of other salts; particularly aluminium, boron, or lead. The aluminosilicate glasses give low noise recordings. They have high melting points (> 900 °C) and so are difficult to pull, but can produce nice recordings (e.g. SM150F-7.5 from Clark Electromedical). More commonly, borosilicate glasses are used for patch clamping (e.g. Clark Electromedical GC150F-7.5 or GC150TF-7.5: 1.5 mm diameter, thick-walled, fibre filled or thin-walled respectively). These glasses provide a good compromise between noise level and ease of pipette fabrication, at least for pipettes with resistances down to a few MΩ. Very low access resistances are more easily achieved using lower melting point lead-based glass. These have higher noise levels but can be pulled and fire-polished down to a very blunt, 'bullet'-shaped tip. The main worry with lead glass, and to some extent with any glass, is that it is possible for chelating agents like EGTA to leach metal ions from the glass and these will affect ionic currents in the cell. This has been shown to be the case with several different glasses (10), and is a possibility which must always be borne in mind.

4.2 Patch pipette pulling procedure

4.2.1 Horizontal pullers

These pullers position the pipette glass horizontally and exert pull either from only one end with one fixed end, or from both ends simultaneously (better).

First stage

Carefully clamp the glass so that the middle of the glass piece is within the puller heating filament. Select a fairly high heater current setting which will soften the glass quickly and set a low pull strength with the puller set to turn off the heating current as soon as the glass begins to soften and has elongated by about 6–10 mm. This should produce an 'hour-glass' shape in the middle of the glass piece.

Second stage

Select a lower heat setting and low pull strength with the pull set to continue until the glass pieces separate.

Multi-stage pulls are made by repeating the first stage, successively 'necking' the glass more and more and so producing extremely blunt tips. The new microprocessor-controlled pullers like that made by Sutter Instruments allow many different pull programs to be stored in memory, avoiding the headaches that occur when someone changes the puller settings just after you have made the perfect patch recording! Some of these pullers incorporate sophisticated features such as a timed gas jet which cools the filament and glass towards the end of the pull, producing a thicker and stubbier tip.

4.2.2 Vertical pullers

These pullers are generally of simpler design and in some ways are less versatile than the horizontal pullers. However, they have a reputation for reliability. Popular examples of these are made by Narashige, List Electronic, and Kopf.

First stage

Clamp the glass in the puller with a slightly greater length of glass above than below the filament (this is because vertical pullers use gravity and so the pull is from the lower end giving a longer lower piece unless this is compensated for). Set spaces to stop the fall of the pipette before the glass separates (about 8–10 mm drop). Set a relatively high filament current (about 15 A) and allow the glass to pull, producing the hour-glass shape.

Second stage

Using a suitable spacer re-set the glass by loosening the top clamp and raising the glass with the lower clamp (by about 4–5 mm) and then tighten the top clamp and remove the spacer. Select a lower heating current (about 11 A) and activate the second pull which should produce two approximately symmetrical pipettes.

NB: When pulling patch pipettes on a vertical puller both stages should be under gravity: no solenoid pull should be used.

4.2.3 Pipette puller filaments

It might be expected that delivering heat to the glass is the most critical aspect of making pipettes. This is definitely the case and the filament, and control of the filament current, are crucial. In addition, it is important that the filament is protected from varying air currents during pulling or consistent pipettes will not be obtained. Unfortunately, the filament is also the most vulnerable part of the instrument because each time a piece of glass is loaded the filament is in danger of being knocked out of shape, which consequently alters the delivery of heat to the glass. In addition, because the filament is undergoing repeated cycles of heating and cooling, it will gradually age and stretch and become deformed.

4.3 Coating with Sylgard

Sylgard coating is used to reduce the pipette capacitance (15). The pipette capacitance is a major source of noise in single channel recordings (12). It also contributes noise in whole-cell recordings and generates instability when using series resistance compensation during whole-cell recordings. The pipette capacitance can be minimized firstly by maintaining a low solution depth around the pipette, and secondly by using thick-walled low-resistivity glass coated with a hydrophobic substance to prevent solution 'creep' up the surface of the glass during a recording. Only relatively few substances are both hydrophobic and will 'wet' glass. Sylgard® (Dow-Corning 184 transparent encapsulating resin, available from, e.g., BDH) has these properties, and has a high specific resistivity which means that coating the outside of the pipette with Sylgard is, in terms of noise levels, almost as beneficial as having thicker glass.

Sylgard comes in the form of a resin and a curing agent which are mixed together in 10:1 proportions before use. The mixture sets on heating but only over a period of 6–10 hours at room temperature. It is quite fluid immediately after mixing and it is important to mix it thoroughly, which unfortunately fills the mixture with air bubbles. The trick is to mix 10 or 20 ml and then leave the mixture sitting on the bench at room temperature for three or four hours. This allows the air bubbles to disperse and the mixture to cure partly and take on a thick consistency. The Sylgard can then be aliquoted in 0.2 ml volumes into Eppendorf tubes and kept for weeks in the freezer, taking a tube out each day it is needed.

The Sylgard is most easily applied under a dissecting microscope (or even magnifying glass) using a piece of pipette glass to paste it on to the patch pipette (as illustrated in ref. 25). Sylgard should be applied from about 100 μm from the tip to several millimetres past the shoulder of the pipette. The patch pipette should be rotated during Sylgard application to obtain an even coat and then the Sylgard immediately heated and cured (e.g. by inserting the pipette into a heated wire coil, or using hot air from an electronics heat gun). It is essential that the Sylgard should not reach the tip of the pipette or it will be impossible to form a seal. In fact, if the cells are not sealing and there is no obvious cause, it is always a good idea to try a few non-coated pipettes. During fire-polishing it is possible that some Sylgard residue can be burnt off the pipette tip but this will not always be successful.

4.4 Pipette fire-polishing

Like Sylgard coating, fire-polishing is not something which is essential for successful patch clamping. It is well known that many people make patch-clamp recordings without fire-polishing the pipette first. However, fire-polishing can be useful as it provides a means of controlling the final tip diameter of the patch pipette and it may help to burn off small dust particles or Sylgard residue from the pipette tip while smoothing out small imperfections in the tip shape.

Pipette microforges are available commercially (e.g. Narashige, List) or can be constructed using any microscope with long-working-distance objectives (to avoid heat damage to the objective the microscope can be mounted on its back). A low power objective is needed to locate the pipette and high power magnification (e.g. × 40 objective with × 15 eyepieces) to observe the pipette during fire-polishing. A standard metallurgy microscope with × 10 and × 100 objectives would make a good basis for a microforge. A loop of thin (e.g. 200 μm diameter) platinum wire with a sharp bend in the middle is used to provide localized heat to soften the pipette tip. The wire is heated by connecting each end to the poles of a variable DC voltage supply capable of providing several amps of current. The platinum wire is covered with glass at the bend (by melting glass from a pipette tip on to the wire at a high heat setting) to prevent the hot wire sputtering platinum on to the pipette tip. The pipette is mounted on a micromanipulator and the pipette tip is moved close to the wire loop to a position where it will be within about 100 μm from the wire when the wire heats up and expands. The current is then briefly switched on and the pipette tip observed to constrict very slightly. Normally this is all that fire-polishing entails. However, particularly with softer glasses, it is possible to fire-polish a pipette which initially has quite a wide tip opening down to 1–2 μm, producing a very steep taper which will minimize the pipette series resistance (this is particularly useful if the experiment involves clamping large currents where series resistance errors can be a problem).

5. Building a patch-clamp set-up

A complete patch-clamp set-up is made from a combination of mechanical equipment, which allows the tip of the patch pipette to make contact with the cell membrane, and electrical equipment which allows the ionic currents flowing through ion channels in the cell membrane to be recorded. In addition, the cells which are being patch clamped must be kept healthy with an adequate supply of oxygen, glucose, etc. The complete set-up can be quite sophisticated or relatively simple depending on the type of experiment intended. A patch-clamper's shopping list for a basic set-up might consist of:

- patch-clamp amplifier
- oscilloscope
- amplifier and filter
- tape recorder
- computer, interface, and analysis software
- vibration isolation system (flotation table)
- microscope
- patch-clamp recording chamber
- coarse and fine micromanipulators
- Faraday cage, ground wiring, and alloy mesh for shielding.

5.1 Patch-clamp amplifiers

The patch-clamp set-up begins with a suitable patch-clamp amplifier. Commercially available patch-clamp amplifiers are of such high quality and so convenient to use that it is money well spent to buy a good amplifier.

Amplifiers are available commercially from a variety of sources (e.g. Axon Instruments Inc., Clark Electromedical, List Electronic, Dagan, Warner). My personal preference is for the Axopatch from Axon Instruments which combines excellent layout features with very low-noise recordings.

5.2 Patch-clamp amplifier headstages

The commercially available patch-clamp amplifiers nearly all have headstages with a similar design philosophy (7). The current–voltage converter electronics in the headstage amplify the very small currents which flow in the patch pipette, and at the same time the headstage electronics allow the pipette voltage to be set at any desired level.

5.2.1 Resistor feedback headstages

These are current–voltage converter amplifiers which use a very high value feedback resistor to produce the high gain needed to measure single channel currents. Typically the feedback resistor (R_F) would have a value of 50 GΩ. Since the standard supply voltage used inside electronic instrumentation is ±12 V, this means (from Ohm's law) that a maximum current of $I = V/R_F = 12\ V/5 \times 10^{10}\ \Omega = \pm 240$ pA can be measured with such a headstage. This is fine for single channel recording. However, during whole-cell recordings, currents of more than 1000 pA may occur. Therefore patch-clamp headstages usually contain two feedback resistors (50 GΩ and 500 MΩ: current ranges of ±240 pA and ±24 nA respectively) with the facility to switch electronically between the two. Of course, the 500 MΩ resistor is associated with a higher inherent noise level (from Equation 1), but since in whole-cell recording mode the cell capacitance and resistance are the dominant noise sources, the 500 MΩ resistor does not usually add significantly to the noise in the recording.

The major disadvantage of headstage designs incorporating a very high value resistor is that these resistors have significantly non-ideal properties. Firstly, the thermal noise produced by these resistors is higher than expected (from Equation 1). Secondly, these resistors have significant distributed capacitance (C_F). Thus, the resistor behaves like a low-pass filter of time constant $\tau = R_F C_F$. For even a small capacitance of around 0.1 pF, a 50 GΩ resistor will have a time constant of 5 ms, giving a bandwidth of $1/2\pi\tau = 31.8$ Hz. In order to see the fine detail in a single channel recording, a bandwidth of at least 1000 Hz is necessary. Therefore all resistor feedback headstages are coupled to a (high) *frequency boost* circuit which is usually tuned to provide a time constant of around 30 μs (equivalent to a bandwidth of greater than 5 kHz).

5.2.2 Capacitor feedback headstages

Capacitor feedback headstages work on an entirely different design philosophy for measuring very small currents (7, 12). They use a small capacitor in the feedback path of the current–voltage converter. Currents flowing in the feedback loop charge up the voltage across the capacitor. The *rate* of charging of the voltage across the capacitor is proportional to the current and is measured by a differentiator circuit to give the patch pipette current. Unlike resistors, capacitors have near-ideal properties and so do not generate noise above their theoretical value. Capacitor-feedback headstages are therefore capable of significantly lower noise levels than the traditional resistor-feedback headstages.

The drawback of using a capacitor is that currents flowing in the feedback loop can cause the voltage across the capacitor to reach the supply voltage. At this point no more currents can be measured unless the capacitor is discharged. To do this the voltage across the capacitor is monitored and discharged whenever the voltage approaches either of the supply rails (e.g. at ±10 V). This resetting of the capacitor voltage is done electronically and takes about 50 μs, during which time the output from the headstage is held constant by a 'sample-and-hold' circuit.

Resetting should be invisible to the experimenter but note that data are lost during the reset time. Loss of 50 μs every second or so is probably quite acceptable (equivalent to a holding current of 10 pA when the feedback capacitor is 1 pF). This would occur when holding a patch with a seal resistance of 10 GΩ at −100 mV. However, a 1 GΩ seal would give a reset every 100 μs which is probably unacceptable. Thus, the capacitor feedback is less useful for whole-cell recording where the holding currents tend to be higher. For whole-cell recordings, capacitor-feedback headstages are switched to a 500 MΩ feedback resistor, since the lower theoretical noise of capacitor feedback will be of little significance in a whole-cell recording.

Despite the added complexity of the differentiator and capacitor discharging circuitry, the capacitor-feedback headstage has several advantages over resistor-feedback technology. The only significant disadvantage is the short reset time during which the headstage does not measure any current. The advantages are significantly lower noise, larger transient current compliance, and wider bandwidth. The value of a lower noise level is obvious. The ability to clamp larger transient currents is valuable to allow faster voltage steps in cells with large capacitance or in experiments with lipid bilayers which typically have a large capacitance (7; see Chapter 2). The wide bandwidth is not of great practical advantage as yet, because most single channel currents would be swamped by the noise seen at bandwidths greater than 5–10 kHz, and this bandwidth is already achievable with resistor-feedback headstages.

5.3 Oscilloscopes

The oscilloscope is still a fundamental part of any electrophysiological recording set-up. It is a cheap and extremely versatile instrument and is the primary means for the experimenter to monitor the progress of the experiment and the quality of the recording. It is neither advisable, nor for single channel recordings even possible, to try to measure data displayed on the oscilloscope during an experiment. It is universally accepted that it is much better to tape record the raw data during an experiment for later review. Some written notes and a comprehensive sound-track commentary of the experiment recorded with the data are then used later as a basis for analysis (Section 5.5, tape recorders).

The best value for money in oscilloscopes are probably the digital storage oscilloscopes which can be bought for less than £1000. Some form of storage oscilloscope is essential so that events which may last for only a few milliseconds can be viewed. If you intend to make computer-controlled on-line experiments it may be tempting to buy some of the software packages that allow a microcomputer to emulate an oscilloscope (e.g. Axotape from Axon Instruments). However, although that sort of software can be useful, it is not a good substitute for an oscilloscope, and the temptation to save money in this way should be avoided.

It is very worthwhile buying a digital storage oscilloscope which has a 'roll' mode to allow slow responses to be viewed 'as they happen'. The old persistence analogue storage oscilloscopes are now too expensive to be justifiable, although in some ways they are preferable because they allow signals to be overlaid on the screen. However, LeCroy make a microprocessor-controlled digital oscilloscope (LeCroy 9310, cost approximately £4000) which mimics the properties of analogue storage. Although expensive, this is probably ideal for a patch-clamp set-up because it allows single channel currents to be observed at high time resolution, yet compared over a relatively long time period by superimposing on the screen. In contrast the standard digital oscilloscopes erase each screenful as the next is produced.

5.4 Amplifiers and filters

A variable amplifier and filter is almost essential to allow flexible analysis of patch-clamp data. Most patch-clamp amplifiers have built-in amplifiers and filters but an additional amplifier/filter combination is very useful in any set-up. Bessel filters are best for single channel data because they do not 'ring' in response to a step input. Useful filter −3 dB (half power) frequencies for patch-clamp data would be 0.2 kHz, 0.5 kHz, 1.0 kHz, 2.0 kHz, 5.0 kHz, and 10 kHz. The final bandwidth of the signal at analysis is important. A useful relationship to know is that analogue Bessel filters and digital Gaussian filters add together to give a final bandwidth f_c according to the relationship in Equation 3.

$$\frac{1}{f_c} = \sqrt{\frac{1}{f_1^2} + \frac{1}{f_2^2} \cdots \frac{1}{f_n^2}}. \qquad [3]$$

Thus, if the Bessel filter on the patch clamp is set to 5 kHz, the Bessel filter on the amplifier is set to 5 kHz and the data are digitally filtered with a

Gaussian at 5 kHz after sampling on to the computer, then the final bandwidth of the data will be approximately 2.89 kHz.

It is generally useful to tape record data with enough gain so that the signal utilizes at least half the dynamic range of the tape recorder and with the filter on a fairly high setting so that no information in the signal is lost. At the same time it may be necessary to filter the data fairly heavily before viewing on the oscilloscope during the experiment. The amplifier should have the facility to apply a variable offset to the data as well as various levels of gain. Useful gain levels might be ×1, ×2, ×5, ×10, ×20, and ×50.

5.5 Tape recorders and data storage

A tape recorder is essential for most types of patch-clamp experiment. The nature of the experiment is that spells of careful preparation are interspersed with sudden periods of intense concentration during data collection. Even for experiments that are controlled by computer it is difficult for the experimenter to concentrate on achieving the best possible recording and ensure the data are captured at the right time and at the best combination of filter and gain settings. Even in the best cases it can be desirable to go back and look at the data several times under different amplifier/filter settings. For all of these reasons a tape recorder is a major aid in making the best use of experimental data. In addition, tape recording is a relatively cheap way to keep a permanent record of all experimental data.

Until about the mid-1980s most patch clampers stored their data in analogue form using an FM tape recorder such as that made by Racal. These tape recorders are quite sophisticated with variable speed (which is proportional to the bandwidth of the recording) and variable input gain. They are also quite expensive, and so a cheaper alternative was developed in the form of a modified video cassette recorder coupled to a digital audio processor (manufactured by Instrutech and also by Vetter, for example). The advantages of this approach are the cheapness of the media and the ability to make long recordings (up to 2 hours) on a single video tape. The machines are also convenient to use with rapid search facilities, remote control, automatic date and time logging on the tape, etc. It should be noted, though, that there is no standard *at all* in the format of the data recorded on video cassette (sampling rates, bandwidth, etc.) and so there is *no* guarantee that a cassette taped on one type of machine can be replayed on any other machine (a good case of 'buyer beware'!).

More recently, modified digital audio tape (DAT) recorders have come into this market. These have advantages similar to that of video recorders for data storage, the main difference being that the tapes are considerably smaller and so easier to store, and data searching is more sophisticated. Both DAT and video recorders adapted for electrophysiological recording can be purchased with computer interfaces which allow direct reading of the tape by computer, usually at a cost of about an extra £1000, but notice that this is comparable to the cost of some purpose-built computer interfaces which, as well as being capable of sampling data from the tape recorder, can provide a variety of options for on-line experimental control.

5.6 Computers and on-line experimental control, data storage, and analysis

By itself the computer is of limited use. It is only when the computer is combined with a suitable *interface* (e.g. Cambridge Electronic Design 1401, Axon Instruments Digidata 1200) and software to drive the interface and analyse the data, that the computer becomes the patch clampers best friend! The computer interface will consist of both analogue-to-digital converters for sampling the data into the computer and digital-to-analogue converters to allow the computer to provide command voltages to the patch clamp, automatically switch solution flow taps, trigger the stimulation of the preparation, and generally communicate with the outside world (see ref. 26 and Chapter 3). Needless to say, this activity can require a prodigious amount of sophisticated software both for experimental control and also for subsequent data analysis. The software can be purchased independently, or along with a computer interface. A good example is the pCLAMP software suite and the Digidata 1200 interface available from Axon Instruments. In addition, there are also a number of different analysis programs available in the academic community for experimental control and analysis of patch-clamp data (26). These software packages can be a quick and convenient way to get started rapidly, but it is often necessary to be able to modify or develop software to suit more specialist needs.

5.7 Vibration isolation system

The mechanical parts of a patch-clamp set-up (microscope, micromanipulators, etc.) usually need to be mounted on some form of vibration isolation system. Unfortunately, there is no escape from environmental vibration wherever one is working. There is no escape from people in other parts of the building banging doors, workmen drilling through the walls, or builders operating a pile-driver down the road. In order to avoid having the tip of the patch electrode shaken (and so losing a whole-cell recording or, at the very least, having the baseline of an isolated patch recording wobbling around) it is necessary to use some form of vibration isolation and damping system. This can be as simple as mounting the microscope and manipulators on a heavy metal plate which rests on tennis balls, squash balls, or inflated bicycle, motor-cycle, or wheelbarrow inner tubes. A more sophisticated approach is to buy an air-damped flotation table available from any supplier of precision optical equipment. These are not cheap (around £2000), but the key is that they feature a heavy ballast plate whose movement is damped using compressed air. Generally it is not necessary to buy the specialist honeycomb table top sold with some flotation tables. Although some applications in optical physics require extremely large and expensive flotation tables, generally the bottom of the specification range is sufficient for patch clamping.

5.8 Microscopes

Nearly all patch clamping involves the use of a microscope to allow visualization of the cell of interest while the pipette tip is manoeuvred on to the cell surface. With isolated cells or tissue-cultured cells an inverted microscope is most commonly used. An inverted microscope also allows easy access to the cells with the patch electrode. For other applications an upright microscope is more suitable and for some, such as patch clamping cells in brain slices (see below), an upright microscope is essential.

5.8.1 Optics

Some form of contrast enhancement is almost essential because living cells are generally of very low contrast. For isolated cells or cells in tissue culture, phase contrast optics may be sufficient. Hoffmann modulation contrast optics are more expensive but will provide sharper resolution, while Nomarski differential interference contrast (DIC) optics is probably ideal for living tissue (27) but adds significantly to the cost of the microscope. Nomarski optics are particularly good for patch clamping because they give a very thin depth of focus. This allows the cell to be viewed in optical slices so that as the patch pipette approaches the cell from above it is very easy to monitor how close the pipette tip is from the cell surface.

5.8.2 Mechanical considerations

The microscope should have a fixed stage so that focusing occurs without moving the preparation relative to the patch pipette. It is important that the microscope is not mounted on rubber feet but instead is firmly fixed to the vibration isolation table or the pipette will crash on to the bottom of the recording chamber every time the focus is adjusted! On a simple system the patch pipette micromanipulator can be mounted directly on the microscope stage. A more elaborate and stable system is to have the stage and manipulators mounted solidly on the vibration isolation table top and entirely separate from the microscope so that manipulation of the microscope or adjusting the focus does not disrupt the recording (see ref. 12).

5.8.3 Patch-clamp recording chamber

The function of the recording chamber is primarily to allow access of the patch pipette to the cells of interest while keeping the cells alive and healthy. A secondary function is usually to allow the controlled application of drugs or other experimental manipulations during a patch-clamp recording. These considerations can mean that the recording chamber is simply a 35 mm culture dish held on the normal microscope stage. At the other end of the sophistication scale the recording chamber may require a flow control system with Peltier temperature control backed up by a complicated array of valves and taps to allow rapid changes of the solution in the bath (28).

The wide diversity of patch-clamp experiments is reflected in the variety of chambers used and in the fact that chambers for special applications are not normally available commercially. This is one of the parts of the system which has to be built, so access to a good mechanical workshop is useful. The recording chamber must be held stably on the microscope stage (e.g. bolted to the stage movement) so that the cells under study can be moved around using the stage movement, but will not move at all once recording has started.

5.9 Micromanipulators

A good quality micromanipulator is essential to allow stable and precise movement of the patch pipette around the cell. The patch-clamp headstage should be mounted directly on top of the manipulator and not on the end of a long rod which will tend to amplify any vibration in the system. Ideally, the fine movement should be done with a remote control: either a hydraulic or a piezoelectric manipulator, and this can be mounted on a coarse manual manipulator for rapid approximate positioning. The coarse manipulator is mounted, in turn, on a solid pillar or platform which should be bolted to the plate of the flotation table for maximum stability. Micromanipulators are available from a variety of sources in the field of optical physics (e.g. Newport has a large selection in their catalogue). Micromanipulators specifically made for electrophysiology are available from Narashige and several other sources. The Narashige water-filler hydraulic micromanipulators are widely used, and Narashige also sell a variety of coarse and fine mechanical manipulators. Fine piezo-electric manipulators are available from Burleigh and a number of optical physics suppliers.

5.10 Grounding and shielding the patch-clamp set-up

It is a fact of life for electrophysiologists that we live in a world full of electromagnetic radiation. Much of this is at high frequencies (such as noise from computer monitors) which are of no significance in the usual bandwidth of patch-clamp recordings (DC to 10 kHz). The biggest problem is line-frequency pick-up (50 Hz or so, and its harmonics: 100 Hz, etc.). Occasional problems arise due to the operation of large electric motors (e.g. lift shafts, air-conditioning systems). To avoid 50 Hz pick-up the patch-clamp set-up must be carefully earthed and shielded (see also refs 12, 8). The theory of grounding and shielding is quite complex. For electrophysiology an empirical approach is most effective, based on the following simple guidelines.

(a) View the patch-clamp current signal on the oscilloscope with the time base on 5 or 10 ms per division and with the time-base trigger set to 'line frequency' (this ensures that 50 Hz line-frequency signals will appear as a standing wave on the oscilloscope screen).

(b) Enclose the microscope, manipulators, and flotation table inside a Faraday cage to shield the patch pipette and headstage from electromagnetic radiation in the room.

(c) Ensure that all metal parts inside the Faraday cage are in electrical continuity by linking each different piece of equipment to a single common ground point (this is known as a star grounding pattern) with reasonably thick earth wire. Ensure that the wire is firmly connected to the equipment (avoid using a maze of spring clips and wires twisted together).

(d) Use a multimeter to measure the resistance between any two metal points in the set-up. This should always be less than 1 Ω.

(e) Connect the single common ground point to the signal ground of the patch-clamp amplifier.

It may or may not be necessary to have a piece of metal shield placed in front of the microscope to remove all trace of 50 Hz interference from the patch-clamp signal. A large piece of cardboard coated with aluminium foil and connected with an alligator clip to the common ground point makes a light shield which is easily removed and replaced during the experiment. With these precautions, no 50 Hz signal should be evident on the current trace when an electrode is in the holder but not in the bath solution (simulating a gigaohm seal). If some 50 Hz interference is still evident, carefully switch on and off each electrical appliance in the room, including any fluorescent lights and attempt to identify the source of the interference. Check that any mains cables are far away from the wire leading the signal from the headstage to the amplifier. A good idea is to insert an extra grounded metal plate (e.g. a 1.5 mm metal plate) between the oscilloscope and the patch-clamp amplifier in the equipment rack (8).

6. Preparation of cells and tissues for patch-clamp recording

Many different cell types have been found to be amenable to patch-clamping after tissue dissociation following enzyme treatment. Cells are either used acutely, following dissociation, or are maintained in primary cell culture. In addition, cell lines have become popular preparations for the study of ion channels and receptor mechanisms using patch-clamp techniques (see Chapters 3 and 5).

In recent years techniques have been developed to allow patch-clamp recordings to be made from cells in tissues which are kept in a state as similar to their *in vivo* state as possible. This has been spectacularly successful in the case of central neurones where techniques for making patch-clamp recordings from cells in brain slices have allowed high resolution studies of synaptic transmission in brain slices to be performed (29, 30).

Factors when considering the advantages and disadvantages of different methods of preparing cells for patch clamping are the following.

(a) Cells that are used acutely are likely to have the same receptors and ion channels as they have *in vivo*.
(b) There is the possibility that proteolytic enzymes used to prepare the tissue for dissociation will damage the receptors or ion channels of interest.
(c) Cells grown in culture may express receptors or ion channels that are not native to the cell *in vivo*, or the distribution or number of receptors and ion channels in the cell membrane may change during cell culture. Receptor transduction mechanisms may be altered compared to the *in vivo* state.
(d) There is the possibility that ion channels expressed by cell lines will suffer genetic mutation.
(e) Whole tissues (e.g. brain slices) may be technically more difficult to deal with than isolated cells, and for most tissues outside the central nervous system it may be almost impossible to make patch-clamp recordings without first using an enzyme treatment to remove connective tissue.
(f) It may be difficult or impossible to identify different cell types once a tissue has been dissociated. This is a particular problem with cells from the central nervous system, where attempts to circumvent this have centred around persuading the cells of interest to take up fluorescent dyes (31) which allow cells to be subsequently identified using a microscope equipped with fluorescence optics.

A large literature exists regarding cell-culture techniques (32–34). A useful discussion of the preparation of autonomic ganglion cells for short-term culture is given by Stansfeld and Mathie (35). Whether cells are to be grown in primary cell culture or used immediately, the basic tissue dissociation procedures are, in principle, similar. However, cell-culture techniques have the additional complexity that all procedures must be performed using sterile techniques.

6.1 Dissociation of peripheral tissues

Peripheral tissues contain varying amounts of extracellular matrix which provides mechanical support for the cells in the tissue. An extreme case would be skin and bone where there is a very large amount of connective tissue (relative to cells). The connective tissue is made up of a variety of glycoproteins, particularly collagen and elastin (important in elastic tissues such as blood vessels). In order to isolate viable cells from the tissue it is necessary to break up the extracellular matrix fibres and this is most often done using collagenase (elastase may also be useful, particularly in tissues which have a dense network of blood vessels).

A good example of the use of collagenase to dissociate cells for patch clamping from a peripheral tissue is the dissociation of bovine adrenal chromaffin cells, a preparation widely used in the study of both secretory mechanisms and receptors and ion channels. The basic principle is to use the blood vessels as a means of giving the enzyme access to as much of the extracellular space as possible. Similar techniques have been used to dissociate heart cells and liver cells. Enzyme solution at 37°C is carried via a fine plastic cannula drawn out to a diameter which just fits the supplying blood vessel (note that in some tissues it is easier to perform a retrograde perfusion via the venous system). Standard cannulae will fit on the end of a plastic disposable 10 ml syringe body (immersed in the water bath) and the perfusion pressure can be regulated by raising or lowering the water bath (20–30 cm is a reasonable pressure head) above the tissue (which can rest inside a sterile plastic disposable 35 mm dish). Monitor the progress of the enzyme action by the rate that the enzyme solution passes through the tissue (the enzyme solution can be recycled to the supply syringe body after collection from the dish in which the tissue is resting).

If it is not practical to perfuse the blood vessels to the tissue and the tissue is larger than 2–3 mm in any dimension then in order to achieve adequate penetration of the tissue with the enzyme, carefully chop the tissue into small fragments (2–3 mm in size) before enzyme treatment. To dissociate small tissues, the whole tissue is immersed in the enzyme, preferably after removal of any connective tissue sheath which may enclose the tissue (35).

Two points to note when using collagenase are:

(a) Collagenase (1 mg/ml, Worthington or Sigma XII) works best when there is a low calcium concentration in the physiological salt solution. A low calcium concentration also helps to dissociate tissues such as liver or heart where the cells are normally joined by tight junctions.

(b) More successful dissociation is often achieved if the collagenase treatment is preceded or followed (opinions vary on which is best) by trypsin treatment (e.g. Sigma XI). If you use trypsin, it is important to follow the trypsin treatment with 10 minutes or so in a trypsin inhibitor (e.g. bovine trypsin inhibitor from Sigma or fetal calf serum from Gibco), or the trypsin will gradually chew up all the cells. There is also some evidence that trypsin is taken up by some cells and can then produce intracellular mayhem too.

Also note that enzyme activity is temperature sensitive. The duration of exposure to the enzyme needed to allow dissociation of healthy cells will thus depend on the temperature. Where the metabolic needs of the tissue for oxygen and glucose can be satisfied at 37°C, then body temperature is probably the most successful temperature for enzyme dissociation. Enzyme exposure time also depends on the type of tissue in use and the age of the animal the tissue came from: some tissues have a much denser connective tissue than others and young animals typically have much less connective tissue.

Where possible, some *gentle* agitation of the tissue during enzyme treatment will almost certainly help. This can be achieved simply by using a very fine plastic tube to bubble 95% O_2/5% CO_2 through the enzyme solution during incubation. Thus, the pH and oxygen supply in a standard bicarbonate-buffered Krebs solution can be maintained during enzyme treatment. The temperature is most easily maintained by keeping the test tube or Eppendorf vial containing the tissue and enzyme solution in a heated water bath. A protocol for enzyme treatment of a peripheral tissue is given in *Protocol 3*.

Protocol 3. Enzyme treatment of a peripheral tissue

Equipment and reagents

- Dissecting instruments to allow tissue isolation
- 250 ml of ice-cold oxygenated physiological saline
- 10 ml of 1 mg/ml trypsin in physiological saline
- 10 ml of 1 mg/ml collagenase + 1 mg/ml trypsin inhibitor
- 10 ml sterile plastic syringe + sterile plastic cannula
- Sterile plastic tubing from 95%/5% O_2/CO_2 supply
- Sterile plastic 35 mm Petri dishes to hold tissue
- Water bath at 37°C (e.g. washing-up basin + aquarium heater)

Method

1. The animal should be killed by a scheduled method and then, if possible, decapitated to drain the blood from the tissues.
2. Quickly expose the tissue of interest and wash the region with cold physiological saline.
3. Carefully dissect the tissue or organ from the body. At this stage it is better to keep the tissue as intact as possible (for example, maintaining the blood vessels if perfusion of the tissue with enzymes is intended).
4. Using fine forceps and scissors remove as much superficial connective tissue as possible without damaging the cells of interest.
5. Perfuse the blood vessels with oxygenated enzyme solutions. Stop perfusion when the tissue shows the first signs of disintegration (10–30 min). At this stage it may be possible to collect the cells in the perfusate. The tissue is then ready for dissociation.
6. Alternatively to step 5, place the whole tissue or chopped tissue in 1 mg/ml trypsin enzyme solution at 37°C and gently agitate with an oxygen bubbler for 15 min. Replace the trypsin enzyme solution with 1 mg/ml collagenase plus 1 mg/ml trypsin inhibitor. Incubate for a further 30–45 min. The tissue is then ready for dissociation.

6.1.1 Tissue dissociation

Dissociation of enzyme-treated tissue is achieved in one or two stages.

Stage 1

Use two pairs of fine watch-makers forceps to gently pull the tissue or chopped tissue fragments apart in 2–5 ml of physiological saline in a 35 ml Petri dish. Gentle shaking of the tissue may release some cells at this point, but usually not enough for experimental purposes.

Stage 2

Use a sterile, fire-polished Pasteur pipette to suck up and then expel repeatedly the tissue pieces in a volume of about 2 ml of physiological saline. The diameter of the Pasteur pipette tip should be as close as possible in size to that of the tissue pieces. *It is critical that this procedure is done gently. Avoid drawing bubbles through the Pasteur pipette.*

Note that the aim is to achieve a small yield of healthy cells (for acute patch-clamp experiments only a few hundred cells may be needed to obtain a reasonable success). Pass the tissue up and down the Pasteur pipette two or three times. If the solution tends to go cloudy, then too much tissue dispersion is occurring. Finally, remove any large lumps of tissue from the cell suspension. The cells are now ready for plating out (see Section 6.3) on coverslips or tissue-culture dishes, prior to immediate use or maintenance in tissue culture. During development of the enzyme treatment and dissociation technique it is useful to examine the tissue under the microscope at each stage of preparation, to monitor the health of the cells and the progress of the dissociation.

6.2 Dissociation of brain tissues

In the central nervous system or in the retina there is little collagen and so collagenase is not used. In fact some areas of the brain, such as cerebellum from young rats, can be dissociated successfully by passing the chopped tissue through a 40 μm nylon mesh without any prior enzyme treatment. Otherwise, when dissociating brain tissue, most procedures use trypsin or papain (both non-specific proteases) with occasionally other enzymes in combination with trypsin such as thermolysin or elastase (particularly in the spinal cord, which is highly vascularized). Papain is preferred to trypsin by some people but great care must be taken to ensure that the enzyme is activated prior to use with cysteine (1 mM) and that the activity (in units/ml of enzyme solution is constant). My personal experience is that trypsin is more reliable.

An effective protocol for enzyme treatment of brain tissue is as given in *Protocol 3*, except that brain slices or chopped tissue pieces of 300–600 μm thickness are necessary to ensure survival of neurones and glia during preparation and usually just one enzyme (e.g. trypsin 1 mg/ml) is used for between

30 and 90 min. A more elaborate slice dissociation method has been published by Kay and Wong (36). Tissue dissociation is achieved by gently passing the tissue pieces up and down a fire-polished Pasteur pipette as described in Section 6.1.1 above.

6.3 Plating out dissociated cells

Healthy cells will stick to collagen-coated (or other treatments: see Section 6.4) 10 mm diameter glass coverslips placed at the bottom of the wells of a multi-well plate. 0.5–1.0 ml of cell suspension is placed in each well and the cells allowed to settle on to the coverslip for 30–90 min. For acute use, cells can be allowed to settle on the bench, while for primary culture, the supernatant should be *carefully* removed after the cells have settled and replaced with cell culture medium: the cells can then be maintained in a standard tissue culture incubator. For cell culture, the dissociation procedure and plating of cells should ideally be done in a laminar flow hood to minimize the chances of infection. However, this is not essential. A quiet corner of the lab with a clean bench surface away from air currents may suffice.

At the start of the experiment, a single coverslip is *carefully* transferred from the multi-well plate to the patch-clamp recording chamber. A good trick here is to use a hypodermic needle which has the tip curved over by tapping on a metal surface to catch the edge of a coverslip in the well and slowly lift it before removing with forceps (avoid ending up with the cells on the underside of the coverslip when this is placed in the recording chamber!).

Growing cells on coverslips in multi-well plates is perhaps the most convenient way to provide cultured cell lines for patch-clamp experiments. Coverslips of around 10 mm diameter will fit easily into the small volume of most patch-clamp recording chambers and can be conveniently stuck to the base of the chamber with a tiny dot of a petroleum jelly such as Vaseline placed at the edge of the coverslip (but take care the jelly does not enter the light path where it will disrupt most contrast-enhancement techniques). Cells grown on coverslips can also be maintained in large cell-culture flasks but this increases the possibility of bacterial infection when the flask is repeatedly accessed each time a new coverslip is needed.

6.4 Immobilizing cells prior to recording

Dissociation procedures provide cells suitable for immediate use. The main problem to be overcome following dissociation is persuading cells to stick to the base of the recording chamber prior to beginning recordings. This is a problem because it is very tedious chasing cells around the recording chamber with a patch electrode and even if cells settle to the bottom of the recording chamber, they will be washed away when perfusion of the recording chamber is begun. Coating coverslips with a variety of positively charged materials is used to encourage cells to stick to coverslips or the base of 35 mm culture dishes. Collagen, laminin, poly-L-lysine, or poly-ornithine are popular substrates.

A simple procedure is to wash a coverslip in a 1% solution of the blue dye alcian blue and then rinse the coverslip with distilled water. The coverslip can then be used immediately (when viewed obliquely the glass takes on a blue tint but otherwise the dye coating is not detectable). The coverslip need not be kept in the bottom of a multi-well plate if the cells are intended to be used immediately. Instead it can be placed directly in the recording chamber and a small 'top hat' of plastic tube (e.g. 1 cm high) sealed over the coverslip with Vaseline (37). The cell suspension is then placed in this chamber and after the cells have settled, the 'top hat' is removed to allow perfusion of the cells in the chamber and recording as usual.

If cells in tissue fragments are to be patch clamped without dissociation, then it is simple to immobilize the tissue (38) in the base of the recording chamber by using a nylon grid (a grid of nylon fibres glued to a piece of platinum wire is laid on top of the tissue as described in Section 7.4 for patch clamping cells in intact brain slices). Note that for any but the thinnest tissue fragments it will be necessary to have an upright microscope to view the cells to be patch clamped on the tissue surface (see also Section 6.5.4)

6.5 Preparation of brain slices for patch-clamp recording

Following the early development of the patch-clamp technique (15) the problem of obtaining a clean access for the patch pipette to the surface of neurones in brain slices seemed insoluble and so the potential advantages of applying patch clamping to brain slices were not achieved immediately. This problem was overcome by the development of techniques to allow patch-clamp recordings to be made from brain slices (38, 39). Brain slices have been widely used for both biochemical and electrophysiological studies and the methods used to prepare brain slices are well documented (for a review see ref. 40). A brief protocol which seems to give good results with a variety of brain areas is given in *Protocol 4* (see also refs 38, 41, 42).

Protocol 4. Preparation of brain slices for patch clamping

Equipment and reagents

- Cylinder of 95%/5% O_2/CO_2
- About 250 ml of ice-cold 'slicing Krebs' solution sitting on crushed ice
- Large scissors
- Small scissors
- Curved, blunt forceps
- No 11 scalpel
- Small spatula (2 mm across)
- Shallow container of ice-cold Krebs solution sitting on crushed ice
- Cyanoacrylate glue
- Large spatula (10 mm across)
- Two fine hypodermic needles or fine dissecting scissors
- Broken and fire-polished Pasteur pipette (opening 3–5 mm)
- Dissecting microscope
- Fibre-optic light source (optional)

Method

1. Kill a rat by a scheduled method and then decapitate the animal.
2. Split the skull along the midline cutting from the base of the skull to forward of the bregma with fine scissors, taking care to avoid damage to the brain underneath.
3. Remove the top half of the skull in two pieces with curved forceps (at this stage it is useful to hemisect the brain using a scalpel or other sharp blade).
4. Remove the brain using the small spatula to cut the optic nerves and cranial nerves and lift the brain out of the skull and into ice-cold physiological solution.

These first four steps should all be done within 60 sec of decapitation (the ice-cold solution should be so cold that it contains many ice crystals and to maintain this temperature the container should be sitting on ice and bubbled with O_2/CO_2.

5. Pause for 3–5 min while the tissue cools down.
6. Trim or block off the tissue using careful clean cuts with a sharp scalpel blade in preparation for glueing to the stage of the tissue slicer. Avoid squeezing or otherwise deforming the tissue at this stage.
7. Apply a *thin* layer of cyanoacrylate glue (Superglue®) to the stage of the tissue slicer and then gently place the tissue at the correct orientation on to the glue (the most common orientation is to have the region of interest near to the blade). Immediately pour ice-cold solution over the tissue until it is submerged.
8. Cut slices (100–300 μm thick) with a vibrating slicer. Fine dissecting scissors or two hypodermic needles can be used to dissect out an area of brain from each brain slice.
9. Using a Pasteur pipette cut and fire-polished to an opening of 3–5 mm across, transfer each slice as it is produced to the slice-holding chamber, which should be in a water bath at 32–35 °C with a good steady flow of O_2/CO_2 bubbling through the solution.
10. Incubate the slices at 35 °C for 30–60 min before beginning recordings.

6.5.1 Notes on making brain slices

i. Time is of the essence

The time from decapitation till immersion of the brain in cold solution is critical (< 1 min). It is perhaps easier to make healthy slices from younger animals (e.g. less than three weeks old) where the skull is soft and can be removed rapidly.

ii. Tissue slicers

The tissue slicer should be able to vibrate at sufficient frequency (around 40 Hz) and with a long enough stroke (1–2 mm) to cut cleanly through the tissue. A simple manual movement is sufficient to move the tissue block towards the oscillating blade (e.g. Vibroslice, Camden Instruments, UK). Some slicers have a Peltier-cooled stage to maintain the tissue close to 0°C during slicing although it is sufficient to use frozen Krebs solution in the bottom of the slicing chamber (or make Krebs ice-cubes) to ensure the tissue stays cold during slicing.

iii. Slicer blades

Blades used for brain slicing need to be very sharp (some companies sell blades specifically for slicing: e.g. Camden Instruments). High-carbon steel blades are preferable to stainless steel razor blades (a high-carbon steel is magnetic and brittle and will break with a sharp snap). Some blades may be coated with oil or wax to prevent corrosion and so it may be worth carefully cleaning the blade with alcohol to avoid possible contamination of the slice.

iv. Slicing solutions

Slices are made with a standard extracellular Krebs solution. For example NaCl, 125 mM; KCl, 2.5 mM; $CaCl_2$, 2 mM; $NaHCO_3$, 26 mM; NaH_2PO_4, 1.25 mM; $MgCl_2$, 1 mM, glucose, 25 mM; pH 7.4 when bubbled with 95% O_2 and 5% CO_2. The exact composition of the Krebs solution varies between laboratories, particularly in the concentrations of Ca^{2+}, $NaHCO_3$, and glucose. To avoid epileptiform activity in the slice, it may be important to keep the K^+ concentration lower than 3 mM and a high glucose concentration may improve neurone survival.

v. Slice incubation chamber

The incubation chamber must be able to provide a good circulation of freshly oxygenated solution around the slices. It should be possible to place or remove a slice from the chamber without disturbing any of the other slices. Preferably the chamber should be simple to make and clean. A simple chamber made from a standard 100 ml beaker is described by Edwards and Konnerth (41) and Gibb and Edwards (42).

6.5.2 Immobilizing the slice in the recording chamber

The slice should be immobilized during recording so that no movements occur as a result of the solution flowing through the bath. Flow rates of between 1 ml and 3 ml per min for bath volumes of less than 1 ml are sufficient. Many people find that a grid (described in ref. 38) made of flattened platinum wire with single nylon strands glued across it with cyanoacrylate glue works well for holding the slice in the recording chamber.

6.5.3 How healthy is the slice?

Visualizing the cells in slices allows an immediate assessment to be made and cells suitable for recording to be carefully picked. A healthy slice under the microscope is a fantastic sight! In a hippocampal slice, for example, lots of bright shiny cells should be visible with a variety of cell dendrites and different cell morphologies evident. If the slice does not contain many bright cells, but instead is uniformly dark with many round opaque cells with clearly visible nuclei evident, then the slice should be discarded (see also ref. 41 for a discussion of assessing the slice visually). Selecting a healthy cell suitable for patch clamping requires experience, as how the cells look depends on the way the microscope is adjusted. Cells should be smooth with a clear outline and have a 'soft' appearance (41). Cells which appear very shiny or 'hard' in appearance do not make seals easily and if observed for some time, appear to die gradually. Dead cells are opaque with visible nuclei and are often swollen and round.

6.5.4 Microscope requirements

An upright microscope is needed for patch clamping cells in slices because the thickness of the slice prevents viewing of the surface of the slice (where the patch electrode can gain access) from below. The microscope should have a high numerical aperture (e.g. 0.75) × 40 water immersion objective. If this is combined with Nomarski differential interference contrast optics then it should be possible to visualize neurones and their dendrites in the slice (looking from above). For Nomarski optics to be effective, however, the maximum slice thickness is around 300 μm. The thicker the slice, the more the light is scattered on passing through the slice and the dimmer and lower the resolution of the image. On the other hand, it is more difficult to obtain healthy thin slices: slices 200–300 μm thick are usually a good compromise.

It should be noted that for best results the numerical aperture of the condenser lens should always be as high or higher (0.9 for example) than that of the objective. This generally means that the working distance of the condenser will allow only a thin glass coverslip or glass base for the recording chamber, if the light from the condenser is to be focused properly on the surface of the slice (plastic chambers, although fine for phase contrast optics, destroy Nomarski imaging). Whatever the precise optical arrangement it is essential that good microscopic practice is followed (27). In particular, good Köhler illumination must be set up with the condenser adjusted to focus the light source diaphragm exactly in the plane of the cells of interest. For DIC optics, the polarizers should be 90° to each other and the analyser adjusted for optimum image quality. Secondary diaphragms in the condenser are then used to cut down the light entering the tissue and so improve the sharpness of the image.

It is often less tiring to view the image using a CCD camera in combination with a standard monitor (27). These are relatively cheap (e.g. from RS Components) and the smaller cameras (<300 g) are light enough to mount directly on top of the microscope trinocular head without applying too much weight to the microscope focusing mechanism. It is generally important to ensure that the camera is insulated from the microscope to avoid conducting interference into the patch clamp signal. Use of a CCD camera also allows either analogue or digital contrast enhancement techniques to be applied to the video signal before viewing (simple and effective contrast enhancement circuits are available from RS Components). These techniques can make a substantial improvement to the quality of the microscope image.

6.5.5 Cleaning cells before patching

Cells are cleaned using a blunt patch pipette (tip diameter 3–10 μm according to the size of cell being cleaned). This pipette is inserted into the bath without filling and will fill with some bath solution by capillary action. The cleaning pipette is brought up close to the surface of the slice and then positive pressure used to produce a gentle stream of solution which will break up the surface of the slice over the cell of interest. After a few seconds of positive pressure, light suction can be used to remove the debris overlying the cell. Numerous investigators have now shown (24, 39) that cleaning cells in a slice in this way is not essential for successful patch clamping. In many situations it may be sufficient to use high positive pressure in the pipette to effect cleaning of the cell as the pipette tip approaches the cell in the slice. Provided the positive pressure is not removed until the pipette tip is against the cell surface, there is a good chance of making a successful recording.

6.5.6 Blind patching

This method was developed in experiments on turtle cortex slices (39). In marked contrast to the method described above, no attempt is made to visualize individual cells. Instead the pipette is slowly advanced ('blindly') into the slice while positive pressure is applied to the back of the pipette (see also ref. 43). Contact of the pipette with the slice produces a change in resistance which is ignored and the pipette is advanced further into the slice. A cell is detected by an increase in pipette resistance (decrease in t-pulse response) and then the positive pressure is removed, suction applied to the pipette, and a seal attempted in the usual way.

The advantages of the blind patching method are that it requires no sophisticated microscopy (a dissecting microscope is used to guide the pipette on to the right area of the slice) and so it is cheaper to set up. No cleaning of the slice is necessary before recording, so saving time. Cells are not disrupted by the cleaning procedure which could conceivably damage synapses on the cell body and cells deep in the slice can be recorded from. The disadvantage of this method is that cells are not identified visually.

6.6 Patch clamping *Xenopus* oocytes

Methods for expressing receptors and ion channels in *Xenopus* oocytes are described in detail in Chapter 5. Methods for making patch-clamp recordings from oocytes were developed by Methfessel *et al.* (44) who give a detailed description of the method. Over the past 10 years or so oocytes have become established as a valuable expression system for the study of receptors and ion channels. The ability to make large macroscopic recordings from the whole oocyte or fine single-channel recordings from excised patches of oocyte membrane (*Figure 4*) means that this preparation will continue to be a popular option for patch-clamp studies directed towards the structure–function aspects of ion channel function.

6.6.1 Removal of the oocyte vitelline membrane

The oocyte follicle cell layer is generally removed by collagenase treatment (1 mg/ml for 1 h at 19°C) followed by manually peeling each oocyte with fine forceps. Oocytes injected with mRNA or cDNA are tested for expression 2–5 days after injection with a two-microelectrode voltage-clamp recording system. Oocytes which have positive responses large enough to suggest a channel density of greater than 1 μm^{-1} are suitable for patch-clamp recording (one of the potentially useful features about both oocytes and transfected cell lines is that the density of channels in a membrane patch can be chosen by varying the time after injection or transfection to begin recording. (Methods for oocyte preparation and injection are described in detail in Chapter 5.)

In order to make patch-clamp recordings, the oocyte vitelline membrane must be removed to allow access of the patch pipette to the oocyte plasma membrane. The vitelline membrane is a tough connective tissue layer which lies between the oocyte follicle cell layer and the plasma membrane. It is extremely useful in supporting the oocyte during injection and during media changes while in culture. Once the vitelline membrane is removed the oocyte is extremely fragile: it will not tolerate exposure to the air and rapidly sticks to plastics and so will tear if sucked off the chamber once it has stuck down.

Removal of the vitelline membrane is achieved by first immersing the oocyte in a hypertonic 'stripping solution' of K-gluconate, 200 mM; EGTA, 10 mM; Hepes–KOH, 10 mM; $MgCl_2$, 1 mM; KC1, 20 mM for 2–3 min to allow the oocyte to shrink inside the vitelline membrane. This is done in a clean 35 mm Petri dish (note that scratches on the dish bottom caused by forceps tips can easily rupture the oocyte membrane once the vitelline membrane has been removed). Oocytes are viewed with a dissecting microscope under ×25 or ×50 magnification and illuminated obliquely from underneath. The vitelline membrane looks like a covering of cling film. It is often difficult to see, but gentle prodding of the oocyte surface usually exposes the membrane. Then gently pinch a fold of vitelline membrane with the forceps. Once the vitelline membrane is firmly held with a pair of forceps, gently slide a second pair of forceps just under the tips of the first pair and gripping the vitelline membrane with both forceps, pull apart to tear the vitelline membrane open. The oocyte will then easily roll out of the membrane. The oocyte is then ready for patch clamping and can be removed from the hypertonic 'stripping' solution and returned to normal frog saline.

6.6.2 Patch-clamp recordings from oocytes

The oocyte can be patched while viewed under a dissecting microscope or on a normal microscope with a low power objective (for example ×2 or ×6). The surface of the oocyte looks something like a partial eclipse. It should look perfectly smooth and undamaged. Once the plasma membrane is broken and the oocyte yolk is spilled, it is very difficult to obtain a gigaohm seal. Patch-clamp recording is done in the usual way, except it is not usually possible to see when the patch pipette is about to touch the surface of the oocyte. Therefore, patch clamping the oocyte is done blind with the patch pipette being very slowly advanced over the last few hundred micrometres to the oocyte. The moment of contact is registered by a reduction in the size of the t-pulse. Release of positive pressure and sealing is then done in the usual way. Methfessel *et al.* (44) note that stretch-activated channel activity is much more common in cell-attached or inside-out than in the outside-out patch configuration when recording from oocytes. Therefore, single-channel recordings will be much cleaner when using outside-out patches.

With oocytes, each patch configuration is formed in the usual way. It is worth noting that with oocytes, breakthrough to the whole-cell recording mode can be disconcerting: the very large size of the oocyte means that it is impossible for the patch-clamp amplifier to clamp the oocyte membrane potential and so the current signal on the amplifier goes into saturation on breakthrough in just the same way as if the seal had been lost. Even when the pipette potential is set at the resting potential, the massive oocyte capacitance means that there is no obvious cell capacity transient. However, slowly pulling the electrode away from the oocyte will nearly always produce the reassuring signs of formation of an outside-out patch (*Figure 4*). In general, provided the oocytes are healthy, they give a high success rate in obtaining seals and provide stable patches which can be used for prolonged recordings.

7. Conclusion

Patch-clamp recording is a technique which has experienced a rapid and sustained rate of development since its discovery almost twenty years ago. This development seems to have been fuelled partly by constant improvements in instrumentation for the recording and analysis of patch-clamp signals but mainly by the imagination of experimenters who continue to apply the technique to an ever-widening range of biological questions. The principles of the technique now seem well established and it should become increasingly easy for students in any biological science to gain experience of the technique and its many possibilities.

References

1. Neher, E. and Sakmann, B. (1976). *Nature* **260**, 799.
2. Neher, E. (1992). *Neuron* **8**, 605.
3. Sakmann, B. (1992). *Neuron* **8**, 613.
4. Sakmann, B. and Neher, E. (ed.) (1983). *Single-channel recording*. Plenum Press, London.
5. Rudy, B. and Iversen, L. E. (ed.) (1992). *Methods in enzymology*, Vol. 207. Academic Press, New York.
6. Gurney, A. M. (1990). In *Receptor–effector coupling: a practical approach* (ed. E. C. Hulme), p. 55. IRL Press, Oxford.
7. Finkel, A. (1991). In *Molecular neurobiology: a practical approach* (ed. J. Chad and H. Wheal), p. 3. IRL Press, Oxford.
8. Sherman-Gold, R. (ed.) (1993). *The axon guide*. Axon Instruments Inc., Foster City, CA.
9. Ogden, D. (ed.) (1994). *Microelectrode techniques: the Plymouth workshop handbook*. Company of Biologists Ltd, Cambridge, UK.
10. Rae, R. A. and Levis, J. L. (1992). In *Methods in enzymology* (ed. B. Rudy and L. E. Iversen), Vol. 207, p. 67. Academic Press, London.
11. Sigworth, F. J. (1983). In *Single-channel recording* (ed. B. Sakmann and E. Neher), p. 3. Plenum, Press, London.
12. Levis, J. L. and Rae, R. A. (1992). In *Methods in enzymology* (ed. B. Rudy and L. E. Iversen), Vol. 207, p. 14. Academic Press, London.
13. Milton, R. L. and Caldwell, J. H. (1990). *Pflüg. Arch. Eur. J. Physiol.* **416**, 758.
14. Horn, R. (1993). *Biophys. J.* **60**, 329.
15. Hamill, O. P., Marty, A., Neher, E., Sakmann, B., and Sigworth, F. J. (1981). *Pflüg. Arch. Eur. J. Physiol.* **391**, 85.
16. Marty, A. and Neher, E. (1983). In *Single-channel recording* (ed. B. Sakmann and E. Neher), p. 107. Plenum Press, London.
17. Penner, R. and Neher, E. (1989). *Trends Neurosci.* **12**, 159.
18. Lindau, M. and Neher, E. (1988). *Pflüg. Arch. Eur. J. Physiol.* **411**, 137.
19. Neher, E. (1992). In *Methods in enzymology* (ed. B. Rudy and L. E. Iversen), Vol. 207, p. 67. Academic Press, London.
20. Pusch, M. and Neher, E. (1988). *Pflüg. Arch. Eur. J. Physiol.* **411**, 204.
21. Horn, R. and Marty, A. (1988). *J. Gen. Physiol.* **92**, 145.
22. Horn, R. and Korn, S. J. (1992). In *Methods in enzymology* (ed. B. Rudy and L. E. Iversen), Vol. 207, p. 149. Academic Press, London.
23. Sakmann, B. and Neher, E. (1983). In *Single-channel recording* (ed. B. Sakmann and E. Neher), p. 37. Plenum Press, London.
24. Stuart, G. J., Dodt, H. U., and Sakmann, B. (1993). *Pflüg. Arch. Eur. J. Physiol.* **433**, 511.
25. Corey, D. and Stevens, C. F. (1983). In *Single-channel recording* (ed. B. Sakmann and E. Neher), p. 53. Plenum Press, London.

26. Dempster, J. (1992). In *Microcomputers in physiology: a practical approach* (ed. P. J. Fraser), p. 51. IRL Press, Oxford.
27. Lacey, A. J. (ed.) (1989). In *Light microscopy in biology: a practical approach.* IRL Press, Oxford.
28. Forsythe, I. D. (1987). *J. Neurosci. Meth.* **25**, 19.
29. Edwards, F. A., Konnerth, A., and Sakmann, B. (1990). *J. Physiol.* **430**, 213.
30. Stern, P., Edwards, F. A., and Sakmann, B. (1992). *J. Physiol.* **449**, 247.
31. Gibb, A. J. and Walmsley, B. (1987). *J. Neurosci. Meth.* **20**, 35.
32. Adams, R. L. P. (1990). *Cell culture for biochemists.* Elsevier, New York.
33. Freshney, R. I. (1992). *Animal cell culture: a practical approach.* IRL Press, Oxford.
34. Freshney, R. I. (1994). *Culture of animal cells: a manual of basic technique*, 3rd edn. Wiley-Liss, New York.
35. Stansfeld, C. and Mathie, A. (1993). In *Electrophysiology: a practical approach* (ed. D. I. Wallis), p. 3. IRL Press, Oxford.
36. Kay, A. R. and Wong, R. K. S. (1986). *J. Neurosci. Meth.* **16**, 227.
37. Gibb, A. J. and Colquhoun, D. (1992). *J. Physiol.* **456**, 143.
38. Edwards, F. A., Konnerth, A., Sakmann, B., and Takahashi, T. (1989). *Pflüg. Arch. Eur. J. Physiol.* **414**, 600.
39. Blanton, M. G., Lo Turco, J. J., and Kriegstein, A. R. (1989). *J. Neurosci. Meth.* **30**, 203.
40. Alger, B. E., Dhanjal, S. S., Dingledine, R., *et al.* (1984). In *Brain slices* (ed. R. Dingledine), p. 381. Plenum Press, New York.
41. Edwards, F. A. and Konnerth, A. (1992). In *Methods in enzymology* (ed. B. Rudy and I. E. Iversen), Vol. 207, p. 208. Academic Press, New York.
42. Gibb, A. J. and Edwards, F. A. (1994). In *Microelectrode techniques: the Plymouth workshop handbook* (ed. D. Ogden). Company of Biologists Ltd, Cambridge, UK.
43. Madison, D. V. (1991). In *Cellular neurobiology: a practical approach* (ed. J. Chad and H. Wheal), p. 137. IRL Press, Oxford.
44. Methfessel, C., Witzemann, V., Takahashi, T., Mishina, M., Numa, S., and Sakmann, B. (1986). *Pflüg. Arch. Eur. J. Physiol.* **407**, 577.
45. Gibb, A. J., Carr, J. A., and Colquhoun, D. (1993). *J. Physiol.* **459**, 404.
46. Edwards, F. A., Gibb, A. J., and Colquhoun, D. (1992). *Nature* **359**, 144.
47. Gibb, A. J., Kojima, H., Carr, A. J., and Colquhoun, D. (1990). *Proc. Roy. Soc. Lond. B* **242**, 108.

2

The measurement of the function of ion channels reconstituted into artificial membranes

ALAN J. WILLIAMS

1. Introduction

As is clear from the first chapter in this book, the function of individual ion channels or populations of ion channels in the cell surface membrane can be monitored in tissues and cells using voltage-clamp techniques. This chapter describes a more biochemical approach to the study of ion channel function. The procedures outlined here are commonly referred to as 'ion channel reconstitution', and involve the incorporation of channel proteins into artificial phospholipid bilayers so permitting the characterization of channel function under voltage-clamp conditions. In practice, this term is applied both to procedures which involve true reconstitution (i.e. studies involving purified channel proteins), and to protocols in which intact isolated membrane vesicles, containing the channel protein of interest, are incorporated into an artificial membrane. For more information on (membrane) protein purification, see refs 1 and 2 in conjunction with Chapter 4.

Ion channel reconstitution provides

- an alternative approach to the measurement of the function of ion channels under voltage clamp conditions;
- experimental strategies which make possible the investigation of channels not accessible to conventional electrophysiological approaches (for example the channels of intracellular membrane systems);
- methods for assessment of the properties of purified ion channel protein molecules.

I will describe a number of methods currently in use for the study of ion channel properties in reconstituted systems. The use of this approach is not as widespread as the more conventional electrophysiological techniques of whole-cell voltage clamp and patch clamping and, although some apparatus is commercially available, it is common for various items of equipment used in these experiments to be 'home made'. For this reason I will provide some details of equipment design together with standard experimental protocols; other protocols and opinions can be found in ref. 3.

2. The production of artificial phospholipid bilayers

The first step in any ion channel reconstitution experiment is the construction of a suitable artificial planar phospholipid bilayer into which the channel protein can be incorporated. All workers who use ion channel reconstitution as a routine technique would probably agree that the formation of a stable phospholipid bilayer is the most important and often the most frustrating step in the entire procedure. As noted by Stephen White (4), the difficulties associated with bilayer formation are often reduced by an understanding of the physicochemical nature of the system. For this reason, any fledgling 'reconsituter' should familiarize himself or herself with these concepts (4).

In all ion channel reconstitution experiments, the bilayer is formed across a hole which connects two fluid-filled chambers. The dimensions of the bilayer and the nature of the support on which it is made vary considerably depending on the particular system employed, but in all cases bilayers are formed using one of two basic protocols.

2.1 Bilayers formed by spreading phospholipid dispersions (painted bilayers)

The formation of planar bilayers by the spreading of phospholipid dispersions was introduced in the early 1960s by Mueller and colleagues (5, 6) and is probably the procedure used most commonly today.

2.1.1 Apparatus and reagents

i. The bilayer block and cup

The apparatus used in our laboratory for the formation of painted bilayers is shown in *Figure 1*. The system consists of two chambers. The block (A) is machined to produce an oblong chamber (*trans*) which connects with a circular well into which the cup (B) fits. The cup contains a well (*cis*). The face of the cup adjacent to the well is machined to produce a thin wall or septum (approximately 200 μm) and the hole (200 μm diameter) on which the bilayer is formed is drilled through this wall. In operation, the cup is positioned in the block with the hole in the cup facing the *trans* chamber so that the *cis* and *trans* chambers are connected via the hole. The volumes of the *cis* and *trans* chambers used in our laboratory are approximately 0.5 ml and 1.0 ml respectively. Clearly, the volume of either chamber can be varied during construction, as can the diameter of the hole on which the bilayer is formed.

The front face of the block is glass and provides an unobstructed view of

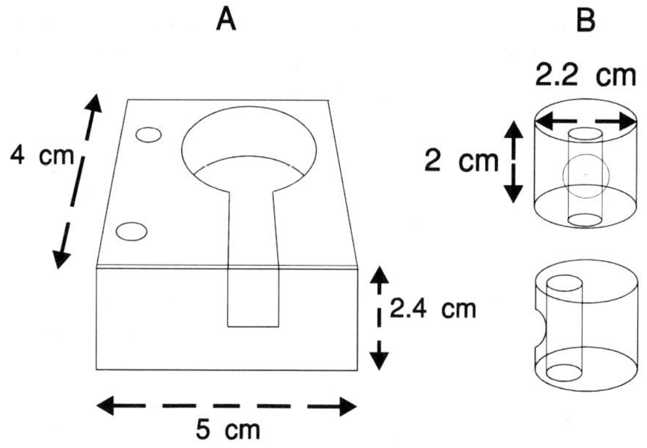

Figure 1. Experimental chambers for the formation of bilayers by spreading dispersions of phospholipids (see Section 2.1 for details). Panel A shows the block containing the *trans* chamber and the circular hole into which the cup containing the *cis* chamber fits. Panel B shows two views of the cup: the lower (side-on) view illustrates the machining of the front wall of the cup to create the septum. The contents of the chambers can be stirred by small Teflon-coated magnetic fleas driven by a magnetic stirrer located beneath the block (Section 5.1). The measurements given on the figure are approximate.

the hole once the cup is positioned in the block. The block may be constructed from any plastic; we use Perspex but materials such as Teflon are equally suitable (7). The material used for the cup can have a major influence on the ease of formation and the stability of the planar phospholipid bilayer. We use a styrene copolymer (Polypenco Q 2005 from Amari Plastics). Cups manufactured from this material provide a durable support on which lipid films thin readily to form mechanically and electrically stable bilayers. Other plastics which have been used as supports for painted bilayers include Teflon (7) and polycarbonate (8).

In addition to the *trans* chamber, the block contains two small wells (*Figure 1*). During use these wells are filled with an electrolyte solution and provide a means of coupling the *cis* and *trans* chambers to the head amplifier via agar bridges and silver–silver chloride electrodes (see Sections 5.1 and 6.4).

ii. The 'painting stick'

To form a planar bilayer, a dispersion of purified phospholipids is drawn across the hole in the cup using some form of implement. The design of the 'painting stick' varies enormously from laboratory to laboratory. Most workers use some form of plastic rod or 'nib' although others use a small, plucked paintbrush or an air bubble at the end of a capillary tube.

iii. The lipid

The quality and condition of the phospholipid used to form the planar bilayer are critical. Painted bilayers can be constructed from either one or a mixture of purified phospholipids. The phospholipid composition of the final bilayer may influence the ease of channel incorporation (see Section 3.1.1) and/or the properties of the channel following reconstitution. Pure native and synthetic phospholipids can be obtained from Avanti Polar Lipids. Phospholipids should be stored refrigerated as stock solutions in chloroform, preferably at −70°C. Painted bilayers are formed from dispersions of phospholipids in a non-polar solvent such as *n*-decane or squalene (*Protocol 1*).

Protocol 1. Making the phospholipid dispersion

Equipment and reagents

- Stock solutions of one or more pure phospholipids in chloroform (100 mM)
- Non-polar solvent, e.g. *n*-decane (D90-1, 99+%, Aldrich Chemical Co.)
- Nitrogen gas cylinder

Method

1. Mix stock phospholipids in small glass or Eppendorf tube to give the desired final composition (e.g. 70% phosphatidyl ethanolamine, 30% phosphatidyl serine). Use sufficient stock solutions to give a final volume of dispersion of 10–50 μl.
2. Evaporate the chloroform under a stream of nitrogen.
3. Add *n*-decane to the dried phospholipids to give a final phospholipid concentration of 20–50 mM.
4. Stir the suspension vigorously (vortex) to create a dispersion. Seal the tube and store it on ice.[a] Discard the dispersion at the end of the day.

[a] Note that some synthetic liquids become 'gel'-like on ice, and should be kept at room temperature.

iv. Solutions

Before bilayer formation, the *cis* and *trans* chambers should contain some form of electrolyte solution, usually one of the eventual experimental solutions. Since the chambers are physically linked, the solution should be the same in both chambers.

2.1.2 Painting the bilayer

Before attempting to form the bilayer it is important that the surface of the cup around the hole is 'primed' with the phospholipid dispersion (produced according to *Protocol 1*). After application, the priming dispersion should be dried with a stream of nitrogen.

A detailed description of the physicochemical process involved in bilayer formation from a dispersion of phospholipids can be found in (4); a brief description is given here. Following priming of the support around the hole, the cup is positioned in the block (see Section 2.1.1) and the *cis* and *trans* chambers are filled with an appropriate electrolyte solution. A quantity of phospholipid dispersion is then drawn across the hole using the painting stick. Initially the film will be several μm thick. The film will be in equilibrium with a much larger annulus or torus of dispersion formed at the perimeter of the hole (the Plateau–Gibbs border). The film should thin spontaneously to form a phospholipid bilayer. Thinning is primarily driven by Plateau–Gibbs border suction but, as the film thins, London–van der Waals attraction between the aqueous phases on either side of the film provide an additional force. The size of the hole can influence the ease with which thinning occurs. We find it difficult to obtain reproducible thinning on holes in styrene copolymer cups with diameters below 100 μm. Alvarez (7) reports similar difficulties with 100 μm diameter holes in Teflon cups.

2.1.3 Monitoring thinning of the bilayer

Bilayers spread from dispersions of phospholipids in non-polar solvents will take some time to form. The success of an ion channel reconstitution experiment depends entirely upon the thinning of the film to form a bilayer. It is not possible to monitor channel function in a film that will not thin. Therefore it is vital that the investigator knows when he or she is in possession of a bilayer. There are two methods for monitoring thinning.

i. Membrane capacitance

An indication of the thickness of the film formed across the hole can be obtained by monitoring capacitance. As the film thins and the area of the hole covered by a bilayer increases, so the observed capacitance will increase. Capacitance can be estimated using the voltage clamp and current monitoring systems described in Section 5.1. A low frequency (1–10 Hz), low amplitude (5–20 mV peak-to-peak) triangular wave is used. Prior to the application of the lipid dispersion to the hole, the electrolyte solutions in the *cis* and *trans* chambers are electrically coupled through the hole; under these conditions application of the oscillating voltage command to the system will produce saturating fluctuations in current. As the phospholipid dispersion is painted across the hole, the hole becomes blocked and small deflections from zero current are seen with each transition of the voltage clamp. With time, as the film thins, a square wave of current will occur with each voltage transition. The size of this current deflection will continue to increase until the maximum area of bilayer is reached. The final capacitance of the bilayer should be approximately 0.4 $\mu F/cm^2$ (7). Quantification of membrane capacitance requires calibration and determination of the surface area of the bilayer (4). In practice, bilayer formation is often determined empirically. The investigator relies upon past experience to judge the magnitude of current fluctuation which corresponds to a capacitance indicating an area of bilayer which will be stable and allow efficient incorporation of channels.

ii. Observation of the film in reflected light

The thinning of the film can be followed by observing light reflected from the structure using an appropriately positioned lens or low-power microscope. When first spread, the phospholipid film is thick and reflects considerable amounts of light, appearing as a multicoloured structure. As a result of the buoyancy of the forming dispersion, the bilayer starts to form at the bottom of the hole and spreads upwards (4). The amount of light reflected from the film decreases as it thins. When a bilayer is produced, the structure is so thin (25–50 Å) that it reflects no light and appears black. This property gives rise to the term black lipid membrane ('BLM').

iii. Problems associated with film formation and thinning

There are two common problems encountered during bilayer formation:

- it is not possible to form a film across the hole;
- the film will not thin to form a bilayer.

Difficulty in forming a film usually indicates inadequate priming of the area of the cup around the hole. Problems with film thinning often arise when the hole is too heavily primed or when repeated painting results in a build up of excess lipid dispersion around the hole. The solution to both of these problems is to remove the cup, wash the original priming dispersion off (Section 6.1), and re-prime the hole (Section 2.1.2).

2.2 Bilayers formed from monolayers (folded bilayers and bilayers on patch pipettes)

Bilayers made by the spreading of phospholipid dispersions inevitably contain non-polar solvent. The second general technique for the formation of planar phospholipid bilayers results in membranes that contain somewhat less solvent and this may be an advantage for some applications. In this approach a bilayer is produced from two phospholipid monolayers formed at an air–water interface.

2.2.1 Formation of a phospholipid monolayer

Under appropriate conditions it is possible to form a monolayer of phospholipid molecules at the interface of an aqueous solution with the air; the phospholipid molecules orientate so that the polar heads of the molecules interact with the aqueous environment and the hydrophobic tail of the molecules stick into the air. Monolayers can be formed using either of the following methods.

i. Application in a volatile solvent

Monolayers can be formed by applying phospholipids in a volatile solvent, such as *n*-pentane, to the surface of an aqueous solution (see *Protocol 2* and refs 7–9). The monolayer forms at the air–water interface as the solvent evaporates.

Protocol 2. Monolayer formation from phospholipids in *n*-pentane

Equipment and reagents

- Stock solutions of one or more pure phospholipids in chloroform (100 mM)
- nitrogen gas cylinder
- *n*-pentane (BDH, Merck)

Method

1. Mix stock phospholipids in small glass tube to provide the required final composition.
2. Evaporate the chloroform under a stream of nitrogen.
3. Add *n*-pentane to give a final phospholipid concentration of 10 mM.
4. Seal the tube and store it on ice.
5. Apply the *n*-pentane solution of phospholipids to the surface of an aqueous solution and allow at least 5 min for the evaporation of the solvent before attempting to form a bilayer (Section 2.2.2). The quantity of phospholipid solution added for monolayer formation will vary with the surface area of the chamber and should be determined empirically.

ii. Monolayers derived from membrane vesicles

Monolayers may also be generated from liposomes, mixtures of liposomes and native membrane vesicles, or liposomes and purified channel proteins. Membrane vesicles are added to an aqueous electrolyte solution and following a period of equilibration, a monolayer derived from the membrane vesicles forms at the surface of the aqueous solution. Bilayer formation requires the surface pressure of the monolayers from which it is formed to be $\geqslant 25$ dyne/cm (12). The complex methodology and specialized equipment required to undertake these experiments may well be responsible for the fairly limited use of the technique. In practice, the protocol has been used by two groups to study the properties of the nicotinic acetylcholine receptor. Specific protocols are not provided here; interested readers are directed to publications from these groups (12–14, 16).

2.2.2 Formation of the bilayer from the monolayer

Phospholipid monolayers created using the protocols described in Section 2.2.1 can be used to form bilayers using one of two approaches.

i. Monolayer folding

Design of the block

In this method, a bilayer is formed as the result of the apposition of the hydrophobic faces of two monolayers made at an air–water interface (see *Protocol 3*). The apparatus used in these experiments is very similar to that described in Section 2.1.1 for the formation of bilayers from dispersions of phospholipids. In both cases the bilayer is created on a hole in a septum between two chambers. For monolayer folding, the septum is not formed from a rigid plastic support but consists of a thin (10–25 μm) sheet of Teflon supported between two thicker Teflon membranes (Teflon sheet can be obtained from Goodfellow Metals). The septum is fixed between two Teflon chambers as shown in *Figure 2*. The Teflon septum is held in position on a central partition using some form of insulating sticky substance such as petroleum jelly. Care must be taken to prevent physical and electrical leakage around the septum. A highly refined version of this basic design is described by Montal *et al.* (16).

Manufacture of the hole

Holes in the septum over which the bilayer is created can be made by simply punching a hole in the septum using a modified hypodermic needle (16). The size of the hole can be varied by using different sized needles. A more reliable method involves melting the Teflon with a spark from a car ignition coil (17). The first step in this procedure is to make a small defect in the Teflon septum using a needle held in a micromanipulator. The septum is placed on a sheet of aluminium and the needle lowered until it presses gently into the Teflon. The hole is then produced by discharging a spark from the needle to the

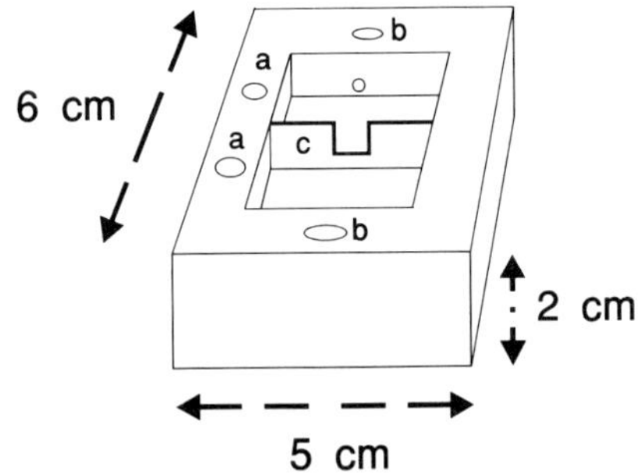

Figure 2. Simple chamber for the formation of bilayers by folding preformed phospholipid monolayers. The chamber is machined from a block of Teflon. Wells (a) in the block are filled with a concentrated electrolyte solution which links the solutions on either side of the membrane to the head amplifier via agar bridges and silver–silver chloride electrodes (Section 6.4). Solution is injected into the chamber via the ports (b). The thin Teflon septum is fixed to the central partition (c) (Section 2.2.2). As with the system shown in *Figure 1*, the solutions in the chambers can be stirred using magnetic fleas. The measurements given on the figure are approximate.

grounded aluminium plate. A power supply or battery is used to supply 12 V to the low tension terminal of a standard car ignition coil. The needle, positioned above the septum, is connected to the high voltage output of the coil. When the 12 V input is uncoupled using a switch, a several-thousand-volt discharge occurs between the point of the needle and the ground plate. This spark melts the Teflon septum and produces the hole. The size of the hole can be varied by applying multiple sparks to the septum. Whichever procedure is used, the quality of the hole should be checked under a microscope. Holes with ragged edges or tears should not be used.

An advantage of the monolayer folding technique is that it should be possible to use smaller holes than is possible with bilayers spread from dispersions of phospholipids. As noted in Section 2.1.2, it is very difficult, if not impossible, to induce film thinning on holes of diameter < 100 μm. This problem should not arise when bilayers are produced by monolayer folding. As the area of the bilayer is decreased so mechanical stability increases and the background noise of the system is lowered. A potential disadvantage of the use of a very small hole is that the small surface area may well reduce the likelihood of channel incorporation (Section 3.1).

Forming the bilayer

The principles involved in creating bilayers by monolayer folding are shown in *Figure 3*.

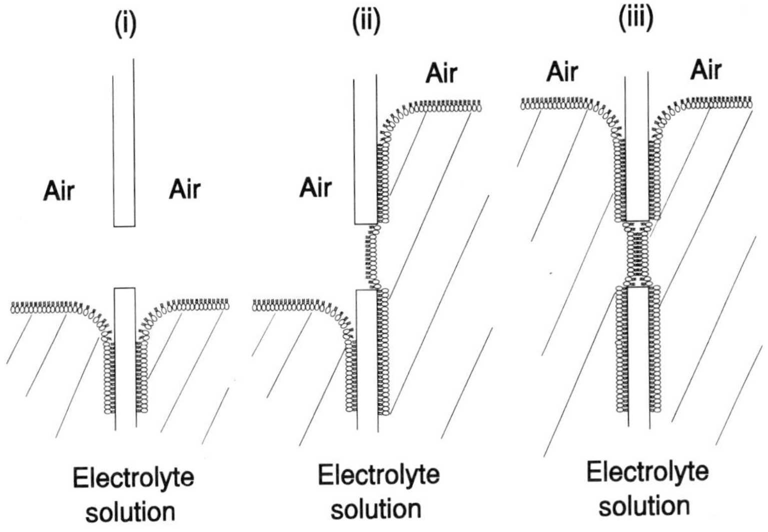

Figure 3. The probable sequence of events during the formation of a bilayer by the folding together of two preformed phospholipid monolayers on a thin Teflon septum: (i) solution in both chambers below hole; (ii) solution on one side raised to form a monolayer across the hole; (iii) solution raised on opposite side to 'appose' another monolayer. (Note that lipid molecules probably repeatedly reorientate during this procedure.) See Section 2.2.2.i and *Protocol 3* for details. The components are not drawn to scale.

Protocol 3. Bilayer formation by folding monolayers

Equipment and reagents

- Experimental chamber complete with a septum containing a hole (*Figure 2*)
- Phospholipids in *n*-pentane (see *Protocol 2*) or membrane vesicles for production of monolayer (Section 2.2.1)
- Priming solution: 0.5% (v/v) hexadecane in hexane (both from BDH)
- Electrolyte solution
- 5 ml syringes connected to fine plastic tubing

Method

1. Treat the septum surrounding the hole with a drop of the priming solution and dry with a stream of nitrogen gas (this procedure provides a hydrophobic environment for bilayer formation).
2. Add electrolyte solutions to both chambers of the block to a level below the hole.
3. Form a phospholipid monolayer at the air–water interface of the solutions in both chambers using either of the methods described in Section 2.2.1.
4. Following monolayer formation, raise one monolayer over the hole in the septum by slowly and steadily increasing the volume of the solution in the chamber. Add solution using a syringe, via a length of plastic tubing, through entry ports at the base of the chamber (*Figure 3*).
5. Raise the second monolayer in a similar manner to form a bilayer across the hole. Monitor formation of the bilayer by measuring capacitance as described in Section 2.1.3. Bilayers formed using this procedure will have a capacitance of approximately 0.8 $\mu F/cm^2$ (4, 16). Failure to form a membrane probably reflects inadequate formation of one or both of the monolayers.

ii. Bilayers on patch pipettes

An alternative procedure for the formation of bilayers from preformed phospholipid monolayers was first described by Hanke and colleagues (18). Using this approach, bilayers are constructed on the tip of a patch-clamp pipette. The potential advantages of this technique are that

(a) the final bilayer will have a very small surface area and hence should be electrically and mechanically stable and

(b) the bilayer will be truly free of solvent.

Formation of the bilayer on the glass surface of the pipette does not require hydrophobic priming. Standard patch-clamp pipettes are used (Chapter 1). Published accounts describe bilayer formation on either fire-polished or un-

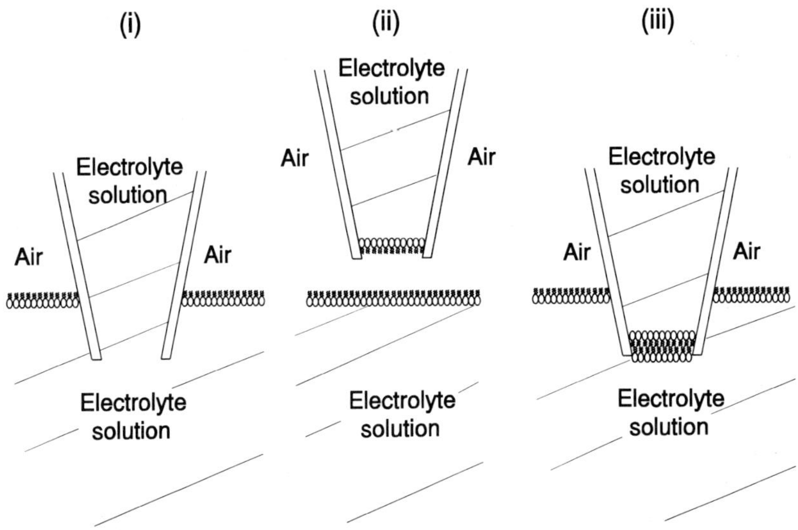

Figure 4. The proposed mechanism for the formation of a bilayer from preformed phospholipid monolayers at the tip of a patch pipette: (i) lipid monolayer surrounding pipette; (ii) pipette pulled up to 'catch' first monolayer; (iii) second dip to press on apposing monolayer to form bilayer. (Again note that the lipid molecules will in fact reorientate twice during this process, not shown.) See Section 2.2.2.ii and *Protocol 4* for details. The components are not drawn to scale.

polished pipettes with tip diameters ranging from 0.5 to 5.0 μm (10, 16). The strategy for bilayer formation is described in *Protocol 4* and illustrated in *Figure 4*.

Protocol 4. Bilayer formation at the tip of a patch pipette

Equipment and reagents

- Multi-well disposable tray (well volume 0.5 ml) (Sterilin)
- Phospholipids in *n*-pentane (see *Protocol 2*) or membrane vesicles, for production of monolayer (Section 2.2.1)
- Patch-clamp pipettes (See Chapter 1)
- Electrolyte solution

Method

1. Mount an electrolyte-filled patch pipette connected to a patch-clamp head amplifier on a micromanipulator so that it can be easily moved up and down.
2. Add electrolyte solution to one well of a disposable tray.
3. Lower the tip of the patch pipette into the solution.
4. Form a phospholipid monolayer at the surface of the solution using either of the methods described in Section 2.2.1.

Protocol 4. *Continued*

5. Following monolayer formation, raise the tip of the pipette through the air–water interface to pick up a monolayer (*Figure 4*).
6. Lower the tip of the pipette back into the solution so picking up a second monolayer and completing the formation of the bilayer (*Figure 4*).
7. If you fail to make a bilayer on your first attempt it is worth repeating this sequence. Apply positive pressure to the pipette as it is lowered back into the solution to prevent blockage. Then follow steps 5 and 6. If this does not work, start again from step 1 with a new well and pipette.

Various phospholipids have been used to form bilayers using this approach. Coronado (19) reported that the likelihood of bilayer formation from monolayers containing negatively charged phospholipids was significantly enhanced by the inclusion of divalent cations in the electrolyte solutions in the well and the pipette.

2.2.3 Channel incorporation during bilayer assembly

Monolayers derived from membrane vesicles that include channel proteins (see Section 2.2.1.ii) will contain those proteins orientated with the polar regions of the protein at the interface with the aqueous phase and the hydrophobic region of the protein in the air. Under these conditions, channel proteins in the monolayer may be transferred directly to a bilayer during assembly by folding on a hole in a septum or at the tip of a patch pipette (13–15, 20).

3. Ion channel insertion into bilayers

With the exception of the cases discussed in Section 2.2.3, the end product of any of the approaches described in Section 2 should be a mechanically and electrically stable, pure phospholipid bilayer. A typical 'good' bilayer will have a leakage current of less than 1 pA at a holding potential of 100 mV (7). If your efforts result in anything other than a 'rock-solid' membrane, there is really little point in attempting to insert channels. Unstable membranes lead to unmanageable offsets, variable baselines, and unpredictable current fluctuations unrelated to channel gating.

Channel proteins studied using the techniques described in this chapter are either sitting comfortably in a native membrane vesicle or have been isolated from that membrane following solubilization with a detergent. This second group of proteins (i.e. purified channel proteins) may either have been reconstituted into proteoliposomes by the removal of detergent in the presence of exogenous phospholipid, or may still be in a solubilized state exposed to detergent. Purified channel proteins in detergent may be incorporated into bilayers by simply adding an aliquot of the sample to the electrolyte solution on one side of the bilayer (21). The success of this approach will depend on a number of factors, including the concentration of the purified protein in the sample (a low concentration may mean that large volumes and hence large quantities of potentially disruptive detergent will need to be added to achieve reasonable rates of channel incorporation) and the stability of the channel protein in the detergent. In general, it is advisable to reconstitute purified channel proteins into proteoliposomes before incorporation into the bilayer. The method adopted for reconstitution will be governed by the particular detergent present in the sample. A comprehensive review of membrane protein solubilization and reconstitution is given in ref. 22.

3.1 Channel insertion by vesicle fusion

Channel proteins in native membrane vesicles or in reconstituted proteoliposomes are routinely inserted into phospholipid bilayers by fusion. This method was developed by Miller in the mid 1970s (23, 24). The approach has subsequently been used by many others using bilayers formed either by spreading from phospholipid dispersions or from monolayers. Whilst there may be minor differences, the general principles controlling vesicle fusion are believed to hold for both forms of bilayer (25).

Channel incorporation by vesicle–bilayer fusion is an extremely useful and flexible approach. Studies of channel gating often require investigations of the properties of a single channel, whilst studies of channel conduction and selectivity can usually be carried out using bilayers containing more than one channel. Alternatively it might be desirable to investigate the properties of several hundred channels at one time. A good working knowledge of the systems governing vesicle–bilayer fusion should allow the investigator to control the process and obtain the desired level of incorporation. *Figure 5* shows a schematic representation of the processes believed to occur during the fusion of an isolated membrane vesicle with a planar phospholipid bilayer.

3.1.1 The pre-fusion state

Prior to fusion, vesicles are thought to associate with, or bind to, the bilayer (25). The likelihood of this association occurring can be influenced by the phospholipid composition of the components of the system. If either the membrane vesicle or the bilayer contain phospholipids bearing a net negative charge, the probability of the pre-fusion state occurring can be increased by the inclusion of millimolar concentrations of either divalent or trivalent cations in the solution (25, 26). The probability of vesicle–bilayer interaction will also be influenced by other, fairly obvious, factors. Thus, stirring the solution containing the vesicles clearly increases the probability of contact between the vesicles and the bilayer. Similarly, increasing the surface area of the bilayer will result in more contacts. Conversely, very small areas of

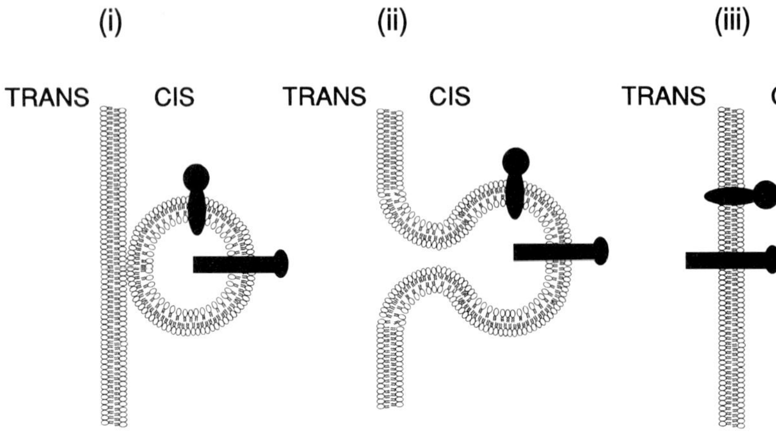

Figure 5. Fusion of a membrane vesicle with a bilayer. Vesicles are initially in a pre-fusion state associated with the bilayer (i). Fusion (ii) is induced as water flows *trans* to *cis* in response to an osmotic gradient (see Section 3.1 and *Protocol 5* for details). The orientation of the channel proteins (filled-in symbols) in the bilayer (iii) is governed by their orientation in the vesicle.

bilayer will decrease the likelihood of contact and hence fusion. This is a particular problem when bilayers are formed on very small holes such as the tips of patch pipettes (Section 2.2.2.ii). To overcome this problem, Hanke *et al.* (18) greatly increased the numbers of vesicles in the vicinity of the bilayer by adding them to the solution inside the pipette rather than the bath.

3.1.2 Fusion

Membrane vesicles in the pre-fusion state can be induced to fuse with the bilayer if they are made to swell (27). It is generally accepted that membrane proteins are transferred to the bilayer as swollen vesicles burst and merge with the bilayer (*Figure 5*). In practice, vesicles are encouraged to swell by the establishment of an osmotic gradient across the bilayer. Following bilayer formation the electrolyte solutions bathing both sides of the bilayer (*cis* and *trans*) are identical. Membrane vesicles are added to the solution on one side of the bilayer (*cis*) and the osmotic pressure of this solution is raised by the addition of an 'osmicant' (25). Under these conditions there will be a net movement of water across the bilayer from *trans* to *cis*. Vesicles in the pre-fusion state will swell if water enters the lumen as it crosses the bilayer.

A wide variety of potential osmicants have been examined. The ideal substance should permeate the vesicle membrane but should not readily cross the bilayer as this would result in the breakdown of the osmotic gradient. Cohen (25) describes urea and glycerol as effective in stimulating vesicle–bilayer fusion. Ethylene glycol and formamide, which are more permeant, are less effective. Since the hydrostatic pressure developed in the vesicle determines fusion, it is helpful if the lumen of the vesicle can be made hyperosmotic to the solution in the *cis* chamber. This can be accomplished by equilibrating and storing the vesicles in a medium containing a high concentration of a sugar (e.g. 0.3–1.0 M sucrose).

Membrane vesicles will fuse with a bilayer if they can be induced to swell. Vesicles will swell under the influence of an osmotic gradient across the bilayer if the vesicle contains a permeability pathway for the osmicant. The osmicants discussed above are inherently membrane permeant. Other osmicants can be used if the vesicle contains suitable permeability pathways. Therefore, osmotic gradients created with salts will induce fusion if the vesicle contains channel proteins permeable to either the cation or anion (25). This observation has formed the basis of the design of a general strategy for vesicle–bilayer fusion reported by Woodbury and Miller (28). The method involves the production of hybrid vesicles that contain native membrane proteins and phospholipids together with ergosterol and nystatin. In conjunction with ergosterol, nystatin functions as a relatively non-selective permeability pathway; therefore its presence in the vesicles renders them likely to swell in the presence of an osmotic gradient generated by a salt such as KCl. Having played its role in promoting swelling and fusion nystatin does not modify the conductance of the bilayer into which it incorporates. On fusion, the small quantities of ergosterol donated to the bilayer diffuse into the massive relative bulk of the bilayer and the permeability properties of nystalin are lost.

The nystatin method is potentially valuable as it allows the investigator to survey the variety of channel species in a membrane fraction and makes possible the determination of the density of channels in a preparation. However, once the channel species present in a membrane fraction have been identified, fusion is usually induced by the establishment of osmotic gradients with salts. The vesicles of interest will, by definition, contain a permeability pathway for an ionic solute.

Protocol 5 is a standard fusion protocol used in my laboratory for the incorporation of a reconstituted purified channel protein (the ryanodine receptor) into a painted bilayer formed on a 200 μm diameter hole. The volume of the *cis* chamber is 0.5 ml and that of the *trans* chamber is 1.0 ml (29). The same principles will apply for bilayers created from preformed monolayers.

Protocol 5. Vesicle–bilayer fusion

Reagents

- Painted bilayer (formed from phosphatidyl ethanolamine in *n*-decane (35 mM), *Protocol 1*, Section 2.1.2) with 200 mM KCl, 20 mM Hepes, KOH to pH 7.4 in *cis* and *trans* chambers
- Membrane vesicles containing reconstituted channel proteins suspended in 0.4 M sucrose
- 3 M KCl

Protocol 5. *Continued*

Method

1. Following bilayer formation, establish an osmotic gradient by adding 100 μl of 3 M KCl to the *cis* chamber.[a] Stir the contents of the chamber (Section 5.1).
2. Add an aliquot of membrane vesicles to the *cis* chamber. The volume added will depend on the channel density; this may vary from preparation to preparation. Stir.
3. Vesicles should fuse with the bilayer (you should see channel openings and closings within 5 min).
4. If fusion does not occur, increase the osmotic gradient by adding an additional 100 μl of 3 M KCl to the *cis* chamber.

[a] NB The solution in the *cis* chamber must be earthed during all additions; see Section 6.3.

3.1.3 Stopping fusion

Different types of channel experiment require different numbers of channels in the bilayer. Therefore, it is important that the investigator can stop the fusion process. In cases where divalent or trivalent cations have been added to the solution to enhance rates of fusion (Section 3.1.1), chelation of the cation with a suitable agent (e.g. EGTA) will slow the process. As fusion is driven by vesicle swelling induced by an osmotic gradient, the rate of vesicle fusion can be dramatically reduced by dissipation of the gradient; this can be achieved in two ways:

- addition of osmicant to the *trans* chamber;
- replacement of the high osmotic strength solution in the *cis* chamber with a solution of osmotic strength equivalent to that in the *trans* chamber.

The latter manoeuvre is best accomplished by perfusion of the *cis* chamber (Section 6.3). An added advantage of *cis* chamber perfusion is that this process also removes the majority of membrane vesicles which have not fused with the bilayer. This is an important point; the unexpected incorporation of additional channels during an experiment is very undesirable as it is likely to give rise to spurious increases in 'channel open' probability.

3.2 Channel orientation in the bilayer

If vesicle–bilayer fusion follows the pattern described in *Figure 5*, channel proteins will be inserted into the bilayer with an orientation that is dependent upon their orientation in the vesicle. Channel proteins are usually asymmetric in terms of function; ligand-binding sites are located on one face of the protein, blockers often only have access to the conduction pathway from one side of the channel. The orientation of the channel protein will therefore have a major influence on the interpretation of data obtained in an experiment. Where possible, the orientation of the channel should be determined using a side-specific agonist or blocker.

4. Patch clamping channels in liposomes

The techniques described above require equipment and experimental approaches which are unfamiliar and may not be readily available to electrophysiologists. However, one approach to the study of ion channel function in reconstituted systems can be carried out using standard patch-clamp technology. Isolated native membrane vesicles and hybrid liposomes containing channel proteins produced by detergent removal are too small to patch clamp (0.1–1.0 μm diameter). However, these membrane vesicles can be transformed into larger structures. Two methods are available, freeze–thaw and dehydration–rehydration. These are considered below.

4.1 Freeze–thaw

Successive freezing and thawing of membrane vesicles to form large, patch-clampable structures was introduced by Tank and Miller in the early 1980s (30). Using this approach, native membrane vesicles containing the channel to be investigated are solubilized and reconstituted into hybrid liposomes with exogenous phospholipids. Repeated cycles of freezing and thawing result in the formation of large multilamellar structures with diameters ranging from 10 to 30 μm. These structures are produced as the formation and removal of ice crystals causes rupture and re-sealing of the liposomes. Therefore, the structures will not form if the solutions contain anything that prevents ice crystal formation. For this reason it is important to wash the membrane vesicles clear of storage cryoprotectant such as sucrose. Larger ice crystals are produced by slow freezing: I have found the best results when samples are frozen slowly in a freezer rather than rapidly in liquid nitrogen.

Protocol 6 is used in my laboratory for the production of patch-clampable liposomes containing the K^+-selective channel of skeletal muscle sarcoplasmic reticulum (SR) membrane vesicles (31).

Protocol 6. The production of patch-clampable liposomes by freeze–thaw

Equipment and reagents

- SR membrane vesicles stored in 0.4 M sucrose at −80°C
- Potassium cholate
- L-α-phosphatidyl choline (Type II-S from Soybean; Sigma)
- Dialysis membrane (1000 dalton cut off, Spectrum/Por 6; Spectrum Medical Industries)
- Solution A: 200 mM KCl, 1 mM EDTA, 5 mM Hepes, pH 7.8
- Solution A containing 10 mg/ml phosphatidyl choline (sonicated to clarity) and 1.4% potassium cholate
- Recording solution: composition to depend on type of channel under investigation

Protocol 6. *Continued*

Method

1. Thaw an aliquot of SR vesicles and dilute to 1.8 mg protein/ml in solution A.
2. Add potassium cholate to give a final concentration of 1.4% (w/v).
3. Leave on ice for 30 min.
4. Sediment unsolubilized material by centrifugation at 100 000 × *g* for 45 min.
5. Collect the supernatant and mix this with an equal volume of solution A containing L-α-phosphatidyl choline (10 mg/ml) and 1.4% potassium cholate.
6. Reconstitute membrane proteins by removing the potassium cholate by dialysis against solution A at 4°C for 1 h at a ratio of 3 ml to 1 litre, followed by 24 h (with two changes of dialysis buffer) at a ratio of 3 ml to 2 litre.
7. Patch-clampable liposomes are produced by freeze–thaw. Place an aliquot of the material obtained in step 6 in a small glass test tube and freeze it by placing it in a freezer at −80°C for 30 min. At the end of this period remove the sample from the freezer and allow it to thaw at room temperature. As large multilamellar structures form, the initially clear sample becomes increasingly turbid. The quantity and size of patch-clampable structures can be increased by repeated cycles of freeze–thaw.
8. Place 50 μl of the sample on to a 3.5 cm diameter tissue-culture dish (Sterilin) and flood the dish with the desired recording solution. Transfer the dish to the patch-clamp set-up.

4.2 Dehydration–rehydration

Criado and Keller (32, 33) have described an alternative method for the production of large membrane structures that can be patch clamped. The initial stages of the protocol are very similar to those used for the freeze–thaw routine described above. Membrane proteins are solubilized and reconstituted into small unilamellar vesicles by detergent removal in the presence of exogenous lipid (33). The size of these vesicles is increased by successive dehydration and rehydration. *Protocol 7* is based upon the method described by these authors (33).

Protocol 7. The production of patch-clampable liposomes by dehydration and rehydration

Equipment and reagents

- CHAPS solution: 1% 3-[(3-cholamidopropyl)-dimethylammonio]-1-propane sulphonate (CHAPS) in 150 mM NaCl, 20 mM MOPS pH 7.5, 0.08% L-α-phosphatidyl choline, 5% glycerol (w/v), 0.1 mM phenylmethylsulphonyl fluoride (PMSF)[a], 1 μM pepstatin A
- Dialysis buffer: 100 mM NaCl, 20 mM Tris–HCl pH 7.4 with and without 1 mM $CaCl_2$
- MOPS solution: 10 mM MOPS pH 7.4 containing 5% (w/v) ethylene glycol
- Airfuge (Beckman Instruments)

Method

Carry out all procedures at 4°C.

1. Solubilize membranes with a CHAPS solution at a CHAPS/protein ratio (w/w) of 2.
2. After intermittent stirring for 5 min, sediment the insoluble material (40–50% of the starting protein) if possible in an airfuge at maximum speed for 15 min.
3. Mix 1 ml of the supernatant with 167 μl (16.7 mg) of L-α-phosphatidylcholine, suspended in water by sonication.
4. Dialyse the mixture for 20 h and 50 h against 400 volumes of dialysis buffer. Include 1 mM $CaCl_2$ in the dialysis buffer during the first dialysis.
5. Sediment 150–200 μl of the dialysed sample in an airfuge (maximum speed) for 1 h.
6. Resuspend the pellet in 10–15 μl of MOPS solution and place the suspension on a glass slide in a compact drop.
7. Dehydrate the drop at 4°C in a desiccator over solid $CaCl_2$.[b] The presence of ethylene glycol prevents complete dehydration.
8. Cover the partially dehydrated film with 20 μl of the required experimental solution and allow the sample to rehydrate by leaving the slide at 4°C overnight in a closed Petri dish containing a wet filter paper pad.
9. Following rehydration large liposomes can be seen along the edges of the film.

[a] AEBSF may be a possible less toxic alternative to PMSF (Editor's note).
[b] Keller *et al.* (33) suggest a 3 h dehydration period for a drop with a starting volume of 20 μl.

4.3 Patch clamping large liposomes

Single-channel current fluctuations of channels incorporated into liposomes are monitored using conventional patch-clamp procedures (Chapter 1). Using fire-polished pipettes with tip resistances in the range 5–20 GΩ it is a relatively simple process to obtain gigaohm seals. Patches should be removed from the surface of the liposome for recordings (inside-out configuration). As with techniques involving the incorporation of channel proteins into planar bilayers, every effort should be made to assess the orientation of the channel in the final patch.

5. Monitoring current flow through ion channels

The final section of this chapter deals with the measurement of channel function following their incorporation into an artificial membrane.

5.1 Electronics

The systems described above can be divided on the basis of the size of the artificial membrane used. The two groups are

- large bilayers on holes in a septum (painted or folded from monolayers);
- small bilayers on patch pipettes (from monolayers or from patching liposomes).

Current flow through channels is monitored using an operational amplifier as a current–voltage converter. Channels incorporated into bilayers on patch pipettes can be monitored using a conventional patch-clamp apparatus (Chapter 1). Whilst the basic principles governing the measurement of current flow through channels in larger bilayers are identical, the size of the bilayer does introduce some differences which should be considered. With large bilayers

- there is no series resistance;
- the large surface area, and hence the much larger capacitance of the bilayer, makes capacitance compensation very difficult.

The system used in my laboratory for the measurement of current flow through single channels incorporated into large planar bilayers was designed and built by Richard Montgomery. An outline of a minimal circuit layout is provided in *Figure 6*: a construction service is available (R.A.P. Montgomery, 26 Meadowcroft Rd, Palmers Green, London N13 4EA, UK).

The solutions in the *cis* and *trans* chambers are linked to the circuit via agar bridges (2% (w/v) in 3 M salt) and silver–silver chloride electrodes (Section 6.4). The system uses a low noise electrometer operational amplifier (A1) (OPA 102BM from Burr-Brown) together with a frequency compensation circuit as head amplifier. The feedback resistor, *R*1, must be of high quality (1% tolerance with a good 'lumped' parasitic capacitance can be obtained from Kobra), to guarantee an accurate output. The rise time of the system is limited as the parasitic capacitance has to be charged through *R*1. For a 10 GΩ resistor the rise time will be in the order of several msec. This

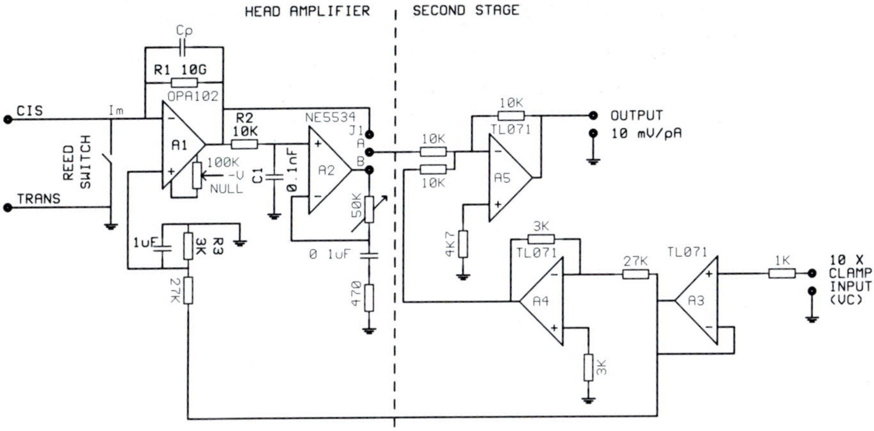

Figure 6. Minimum head amplifier and second-stage circuit layout for the measurement of channel current fluctuations. The solution at the *trans* side of the bilayer is held at virtual ground and the solution in the *cis* chamber is clamped at a potential relative to ground.

is shortened to 200 μsec by the frequency compensation circuit. The frequency compensation circuit is a frequency-sensitive amplifier producing greater amplification as the speed of change of the signal is increased, with unity gain at DC. Therefore noise from the bilayer will be amplified preferentially. High frequency bilayer noise is cut off prior to frequency compensation by *R*2 and *C*1 acting as a low-pass single-pole filter, although the frequency-compensated signal will always be noisier. To reduce noise, the command voltage to the head amplifier is ten-fold greater than necessary and is scaled down at the head amplifier. The signal is filtered by a 1 μF capacitor across the 3KΩ resistor (*R*3), which will limit the rise time of the command signal.

The bilayer capacitance has to be charged through *R*1 which results in saturation of the amplifier when a step change in clamp input is applied. The relatively large capacitance associated with these bilayers (typically approximately 50 pF for a bilayer on a 200 μm diameter hole) means that capacity compensation for the bilayer is difficult to achieve using the methods applied in patch-clamp experiments. Head amplifier output will be a voltage equivalent to $-(I_m \times R1) + (V/10)$ where I_m is the membrane current. The contribution of V is removed by the second-stage-amplifier which sums the head amplifier output with an inverted signal of $V/10$. The head amplifier is supplied with a low-noise power source from the second-stage amplifier, which also provides a power supply for a small motor which rotates a magnet under the bilayer chamber (*Figure 1*) and stirs the solutions with small magnetic fleas. Both leads to the motor must be grounded when the stirrer is switched off. Star earthing is used throughout to minimize noise. Bilayer amplifiers are also available commercially from Axon Instruments and

Biologic or a conventional patch-clamp amplifier can be employed with appropriate modifications to route the electrodes.

5.2 Noise

Channel current signals will be corrupted by both electrical and mechanical noise. The apparent noise on the signal can be reduced by low-pass filtering, but this will limit the resolution of channel open and closed lifetimes. In practice some degree of low-pass filtering of the signal is essential. The cut-off used will depend on the size of the channel current fluctuations under investigation and the level of noise. For this reason it is essential that steps are taken to reduce noise in the system.

5.2.1 Electrical noise

Electrical noise inherent in the systems required to monitor channel activity is discussed above. Shot noise at the bilayer will rise as the surface area of the bilayer is increased; large bilayers will be 'noisier' than bilayers on patch pipettes. The system will also be prone to electrical interference and for this reason the bilayer chamber should be 'screened'. This is easily accomplished by placing the bilayer block and the head amplifier in an openable metal box. Mains noise can often be reduced by earthing. Optimal earthing of the apparatus can only be determined empirically.

5.2.2 Mechanical noise

Once again, the larger the surface area of the bilayer, the more likely is the system to suffer from mechanical disturbance. This can be a serious problem with large planar bilayers. We have countered this problem by fixing the bilayer block in a very heavy base (a lead block) which damps vibration. As with patch-clamp experiments, mechanical disturbance can be significantly reduced by placing the apparatus on a very sturdy table, or even better on some form of vibration isolation system.

6. General hints

Ion channel reconstitution can be an extremely frustrating business. While I have attempted to describe the various approaches in some detail, it would take much more space than is available here to list all the minor problems that one may encounter in each technique, let alone the possible solutions. There is no substitute for experience. However, some general points which should prove helpful are discussed below.

6.1 Cleaning the apparatus

The apparatus used for the insertion of channels into membranes on patch pipettes is generally disposable. However, the formation of bilayers on holes either by spreading dispersions of phospholipids or by folding from monolayers involves plastic chambers which will be used repeatedly. As with all procedures involving the measurement of single-channel events, it is important that these chambers be cleaned well between experiments. We clean and store bilayer chambers and painting sticks (*Figures 1* and *2*) in solutions of commercial dishwashing detergent (a good squirt in a 500 ml beaker of water). Before use, the chambers are washed in running water for 30 min. Other cleaning procedures are described in refs 7 and 16.

6.2 Solutions

All solutions used in single-channel experiments should be filtered before use. We use 0.45 μm diameter pore Millipore filters.

6.3 Changing solutions

A major advantage of the insertion of ion channels into a planar bilayer formed on a hole in a septum is that the investigator has free access to, and can change, the solutions on both sides of the membrane. This is achieved by perfusion of the chamber. Fresh solution is pumped into the chamber whilst solution is pumped out. If the delivery and removal rates are the same, the solutions will eventually be exchanged with no change in volume. The volume of perfusate required to give complete exchange must be determined empirically (this can be done by monitoring the concentration of a dye by measuring absorbance) and will be influenced by the relative densities of the two solutions and stirring. Equal rates of delivery and removal can be obtained by using two syringes with their plungers connected back to back. If the barrels of the syringes are fixed so that the plunger of one is fully depressed whilst the other barrel is full of the new solution, movement of the plungers will draw solution from the chamber at the same rate as new solution is delivered. Alternatively, perfusion can be carried out using small, two-way peristaltic pumps (Schuco International model MV2).

During perfusion or additions to the *cis* chamber, it is important to short out the chambers, as this prevents the development of large voltages across the bilayer. This can be accomplished by placing an agar bridge across the bilayer or, as in the layout shown in *Figure 6*, by locating a reed switch between the input sockets of the head amplifier. The switch is activated with a small bar magnet.

6.4 Agar bridges and silver–silver chloride electrodes

Agar bridges are made by heating agar in a concentrated electrolyte solution (e.g. 3M KCl) to give a final concentration of 2% (w/v). With the agar in solution, the hot mixture is drawn into either preformed glass tubing or flexible plastic tubing, using a syringe. The agar in the bridges should be allowed to cool before use and the bridges should be stored, refrigerated, in the high salt solution. High ionic strength bridges minimize liquid-junction potentials when measurements are made under mixed ionic conditions.

Silver–silver chloride electrodes are made by chloriding silver wire. This can be done by immersing freshly cleaned silver wire in a small volume of domestic bleach.

Acknowledgements

I am grateful to the British Heart Foundation, The Medical Research Council, and The Wellcome Trust for support. I would also like to thank Richard Montgomery for his innovative approach to the design and construction of equipment and for his comments on this chapter.

References

1. Findlay, J. B. C. and Evans, W. H. (ed.) (1990). *Biological membranes: a practical approach*. IRL Press, Oxford.
2. Harris, E. L. V. and Angal, S. (ed.) (1990). *Protein purification applications: a practical approach*. IRL Press, Oxford.
3. Miller, C. (ed.) (1986). *Ion channel reconstitution*. Plenum, New York.
4. White, S. H. (1986). In *Ion channel reconstitution* (ed. C. Miller), p. 3. Plenum, New York.
5. Mueller, P., Rudin, D. O., Tien, H. T., and Wescott, W. C. (1962). *Circulation* **26**, 1167.
6. Mueller, P. and Rudin, D. O. (1969). In *Laboratory techniques in membrane biophysics* (ed. H. Passow and R. Stampfli), p. 141. Springer, Berlin.
7. Alvarez, O. (1986). In *Ion channel reconstitution* (ed. C. Miller), p. 115. Plenum, New York.
8. Nelson, M. T., French, R. J., and Krueger, B. K. (1984). *Nature* **308**, 77.
9. Montal, M. and Mueller, P. (1972). *Proc. Natl Acad. Sci. USA* **69**, 3561.
10. Coronado, R. and Latorre, R. (1983). *Biophys. J.* **43**, 231.
11. Hanke, W. and Miller, C. (1983). *J. Gen. Physiol.* **82**, 25.
12. Schindler, H. (1980). *FEBS Lett.* **122**, 77.
13. Schindler, H. and Quast, U. (1980). *Proc. Natl Acad. Sci. USA* **77**, 3052.
14. Nelson, N., Anholt, R., Lindstrom, J., and Montal, M. (1980). *Proc. Natl Acad. Sci. USA* **77**, 3057.
15. Labarca, P., Lindstrom, J., and Montal, M. (1984). *J. Gen. Physiol.* **83**, 473.
16. Montal, M., Anholt, R., and Labarca, P. (1986). In *Ion channel reconstitution* (ed. C. Miller), p. 157. Plenum, New York.
17. Hartshorne, R., Tamkun, M., and Montal, M. (1986). In *Ion channel reconstitution* (ed. C. Miller), p. 337. Plenum, New York.
18. Hanke, W., Methfessel, C., Wilmsen, U., and Boheim, G. (1984). *Bioelectrochemistry and Bioenergetics* **12**, 329.
19. Coronado, R. (1985). *Biophys. J.* **47**, 851.
20. Suarez-Isla, B. A., Wan, K., Lindstrom, J., and Montal, M. (1983). *Biochemistry* **22**, 2319.

21. Lai, F. A., Erickson, H. P., Rousseau, E., Liu, Q.-Y., and Meissner, G. (1988). *Nature* **331**, 315.
22. Jones, O. T., Earnest, J. P., and McNamee, M. G. (1990). In *Biological membranes: a practical approach* (ed. J. B. C. Findlay and W. H. Evans), p. 139. IRL Press, Oxford.
23. Miller, C. and Racker, E. (1976). *J. Memb. Biol.* **30**, 283.
24. Miller, C. (1978). *J. Memb. Biol.* **40**, 1.
25. Cohen, F. S. (1986). In *Ion channel reconstitution* (ed. C. Miller), p. 131. Plenum, New York.
26. Hanke, W. (1986). In *Ion channel reconstitution* (ed. C. Miller), p. 141. Plenum, New York.
27. Finkelstein, A., Zimmerberg, J., and Cohen, F. S. (1986). *Ann. Rev. Physiol.* **48**, 163.
28. Woodbury, D. J. and Miller, C. (1990). *Biophys. J.* **58**, 833.
29. Lindsay, A. R. G. and Williams, A. J. (1991). *Biochim. Biophys. Acta* **1064**, 89.
30. Tank, D. W. and Miller, C. (1983). In *Single-channel recording* (ed. B. Sakmann and E. Neher), p. 91. Plenum, New York.
31. Tomlins, B. and Williams, A. J. (1986). *Pflügers Arch. Eur. J. Physiol.* **407**, 341.
32. Criado, M. and Keller, B. U. (1987). *FEBS Lett.* **224**, 172.
33. Keller, B. U., Hedrich, R., Vaz, W. L. C., and Criado, M. (1988). *Pflügers Arch. Eur. J. Physiol.* **411**, 94.

3

Concepts of single-channel analysis: inferring function from fluctuations

GUY W. J. MOSS and EDWARD MOCZYDLOWSKI

1. Introduction

1.1 Single channels in practice and theory

The defining act of an ion channel protein is to allow an electrical current of ions to flow passively across a cell membrane. Thus, among the variety of approaches discussed in this text, perhaps the most basic measurement of channel function is its unitary current. As described in Chapters 1 and 2, single channel currents can be recorded with techniques that involve patch pipettes or artificial bilayers. Depending on the noise characteristics of the system, a patch-clamp amplifier allows one to measure unitary currents as small as ~0.1 pA. Regardless of how the single-channel recording is obtained, the primary objective is to interpret or make sense of it in terms of channel function. For this formidable task, the record supplies only two basic types of information: current amplitudes and dwell times. These pieces of information are illustrated for a recording of a Ca^{2+}-activated K^{+} channel in *Figure 1*. The first trace in *Figure 1* shows that this particular channel fluctuates between a well-defined closed (down) and open (up) state. The unitary current, *i*, of the channel is measured as the difference between the closed and open levels. A dwell time in the open state is defined as the duration of an event in the open state between two consecutive closing events. Likewise, a dwell time in the closed state is that between two consecutive openings.

In order to elucidate some aspect of channel function with this information, a theoretical framework is needed to guide experimentation and analysis. Since unitary current is a direct measure of the rate of ions moving through the open channel, various theories of ion permeation or conduction have been developed to explain or model this process. Similarly, the dwell times of channel openings and closings reflect the underlying kinetic pathway of channel gating. Analysis of channel gating makes use of probability theory and the mathematics of stochastic processes. Since the introduction in 1981 of improvements in patch recording (1), the science of single-channel analysis

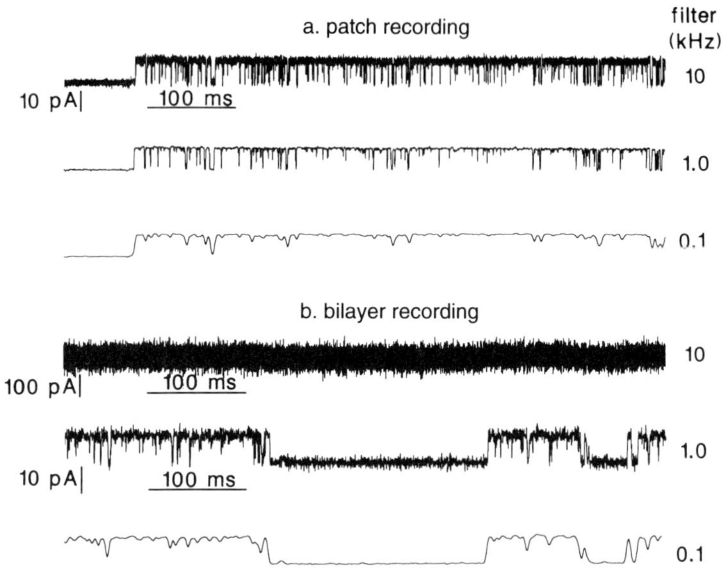

Figure 1. Comparison of records of a single Ca^{2+}-activated K-channel from a cell-excised patch and a planar bilayer. (a) Cell-excised patch pipette recording from a cultured bovine vascular smooth muscle cell. Conditions: 100 mM symmetrical KCl, 10 mM MOPS-KOH, pH 7.4, 40 μM internal Ca^{2+}, holding voltage = +40 mV. (b) Recording of a Ca^{2+}-activated K-channel incorporated into a planar bilayer from membrane vesicles prepared from rat skeletal muscle. Conditions: 50 mM symmetrical KCl, 10 mM MOPS-KOH, pH 7.4, 100 μM internal Ca^{2+}, holding voltage = +40 mV. In both (a) and (b) the same record is shown after low-pass filtering at 10, 1.0, and 0.1 kHz as indicated. The records at 10 kHz were filtered mainly at the amplifier head stage and the records at 1.0 and 0.1 kHz were filtered with a digital Gaussian filter. Channel opening is upward. Note that the bilayer recording at 10 Hz is overwhelmed by high frequency noise and is shown at a reduced current scale.

has grown rapidly into a vast field that cannot be adequately covered in a single chapter. In addition to this volume, a number of excellent books and reviews (2–9) are recommended as reference sources to the extensive theory of channel biophysics and methods for single-channel analysis.

It is often noted that an ion channel is simply an enzyme that has evolved for the purpose of selectively overcoming the extremely high energy barrier that prevents the diffusion of hydrated ions through a lipid bilayer. Yet by comparison to enzyme kinetics, the formal principles of channel analysis have not yet become part of the common language of biology. This is partly due to differences in terminology and is partly the legacy of complex mathematical derivations. In contributing to a 'practical approach' to ion channels, this chapter addresses basic concepts of channel function with a view toward lowering some of the energy barriers that hinder general accessibility. The first half of the chapter attempts to draw a direct connection between channel conduction theory and enzymology. The second part covers some fundamentals of probability distributions relevant to channel analysis and introduces the matrix approach to the analysis of channel gating. We hope that this chapter will provide a basic overview and will prove to be useful in opening the door to more advanced texts and treatises that form the foundation of ion channel research.

1.2 Basic considerations for recording and analysing data

Analysis of single-channel data always involves some type of filtering which excludes high frequency information. Filtering may either occur at the limiting level of the patch-clamp headstage (currently in the range of 10–20 kHz), or at a lower frequency determined by a user-selected low-pass filter. *Figure 1* compares the single-channel activity of a maxi Ca^{2+}-activated K^+ channel as recorded in an excised patch (from a bovine vascular smooth muscle cell) versus a similar channel (from rat skeletal muscle plasma membranes) after incorporation into a planar bilayer. The diameter of the cell membrane patch recorded with the patch pipette was ~1 μm and that of the planar bilayer was ~200 μm. Both of these recordings are shown at three different levels of filtering (10, 1.0, and 0.1 kHz) to illustrate that information derived from such records is highly dependent on the filter cut-off frequency.

An important purpose of filtering is to exclude random high frequency background noise that is not due to channel activity. The major component of this background noise arises from charging currents inherent to the voltage clamping of a membrane (10). The direct dependence of this particular noise component on membrane capacitance or area can be readily appreciated by comparison of the small diameter patch recording (*Figure 1a*) with the bilayer recording (*Figure 1b*) filtered at 10 kHz. In the 10 kHz bilayer record, the underlying opening and closing events of the channel are completely obscured due to large amplitude background noise. However, as shown in the second trace of *Figure 1b*, the channel activity can be revealed by filtering at 1 kHz. As a single-channel record is filtered more heavily, longer dwell times of channel activity are fully resolved from background noise, but some brief channel transitions are also lost along with the unwanted noise component. When recording and analysing single-channel data, it is important to consider that one is always looking at a particular window of time. Fast processes are lost due to low-pass filtering and slow processes may not be recognized because the recording has not lasted long enough. The reality of this situation is that, despite the wonderful details that can be uncovered within an observable time-window, some important ones may lie just beyond view (e.g. brief gating events, open-channel noise, gating mode shifts, modulation by phosphorylation, etc.). Thus, the sampling and filtering conditions of the recording must be appropriately set to match the time-scale of interest.

Single-channel recording with small diameter patch pipettes affords lower noise and superior time resolution; however, the planar bilayer technique is useful for many applications that would be difficult to carry out with patch technology. As discussed in Chapter 2, some channels are readily incorporated into bilayers and form the basis of reliable assay systems. Bilayer resolution is often fully adequate for many purposes, such as studies of slow toxin blocking reactions. Simultaneous access to both sides of a bilayer channel is advantageous for some experiments and an *in vitro* reconstitution system allows certain variables to be manipulated more readily (e.g. lipid composition of the membrane).

2. Measurement of ionic selectivity

2.1 What is selectivity?

A remarkable feature of the excitable cell membrane is the different electrical behaviour that is produced by substitution of simple inorganic ions such as Na^+, K^+, Ca^{2+}, and Cl^-. Physiologists have long sought to understand the mechanism underlying this selectivity. While the concept of ionic selectivity of a channel seems intuitively obvious (some ions flux better than others), it is very difficult to provide a universal definition of this phenomenon or fully deduce its molecular basis. In the first case, this is because selectivity depends on how it is measured. The relative ability of an open channel to pass current of different ions may strongly depend on the absolute voltage, the concentrations, and the species of ions present on each side of the membrane. Thus, there has to be some agreement on the basis for comparison and the conditions used for comparing different ions and channels. Secondly, we now know that very selective channels use a variety of fundamental mechanisms to discriminate between ions. Measurement of pore selectivity often reflects many different aspects of channel structure. This is illustrated in the schematic view of an ion channel given in *Figure 2*.

As indicated in *Figure 2*, evidence suggests that many channel proteins form an ion conduction pathway at the central interface of several homologous subunits or domains. The passageway through the membrane is not necessarily cylindrical, but may have inner and outer funnel-shaped regions called vestibules that are connected by a narrow tunnel. In highly selective channels, the small-diameter tunnel region is where ions and water molecules are constrained to move in single file. Each of these distinct regions makes a special contribution to ion selectivity and the rate of transport. Wide vestibules enlarge the capture radius for collecting ions that bombard the pore by diffusional collision (6). A large capture radius increases the encounter rate of ions with the channel, but diffusional limitation may actually result in current saturation at high voltage (11). The two vestibules of the channel may contain fixed charges due to the functional groups of negatively charged acidic (aspartate, glutamate) and positively charged basic (lysine, arginine, histidine)

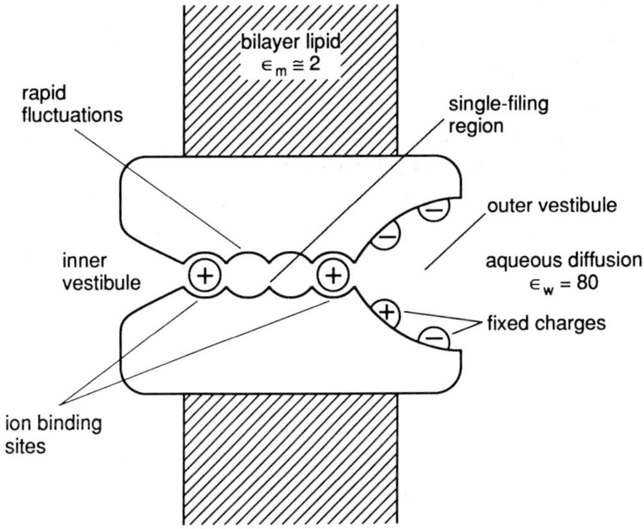

Figure 2. Schematic diagram of an ion channel protein. The drawing summarizes various features of channel proteins that are considered to be important for ion permeation and selectivity, as discussed in the text.

residues of amino acids. These charges establish an electrostatic surface potential that is expected to produce local concentration gradients of cations and anions near the protein surface by attraction and repulsion (12). For example, negative surface potential has been proposed to be important in increasing the local concentration of permeant cations at the entrance of cation-selective channels (13). In addition to the cation-concentrating effect produced by a smeared distribution of negative surface charges, it is possible that cation-selective channels exploit the discrete zwitterionic nature of protein surfaces. For example, mutagenesis work has shown that there is a conserved lysine residue in the pore region of sodium channels which allows permeation of monovalent cations (Na^+, K^+) but apparently prevents the permeation of divalent cations such as Ca^{2+} (14). Thus, fixed charges at pore entrances may contribute to selectivity by attracting either cations or anions into a vestibule depending on the sign and magnitude of the surface potential. In addition, a more localized discrimination of ions at particular sites in the channel may take place by mechanisms based on ion–ligand interactions and steric restrictions.

The single-filing region of the channel is where polar residues of the protein can make direct chemical interactions with ions by substituting for the water molecules that normally hydrate ions in aqueous solution. For alkali cations such as K^+, such interactions might include complex formation with oxygen atoms of peptide carbonyl groups, hydroxyl groups of serine or threonine, or even formation of π-complexes with aromatic groups such as tyrosine (15).

These complexes can be stable enough to form ion binding sites which may be regarded as the most energetically stable positions of the ions in the pore. For many, if not all channels, the single-filing region is apparently long enough to accommodate more than one ion simultaneously. Such channels are termed multi-ion channels. Ca^{2+} channels appear to have two such high-affinity binding sites with effective equilibrium dissociation constants in the range of 2 μM when the channel is singly occupied (16). However, in the case of double occupancy, repulsion between the two Ca^{2+} ions results in a mutually lower binding affinity and faster exit from the channel, giving rise to a higher conductance (17). Thus, chemical selectivity can take place at several sites in the channel and ion–ion interactions can be an important determinant of which ions enter and which ions leave. Finally, the preceding picture was presented as if a channel is analogous to a rigid sewer pipe. In reality we know that protein molecules are capable of many kinds of dynamic motions on widely different timescales. If the pore region exhibits structural fluctuations on a time-scale comparable to transport, these rapid fluctuations may be coupled to ion occupancy and also enter into the determination of flux rate (18, 19).

These important features of channel proteins defy attempts to construct 'minimal' models of ion conduction that can in any way accurately represent the whole process in predicting current and selectivity. Consideration of the cartoon channel in *Figure 2* also leads to the realization that a measurement of unitary current carried by ions through a channel is quite unlike measuring equilibrium binding of a ligand to a protein. An equilibrium binding experiment aimed at determining a K_D value reports on the interaction of a ligand with one or a small number of sites. K_D can be directly related to the free energy of binding which can be compared for different homologues at that bind at the same site. In some cases, when ions are studied as blockers of a channel, it is possible to measure an equivalent inhibition constant or K_D of binding (20). These values of binding affinity can be interpreted in terms of specificity of binding to the blocking site, but they are poor predictors of the relative ease of permeation. In contrast, a current measurement is a non-equilibrium measurement that reports on the net process of ions diffusing into the vestibules from both directions, traversing the channel and exiting it. Thus, the ionic selectivity of a channel current is globally reflective of all the processes summarized in *Figure 2*.

2.2 Continuum versus discrete-state models

Channel biophysicists generally rely on two types of measurements as indicators of ion selectivity. The first is the permeability ratio of two ions, P_A/P_B. This parameter is derived from the reversal potential, V_R, measured in the presence of a known concentration gradient of the two ions across the channel-containing membrane. V_R obtained from the current–voltage (I–V)

relation is simply the voltage at which the measured current through the channel is zero. The second type of measurement is the conductance ratio of two ions, g_A/g_B. This parameter is best expressed as the ratio of the slopes of the single-channel I–V relation independently measured for symmetrical solutions of ion A or B. These two measurements of selectivity generally reflect different aspects of the conduction process. In fact, it is possible for an ion to have a high relative permeability (P_A) but a low or virtually zero conductance (g_A). To understand this extraordinary statement and the meaning of these two measurements of selectivity, it is necessary to examine the behaviour of predictive models.

The development of quantitative models designed to explain ion permeation through membrane pores is an old science that is still a subject of serious biophysical pursuit and ongoing revision. There is still disagreement about which theoretical approach is best. In practical terms, it is important to know that the two major classes of models are continuum models and discrete-state models (21). In the foreseeable future, it may be possible to predict ionic currents through channels by carrying out molecular dynamics simulations with powerful computers that take into account the three-dimensional structure of a channel protein and all the atomic forces between the ions, water, and the protein (22). However, until then, a basic understanding of heuristic models is necessary.

In one major class of continuum models, permeation is described by a diffusion-like process that is based on Nernst–Planck electrodiffusion theory (21, 23). The structure of the channel may be represented by a region of high dielectric constant (aqueous) surrounded by a region of low dielectric constant (lipid, protein) which is used to calculate the electrochemical potential of an ion at a point in the channel. The important variables of the model are the diffusion coefficient and the profile of electrical potential across the pore as calculated from an assumed geometry. This approach is attractive because ionic flux through a channel is actually a modified form of passive diffusion. Until recently, the continuum approach faced difficulties in describing ion binding and interactions between ions in the channel. However, new extensions of the theory have made it possible to analyse the behaviour of multi-ion channels with this model (24).

In contrast to simulations of diffusion, discrete-state models utilize Eyring reaction rate theory to decompose the conduction process into a series of distinct steps, such as entry of an ion into the pore, binding to the first site, hopping to the second site, and exit to the other side of the membrane (4, 6, 25). This approach allows transport kinetics to be represented by a limited number of rate constants that can be related to a series of transition states (energy barriers) and stable intermediates (ion binding sites). Eyring models are useful because they facilitate incorporation of many of the particular features of channels depicted in *Figure 2*. For example, it is possible to use Eyring models to predict how a multi-ion channel behaves in comparison to a single-site channel (26) or to examine how ion–ion interactions, charged vestibules (12), or rapid fluctuations (18) might influence conduction. Despite the usefulness of Eyring models, numerous uncertainties make it impossible to establish a direct relationship between derived rate constants or energies of the barrier profile and the actual structural features of a channel (21). Thus, discrete-state models are primarily useful in determining whether a particular mechanism can account for a particular phenomenon. As a general rule, kinetic analysis can be used to eliminate particular classes of mechanisms but cannot be used to prove directly any particular mechanism.

Figure 3 illustrates two different conceptions of ion permeation envisaged by a classical continuum model versus a discrete-state model. In the basic Nernst–Planck electrodiffusion theory originally derived to account for macroscopic currents, the channel-containing membrane is treated as a homogeneous slab of thickness L. Ion permeation across this partition is governed by a partition coefficient, β, and a diffusion coefficient, D. β is the equilibrium ratio of ions in the membrane phase to that of the aqueous phase, while D characterizes the ability of an ion to move within the membrane. Note that this model (*Figure 3a*) does not include the notion of an ion binding site. Any given ion is assumed to move independently of other ions and the electric field is assumed to be constant within the membrane.

Figure 3b is a representation of a commonly used Eyring barrier model. The particular model shown is called a 3-barrier-2-site (3B2S) model according to the number of energy peaks (barriers or transition states) and wells (binding sites). The passage of ions through the channel occurs in distinct steps in both directions by moving across one of the three barriers. Although

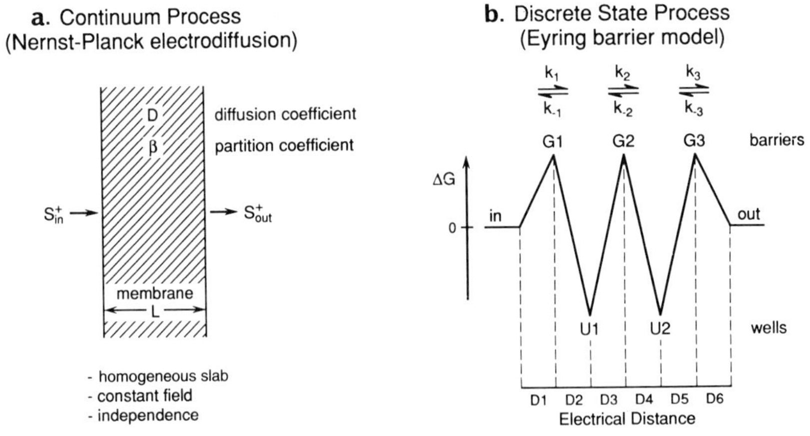

Figure 3. Two different types of models used to analyse ionic conduction and selectivity of channels. (a) Parameters and key assumptions used in the derivation of the classical Nernst–Planck treatment of ion permeation as a continuum process. (b) Description of ion permeation according to a 3-barrier-2-site model using Eyring transition-state theory.

the 3B2S model channel includes two ion binding sites, it can be specified to contain a maximum of either one or two ions at a time depending on the strength of ion–ion interactions. (This will be discussed more thoroughly in Section 3.2.) A probability state of the model represents one of the various possible conditions of occupancy of the two sites. The discrete-state approach allows a transition diagram of routes between empty and occupied sites to be written for very simple (one site, one type of ion) or very complex situations (multiple sites, multiple types of ions). The basic 3B2S model in *Figure 3b* has five energy parameters which correspond to the three peak energies (*G1, G2, G3*) and two well energies (*U1, U2*) at zero voltage. Energy differences are converted to a voltage-dependent rate constant, $k_i(V)$, for crossing a particular energy barrier according to the Eyring rate theory expression:

$$k_i(V) = (k_B T/h)\exp[-\Delta G_i^{\ddagger}(V)/RT]. \qquad [1]$$

In this equation, $\Delta G_i^{\ddagger}(V)$ is the difference in energy (e.g. in kcal/mole or RT units) between a transition state and a neighbouring energy well (or the solution reference state), k_B is the Boltzmann constant, T is absolute temperature, h is Planck's constant, and R is the molar gas constant. At 23°C, $k_BT/h = 6.21 \times 10^{12}$/sec, and $RT = 0.59$ kcal/mole. When calculating $\Delta G_i^{\ddagger}(V)$, the zero-voltage energy of each peak or well $G(0)$ is added to a term, $z\delta FV$, which is the electrical work involved in moving an ion of valence z by a fractional distance δ across the transmembrane electric field:

$$G(V) = G(0) + z\delta FV \qquad [2]$$

where F is Faraday's constant and V is the transmembrane voltage. In *Figure 3*, the parameters *D*1–*D*6 indicate the fraction of the electric field (δ) attributed to each segment of the energy profile, with the sum of *D*1 through *D*6 equal to 1. The voltage dependence of $\Delta G_i^{\ddagger}(V)$ is determined by the direction of ion movement, with the rate constant ($k_i(V)$ of Equation 1) for outward movement of a cation being increased by positive voltage (depolarization). This method for handling the voltage dependence of ion movement is similar to that used in the classic Woodhull model of a voltage-dependent blocking reaction (27). *Figure 3* shows that even for relatively simple energy profiles, Eyring models of ion conduction require a large number of independent energy and distance parameters to be specified. This feature allows such models great flexibility in fitting current–voltage observations (28) but generally means that more than one combination of parameters will provide practically indistinguishable fits. Again, this stresses that barrier models of channel permeation are not meant to be taken literally. They are primarily useful as tools to investigate the theoretical behaviour of particular classes of mechanisms.

2.3 Classical definition of a permeability ratio

The concept of a permeability ratio is usually first encountered by nascent neuroscientists in connection with the Goldman–Hodgkin–Katz (GHK) equations (29) used to describe macroscopic currents and the ionic basis of membrane potential. This treatment, developed before there was any knowledge of channel structure, is an example of a continuum model. Today, it still serves as the basis of the most commonly used method to evaluate the ionic selectivity of a channel. In a typical experiment, the instantaneous I–V relation of a macroscopic current or the unitary I–V curve at the single-channel level is measured in the presence of a gradient of two salts (such as NaCl and KCl) on opposite sides of the membrane. For a purely cation-selective channel, the following GHK voltage equation is used to evaluate the permeability ratio, P_A/P_B, of cations A and B present on the inside (i) or outside (o):

$$V_R = \frac{RT}{F} \ln \left\{ \frac{(P_A/P_B)\,[A]_o + [B]_o}{(P_A/P_B)\,[A]_i + [B]_i} \right\}. \qquad [3]$$

With known concentrations of the two cations and the measured reversal potential, V_R, Equation 3 can be solved to obtain P_A/P_B for the current under investigation. To be exact, concentration terms such as [A] and [B] in the equations discussed in this article represent thermodynamic activities. In practice, activities are calculated as the product of ionic concentration and an appropriate activity coefficient. For ease of discussion, we omit activity coefficients and use the terms concentration and activity interchangeably. The measurement of permeability ratios is generally used to characterize and compare the relative ionic selectivity of various types of channels. It also provides an important basis for identifying and functionally discriminating different families of channels. For example, it has been determined that $P_K/P_{Na} \simeq 0.08$ for voltage-sensitive sodium channels, $P_{Rb}/P_K \simeq 0.7$ for certain types of potassium channels, $P_{Ba}/P_{Ca} \simeq 0.4$ for calcium channels, etc.

Determining such permeability ratios in binary mixtures of salt solutions can be useful for establishing the basic trend of ion selectivity. However, for mechanistic work, it is preferable to measure what is called the bi-ionic permeability ratio. The bi-ionic permeability ratio is determined from the reversal potential with only one type of permeant ion on either side. This provides a better measurement of intrinsic selectivity since, in many cases, permeability ratios depend on the details of the measurement (25). For example, in some studies P_K/P_{Na} for sodium channels has been found to vary with the concentration of internal K^+ (30) and the orientation of the test cations (31). Bi-ionic permeability ratios provide a consistent way of comparing the ionic selectivity of different channels and are more readily interpreted within the context of ion conduction models. The following three equations give the relationship between V_R, ionic activity outside (o) and inside (i), and bi-ionic permeability ratios for the respective selectivity comparison of monovalent/monovalent (Equation 4A), divalent/divalent (Equation 4B), and divalent/monovalent (Equation 4C) ions.

$$V_R = \frac{RT}{F} \ln \left\{ \frac{P_A[A^{1+}]_o}{P_B[B^{1+}]_i} \right\}, \qquad [4A]$$

$$V_R = \frac{RT}{2F} \ln \left\{ \frac{P_A[A^{2+}]_o}{P_B[B^{2+}]_i} \right\}, \qquad [4B]$$

$$V_R = \frac{RT}{F} \ln \left\{ \frac{4P_A[A^{2+}]_o/P_B[B^{1+}]_i}{\exp(FV_R/RT) + 1} \right\}. \qquad [4C]$$

From Equation 4A, if two monovalent cations are present on opposite sides of the channel at the same activity, using RT/F = 25.5 mV (at 23°C), the reversal potential is expected to be +58.7 mV for $P_A/P_B = 10$ and −58.7 mV for $P_A/P_B = 0.1$. The measurement of a bi-ionic reversal potential thus allows the relative ionic selectivity of a channel current to be established quickly. Equation 4C is a modified form of the GHK voltage equation as derived for the case of a divalent cation, A^{2+} outside, and a monovalent cation, B^{1+} inside (32). This latter equation has been used to estimate a permeability ratio of 4200 for Ca^{2+}/Cs^{+} for a cardiac L-type calcium channel (17, 33).

Despite the usefulness and broad application of permeability ratios, it is important to recognize that the continuum model upon which the concept is based is a rather poor approximation of most ion channels. However, the behaviour of a permeability ratio can be used to infer important aspects of the actual conduction mechanism. To pursue these insights, we note that the GHK voltage equation that provides us with the classic definition of the permeability ratio is based on the GHK current equation which is shown below.

$$I_S = \frac{z^2F^2}{RT} P_S V \left\{ \frac{[S]_i - [S]_o \exp(-zFV/RT)}{1 - \exp(-zFV/RT)} \right\}. \qquad [5]$$

This equation describes the predicted I–V relationship for the continuum electrodiffusion model of *Figure 3a*. It describes the current (I_S in A/cm^2) carried by an ion, S, of valence, z, in terms of permeability (P_S), voltage (V), and ionic activities on both sides of the membrane ($[S]_i$, $[S]_o$). Equation 5 is obtained from the sum of the current due to diffusive flux (Fick's law of diffusion) and electrical flux (Ohm's law). The absolute permeability, P (in cm/sec) for an area of membrane is ultimately related to the product of the diffusion coefficient, D (in cm^2/sec) and the partition coefficient, β, as follows:

$$P = D\beta/L \qquad [6]$$

where L is the membrane thickness. Note that under symmetrical conditions, $[S]_i = [S]_o = [S]$, and Equation 5 reduces to

$$I_S = \frac{z^2F^2}{RT} P_S V[S]. \qquad [7]$$

Equation 7 predicts an ohmic (linear I–V) relationship under symmetrical conditions with the membrane conductance, G (in reciprocal ohms or siemens per cm^2), equal to

$$G = \frac{I}{V} = \frac{z^2F^2}{RT} P_S[S]. \qquad [8]$$

Thus, according to the GHK model upon which the classic definition of a permeability ratio is based, a channel should exhibit a purely linear relationship between its conductance and the symmetrical concentration of a permeant ion, [S]. The GHK model is said to exhibit 'independence', which means that the interaction between the ion and the channel walls is not saturable, and there are no interactions between ions of the same species or different species. In other words, there are no binding phenomena and no ion–ion interactions. According to this view of a channel mechanism, permeability ratios for different ions depend on how readily different ions are adsorbed via a partitioning process (β) and how readily they move via a diffusional process (D). In such a hypothetical channel, the permeability ratio for a given pair of ions is an absolute constant that does not vary with the ionic conditions. In reality, many channels have been found to violate this rule.

3. The rate-theory perspective on ion conduction

3.1 A one-site channel compared to a one-site enzyme

As discussed earlier, according to an Eyring-type model, the process of ion conduction may be summarized by an energy barrier diagram. *Figure 4b* illustrates the simplest possible energy profile, a channel that is described by two barriers and one site (2B1S). Such a channel is constrained to contain a maximum of only one ion at a time. *Figure 4b* indicates that ion conduction through a 2B1S channel is governed by two types of rate constants: a pair of association rate constants (k_1, k_2) for entering the site from the inner or outer solutions and a pair of dissociation rate constants (k_{-1}, k_{-2}) for exiting the channel to either side. The kinetic scheme for such a channel may be reduced to the following cyclic form:

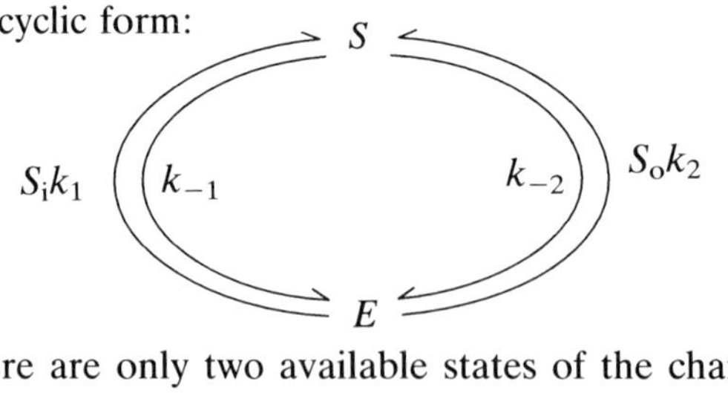

Scheme 1

In Scheme 1 there are only two available states of the channel, empty (E) and occupied (S), and each state is characterized by a corresponding probability, P_E and P_S. Both states can be entered by two pathways and exited by two

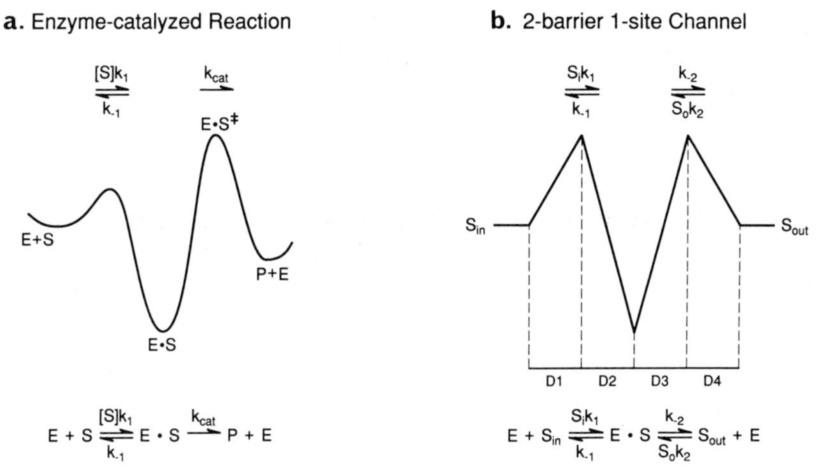

Figure 4. Comparison of energy barrier diagrams and reaction schemes for an enzyme-catalysed process (a) and a 2-barrier-1-site channel (b).

pathways. In the steady-state condition, the rate of change of any state is zero. As a particular example, the rate of change of the probability of the unoccupied state is equal to the sum of the rates for entering state E minus the sum of the paths for leaving state E as follows:

$$\frac{dP_E}{dt} = k_{-2}P_S + k_{-1}P_S - S_o k_2 P_E - S_i k_1 P_E = 0. \qquad [9]$$

The probabilities of states E and S must also sum to unity according to the conservation equation:

$$P_E + P_S = 1. \qquad [10]$$

Equations 9 and 10 may be used to solve for P_E and P_S by substitution to obtain the following expressions for the steady-state probabilities:

$$P_E = (k_{-1} + k_{-2})/(S_i k_1 + S_o k_2 + k_{-1} + k_{-2}) \qquad [11A]$$

$$P_S = (S_i k_1 + S_O k_2)/(S_i k_1 + S_o k_2 + k_{-1} + k_{-2}). \qquad [11B]$$

Equations 11A and 11B may further be used to obtain a solution for the steady-state ionic flux for the system which is equal to the net rate across any barrier. For example, the expression for the outward flux across the outer barrier is

$$J_{out} = P_S k_{-2} - P_E S_o k_2. \qquad [12]$$

When P_E and P_S (Equations 11A, B) are substituted into Equation 12, one obtains

$$J_{out} = (S_i k_1 k_{-2} - S_o k_{-1} k_2)/(S_i k_1 + S_o k_2 + k_{-1} + k_{-2}). \qquad [13]$$

The flux, J_{out} in ions/sec may be converted to current (coulombs/sec) by multiplying by the valence, z, and the elementary charge e_0 (1.602×10^{-12} coulombs/e$^-$):

$$I_{out} = ze_0\left\{\frac{S_i k_1 k_{-2} - S_o k_{-1} k_2}{(S_i k_1 + S_o k_2) + (k_{-1} + k_{-2})}\right\}. \qquad [14]$$

Equation 14 with each of its rate constants expressed as a function of voltage (according to Equations 1 and 2) is a discrete-state analogue of the GHK current equation (Equation 5). To consider its implications, it is useful to first examine the predicted behaviour of unidirectional current. In the case of $S_0 = 0$, the unidirectional outward current is given by

$$I_{out} = ze_0\left\{\frac{S_i k_{-2}}{S_i + (k_{-1} + k_{-2})/k_1}\right\}. \qquad [15]$$

Equation 15 has the same form as the classic Michaelis–Menten equation of enzyme kinetics. Specifically, this particular model of a channel-mediated current may be compared directly to a two-step enzyme-catalysed reaction shown in *Figure 4a*. The rate law for an enzymatic reaction that involves formation of an enzyme–substrate complex (E–S) followed by direct conversion to product (P) is given by the following steady-state solution (34), where $[E]_0$ is equal to the total enzyme concentration:

$$v = \frac{[S]\,[E]_0 k_{cat}}{[S] + (k_{-1} + k_{cat})/k_1} = \frac{[S]V_{max}}{[S] + K_M}. \qquad [16]$$

Comparison of Equations 15 and 16 shows that a one-site channel is expected to follow essentially the same saturating behaviour with respect to internal ion concentration, S_i, as the enzyme-mediated reaction with respect to substrate concentration, [S]. Just as the maximal rate of an enzyme, V_{max}, is determined by the first order rate constant, k_{cat}, for conversion of E–S to P, the maximal unidirectional current through a channel, I_{max}, is determined by k_{-2}, the rate constant for exit from the binding site. We also deduce for this simple mechanism that the one-sided K_M of a channel is

$$K_M = \frac{k_{-1} + k_{-2}}{k_1} = K_S + \frac{k_{-2}}{k_1} \qquad [17]$$

where K_S is the intrinsic equilibrium dissociation constant for ion binding, k_{-1}/k_1, with respect to the inside solution. Analogous to the enzyme-catalysed reaction, the K_M or apparent affinity exhibited by the channel for the ion is larger than K_S by the term k_{-2}/k_1. What determines the channel current or the rate of the enzyme-catalysed reaction at low ion or substrate concentration? In this case, $S_i \ll K_M$ and Equation 15 reduces to

$$I_{out} = ze_0 S_i k_{-2}/K_M. \qquad [18]$$

This equation indicates that in the limit of low substrate concentration, the current is essentially a linear function of S_i with a slope governed by k_{-2}/K_M. This is significant because it relates directly to the concept of catalytic efficiency of an enzyme which is similarly measured by the term k_{cat}/K_M. This latter term is in the form of an apparent second order rate constant (with units of $sec^{-1}\ M^{-1}$) and is a general measure of the efficiency at which an enzyme can process a substrate (34). It can be shown that k_{cat}/K_M cannot be greater than any individual microscopic association rate constant in an enzyme's reaction pathway. For the most efficient enzymes, k_{cat}/K_M values are within an order of magnitude of the bimolecular diffusion limit, approximately $10^9\ sec^{-1}\ M^{-1}$. Another important point borrowed from enzyme kinetics is that, in the case of two substrates competing for the same enzyme, substrate discrimination or selectivity is determined by the ratio of k_{cat}/K_M for the two substrates (34). Likewise, in the Eyring formulation of channel permeation, the analogous ratio may be considered to be a measure of selectivity for different permeant ions. Ion selectivity may thus be considered to be determined by the rate at which an ion can exit the channel (k_{-2}) and the apparent binding affinity for the ion (reciprocal of K_M). In the competition for association at low concentration, a tightly binding ion will be more successful in gaining access to the channel. Ultimately, however, such an ion may carry less current in a higher concentration range because it may have a lower exit rate.

The preceding analysis of unidirectional current is useful in understanding the contribution of energy peaks and wells in the discrete-state approach, but it is somewhat artificial. Channels generally operate under conditions in which permeant ions are present on both sides, meaning that ions may enter and depart the channel from both directions. A good method for evaluating the intrinsic conduction properties of different ions is to characterize the dependence of unitary conductance on symmetrical ion concentration. This would be comparable to a study of velocity versus [substrate] curves for several substrate analogues of a particular enzyme. For the case of symmetrical solutions, $S_i = S_o = S$, and Equation 14 becomes

$$I_{out} = ze_0\left\{\frac{S(k_1k_{-2} - k_{-1}k_2)}{S(k_1 + k_2) + (k_{-1} + k_{-2})}\right\}. \qquad [19]$$

Since the conduction mechanism of Scheme 1 is a reversible process, it is expected to obey the principle of microscopic reversibility, meaning that the product of rate constants obtained by traversing a cycle in one direction must equal that for the other direction. When the voltage dependence of the rate constants is taken into account, microscopic reversibility leads to the following relationship (4, 35):

$$k_1(0)k_{-2}(0) = k_{-1}(0)k_2(0)\exp\{zFV/RT\} \qquad [20]$$

where $k_1(0)$, $k_{-2}(0)$, $k_{-1}(0)$, and $k_2(0)$ are zero-voltage rate constants. This relationship is automatically described by the energy diagram of *Figure 4b* since the calculated change in free energy for binding to the central well is the same from either direction.

In order to obtain an expression for the conductance at zero voltage, g_0, the various rate constants in Equation 19 are expressed as their respective functions of voltage, and the derivative dI_{out}/dV is evaluated at $V = 0$. If this manipulation is performed after simplification by Equation 20, the following result is obtained (4, 35):

$$g_0 = \frac{g_{max}S}{S + K_M} \qquad [21A]$$

where $$g_{max} = (z^2e_0^2/k_BT)k_{-1}k_{-2}/(k_{-1} + k_{-2}) \qquad [21B]$$

and $$K_M = k_{-1}/k_1 = k_{-2}/k_2. \qquad [21C]$$

This result may be compared to that discussed above for unidirectional current. The zero-voltage g_{max} of the symmetrically bathed channel is limited by the slowest step for leaving the channel in either direction (k_{-1} or k_{-2}) and the observed K_M is now simply equal to the equilibrium dissociation constant for binding to the site from either direction.

Equation 21 is now in a convenient form that can be used to extract the definition of permeability according to the discrete-state perspective. This result is obtained by again noting that in the limit of low substrate concentration ($S \ll K_M$) Equation 21A reduces to

$$g_0 = g_{max}S/K_M. \qquad [22]$$

By setting Equation 22 equal to Equation 8, solving for P_S and making use of $k_{-1}k_2 = k_1k_{-2}$ at $V = 0$ (Equation 20), we find:

$$P_S = \frac{RT}{z^2F^2}\frac{g_{max}}{K_M} = \left(\frac{1}{N_0}\right)\frac{k_1k_2}{k_1 + k_2}. \qquad [23]$$

In this equation, Avogadro's number, N_0, is introduced to convert from macroscopic to single-channel current. P_S is in units of cm^3/sec which would be equivalent to multiplying the macroscopic P_S (in cm/sec) of Equation 1 by the area occupied per channel molecule to obtain the unitary permeability of a single channel. Equation 23 shows that the permeability of a single-site channel as interpreted by reaction rate theory is effectively determined only by association rate constants (k_1, k_2), or the peak heights of *Figure 4b*. Furthermore this definition of permeability based on g_{max}/K_M is analogous to the catalytic efficiency of enzyme kinetics, k_{cat}/K_M, which we described previously as an apparent second-order rate constant.

In summary, unifying principles underlie the measurement of substrate/ion selectivity for a one-site enzyme, a one-site channel, and a Nernst–Planck continuum. In the limit of low concentration, a single-ion channel exhibits a linear dependence on ion concentration (*Figure 5*) which mimics the Nernst–

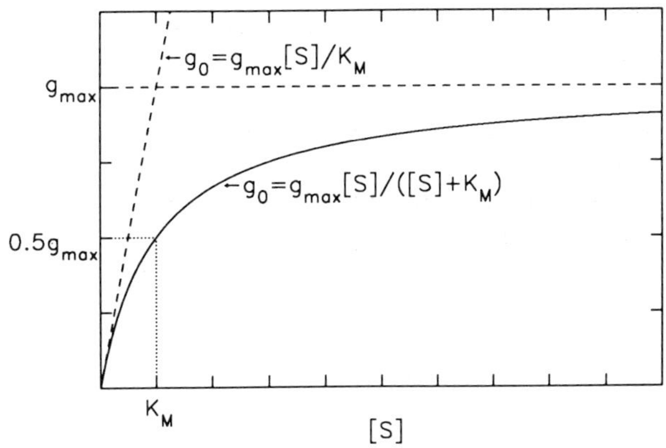

Figure 5. Theoretical dependence of zero-voltage conductance on symmetrical ion concentration for a single-ion channel. The solid line is plotted according to Equation 21A which follows Michaelis–Menten behaviour with respect to symmetrical ion concentration, [*S*]. The dashed lines indicate conductance in the limits of zero and infinite [*S*]. The limiting slope at low [*S*] is related to ion permeability, P_s, as discussed in the text.

Planck model. In this limit, the slope of the conductance–concentration relation is equal to g_{max}/K_M, analogous to k_{cat}/K_M, the specificity constant of an enzyme. Furthermore, g_{max}/K_M of a single-ion channel is analogous to permeability, P_S, in the GHK equations. Just as the ratio of k_{cat}/K_M for different substrates can be used to assess substrate discrimination by an enzyme, permeability ratios provide a good description of ion selectivity in channels. By generalizing an Eyring model to the case of a channel with n wells and $n+1$ barriers, Läuger (35) showed that for any channel that contains only one ion at a time, the bi-ionic permeability ratio, P_A/P_B (or the bi-ionic reversal potential), is independent of ion concentration. In contrast, the conductance ratios for two different ions (g_A/g_B) vary with concentration according to the respective Michaelis–Menten functions for the two ions. Thus, it is possible for an ion with a high-affinity K_M to exhibit a high permeability ratio relative to an ion with lower affinity but a higher g_{max}, *Figure 5* illustrates the conductance–concentration relation of a single-ion channel in the manner typically shown for the Michaelis–Menten function of an enzyme. While most ion-selective channels display conductance saturation indicative of ion binding and deviation from independence, the majority of ion channels studied to date do not conform *exactly* to Michaelis–Menten behaviour. One channel that does behave much like a single-ion channel is a K^+ channel from sarcoplasmic reticulum of skeletal muscle (36). However, the cardiac version of this channel exhibits more complex behaviour (37). Deviations from single-ion behaviour can provide important evidence for multiple-ion occupancy and ion–ion interactions. Such phenomena can be explored by considering the theoretical implications of a multi-ion channel using an Eyring-type model.

3.2 Multi-ion channels and ion–ion interactions

Despite the serious lack of tertiary structural information on channel proteins, much evidence indicates that the pores of highly selective channels can bind simultaneously more than one ion at a time. In the particular case of gramicidin, structural analysis reveals this directly. Gramicidin is an unusual 15-residue bacterial peptide with alternating D- and L-amino acids which forms a cation-selective channel across a lipid bilayer. The water-filled channel is formed through the centre of two single-stranded β-helices joined together by N-terminal dimerization of two gramicidin monomers. X-ray diffraction data (38) indicate that this 26 Å long structure contains two binding sites for Tl^+ which are 19.2 Å apart (38) in multilayer samples of lipid membranes containing the peptide (38).

The existence of a doubly occupied state is also reflected in the conductance behaviour of the gramicidin channel (39). Instead of concentration-independent permeability ratios characteristic of a single-ion channel, gramicidin exhibits permeability ratios that vary with ion activity (40, 41). Also, when the unitary conductance of gramicidin is measured at fixed total ionic concentration, in mixtures of K^+ and Tl^+, the conductance is observed to pass through a minimum (42). This phenomenon is termed 'anomalous mole-fraction behaviour' and is predicted by Eyring models that include double occupancy, but not by models limited to single occupancy. As another indication, instead of Michaelis–Menten behaviour, the conductance–concentration relation of certain ions rises to a maximum and exhibits a decrease in conductance at the high concentration limit (40, 41). While such behaviour is not compatible with Eyring-type models of a single-ion channel, it can be readily explained by a 3B2S model that includes double occupancy.

The state diagram of a 3B2S channel shown in *Figure 6a* indicates that such a channel has four possible occupancy states. The four states correspond to $\underline{EE}$ for an unoccupied channel, $\underline{SE}$ and $\underline{ES}$ for a singly occupied channel with an ion in the outer site and inner site respectively, and $\underline{SS}$ for a doubly occupied channel. Traversing clockwise around the lower cycle of $\underline{EE} \rightarrow \underline{SE} \rightarrow \underline{ES} \rightarrow \underline{EE}$ states results in translocation of an ion from the outside to the inside of the channel using only singly occupied states. As shown in *Figure 6c* and *d* (curve 2), a 3B2S channel restricted to these states would exhibit strict Michaelis–Menten behaviour like a 2B1S model (35). However, a new phenomenon is observed if the doubly occupied state $\underline{SS}$ is allowed. If there is no interaction between the two sites; i.e. if the binding constants of the two sites are not affected by occupancy of the adjacent site, then the conductance (at zero voltage) will rise to a maximum and decrease to zero in the limit of infinite symmetrical ion concentration (*Figure 6c* and *d*, curve 1). This

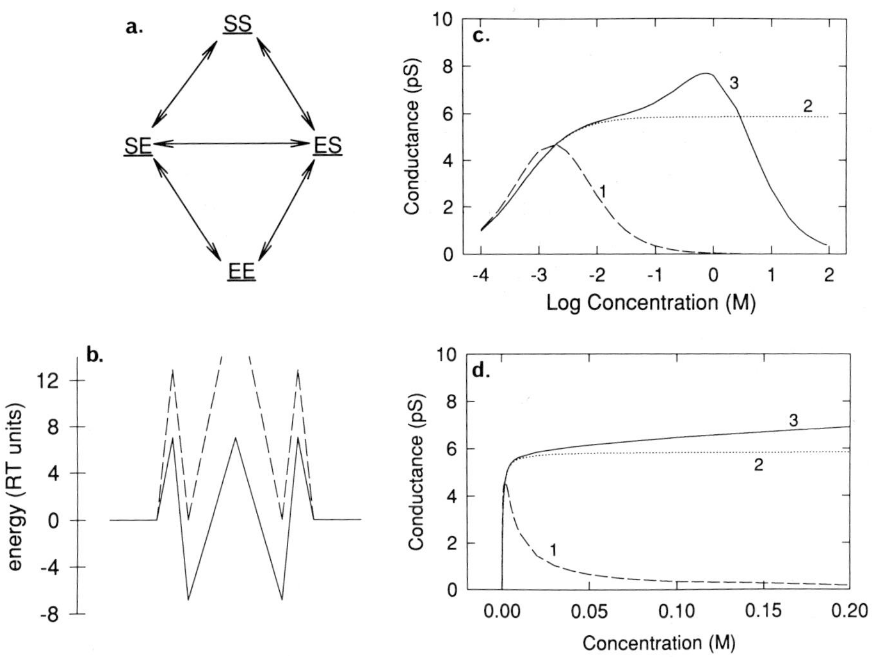

Figure 6. Discrete-state model and theoretical behaviour of a two-site channel. (a) State diagram showing routes of interconversion for a two-site channel. EE, unoccupied state; SE and ES, singly occupied states with an ion (*S*) bound to the inner site or outer site, respectively; SS, doubly occupied state. (b) Energy-barrier profile used to simulate conductance against concentration behaviour shown in (c) and (d). (c) Theoretical prediction of zero-voltage conductance against symmetrical ion concentration (log scale) for various types of ion–ion interactions. Curve 1: no ion–ion interaction is assumed. Ions bind with equal affinity to both sites and all states. Curve 2: the repulsive interaction between ions in the two sites is assumed to be infinitely strong. This eliminates the SS state by preventing binding of a second ion and thus, the channel behaves as a single-ion channel. Curve 3: the repulsion between the two sites is fixed such that affinity for binding of a second ion to a singly occupied channel is 1000-fold lower than that of the unoccupied channel. In this case, the broken line profile in (b) indicates the energy profile for binding the second ion. (d) The same plots in (c) are shown in the range of 0 to 0.2 M using a linear concentration scale. Theoretical curves were calculated using the AJUSTE program described in ref. 28.

phenomenon occurs because at very high *symmetrical* ion concentration, the probability of the doubly occupied SS state will approach 1.0 and no net transport will take place. For this effect to be observed, the K_D of the doubly occupied state must be within the allowable range of experimentation as limited by solubility; i.e. lower than ~2–10 M.

If there is ion–ion repulsion or a negative allosteric interaction between the two sites, then binding of a second ion to a singly occupied channel will exhibit lower affinity than ion binding to the completely unoccupied channel.

Furthermore, if the corresponding K_Ds are widely different, a biphasic conductance–concentration dependence is predicted (26). This is illustrated by the energy diagram in *Figure 6b*, which shows a hypothetical energy barrier profile for an unoccupied channel (solid lines) and a higher energy profile (dashed line) which corresponds to ion binding to an occupied channel. The energy profiles in *Figure 6b* are drawn such that the unoccupied channel has two sites that each bind an ion with a K_D of 1 mM, whereas binding of a second ion to the unoccupied site of a singly occupied channel has a K_D of 1000 mM. If the exit rate of an ion from the channel is enhanced by such an ion–ion interaction, then binding of the second ion will increase conductance and result in a biphasic conductance–concentration relation. This behaviour can be recognized graphically by an inflection in the log–log plot or a creeping increase in conductance at high concentration as observed in the linear plot (see *Figure 6c,d*, curve 3). Behaviour similar to this has been found in the conductance–concentration relations of various potassium and sodium channels where the possibility of multi-ion conduction mechanisms are under active consideration (37, 43–44). The hypothetical model of *Figure 6* emphasizes the importance of ion–ion interactions in single-file, multi-ion pores. If more than one ion can bind simultaneously with high affinity, the channel could easily become occluded. The fact that strong self-block behaviour (illustrated by curve 1 in *Figure 6*) is rare suggests that ion–ion repulsion (or 'negative co-operativity') is the rule rather than the exception in multi-ion channels.

Just as multi-site, allosteric enzymes are known to exhibit an array of complex kinetic behaviour, multi-ion conduction mechanisms lead to numerous complexities that have yet to be explored exhaustively. Aside from anomalous conductance–concentration behaviour, other hallmarks of multi-ion channels include an anomalously high voltage dependence of block ($z\delta > 1.0$), large Hill coefficients for blocker titrations ($n > 1.0$), and '*trans* effects' such as an increase in the dissociation rate of a blocker from one side of the channel caused by increasing permeant ion concentration on the other side. Further information on the extraordinary behaviour of multi-ion channels can be found in a variety of sources (7, 17, 26, 45, 46).

4. Probability distributions used to describe single-channel activity

Biochemists ordinarily work with large populations of molecules ($> 10^{-15}$ mole or 10^8 molecules). Ion channel research is a unique enterprise where the activity of a single protein molecule is a routine observation. Accordingly, different mathematical approaches appropriate to the macroscopic behaviour of a population versus the microscopic behaviour of single molecules are suited to these respective applications. The principles of chemical kinetics and enzyme kinetics are normally derived in terms of the rate of change of a bulk variable such as substrate or product concentration. At the single-channel level, the behaviour of the protein must be inferred from the measurement of many unitary events. In such a system, probability and statistical theory are used to analyse populations of events. The comparison of macroscopic and single-channel perspectives is interesting because it illuminates the underlying probabilistic basis of abstract mathematical concepts such as 'rate constant' and 'exponential'. Of the many theoreticians who have contributed to the foundation of single-channel analysis, D. Colquhoun and A. G. Hawkes have provided an encompassing body of work that is recommended to serious students of this subject (47–51). To introduce this topic using methods previously described by Colquhoun and others, we will first review the basic applications and inter-relationships of various types of probability distributions.

Probability distributions are used to describe the statistical behaviour of a random variable. These distributions can either be discrete or continuous. Discrete distributions are used for variables that only assume certain integral values such the number of channel closures in a burst of openings. Continuous distributions are used for variables, such as time, that can have any real value greater than zero, or current, that can have any negative or positive value.

The most familiar probability distribution is the normal or Gaussian distribution. It describes the continuous distribution of a population of measurements that vary by positive or negative random errors about the most probable or 'true' value called the mean. For example, the distribution of many repeated measurements of the unit current from opening events of the same channel is expected to exhibit a Gaussian distribution, because the background noise of the recording system introduces a small random error in every measurement. The Gaussian distribution has the form:

$$f(x) = \frac{1}{\sigma\sqrt{2\pi}}\exp\left\{\frac{-(x-\mu)^2}{2\sigma^2}\right\} \qquad [24]$$

where the μ is the mean and σ^2 is the variance, as defined by

$$\mu = \int_{-\infty}^{\infty} x f(x)\mathrm{d}x \qquad [25A]$$

$$\sigma^2 = \int_{-\infty}^{\infty} (x-\mu)^2 f(x)\mathrm{d}x = \int_{-\infty}^{\infty} x^2 f(x)\mathrm{d}x - \mu^2. \qquad [25B]$$

The Gaussian distribution is used widely to describe and analyse amplitude histograms of single-channel data. Such histograms may be compiled simply by digitizing a current record at a certain sampling rate (e.g. 1000 points per second) and plotting the number of current amplitudes observed as a function of bin width (e.g. −0.1 to 0 pA, 0 to 0.1 pA, etc.). Generally, for each

Gaussian component detected in the histogram, the mean (μ) and standard deviation ($\sigma = \sqrt{\sigma^2}$) are interpreted as the unitary current of a conductance state (such as open, closed, or sublevel) and a measure of the background noise, respectively. By definition, the amplitude histogram of the closed state should exhibit the same standard deviation as the zero-current record, unless excess noise is superimposed on the closed state from another source such as a leak conductance. In comparison to amplitude histograms of the closed state, those of the open state often exhibit a larger standard deviation or are skewed instead of symmetrical. Skewness toward the closed state can result from brief unresolved channel closures that do not reach baseline level. Open state noise in excess of background can be produced by any fluctuating process that is not completely eliminated by filtering. For example, a very fast channel blocker can produce a noisier open-state level and/or an apparent reduction in unitary conductance. The unitary conductance can appear to be reduced because filtering effectively averages rapid blocking events to produce a lower mean current level. In the example of rapid block, the amplitude histograms can be described adequately by a beta function which takes into account the kinetics of the rapid blocking process and the filter response function (52). A variety of evidence also indicates that excess open-state noise may result from inherent fluctuations of protein structure in the open-pore conformation (19).

Another type of probability distribution encountered in channel research is the binomial distribution. This discrete distribution describes the probability of observing r simultaneous openings from a membrane containing n independent channels, each of which has an individual probability of opening, p. If each channel gates independently, the probability of r channels opening simultaneously, $P(r)$, is equal to the product of three terms: the probability of a group of r channels being open, p^r; the probability of a group of $n - r$ channels not being open, $(1 - p)^{n-r}$; and the number of ways a group of r channels out of n channels can be open. This last term is often written as ^nC_r, a function of combinatorial analysis which gives the number of possible combinations of n things when taking r at a time. Thus, the binomial probability distribution is given by

$$P(r) = {^nC_r}p^r(1 - p)^{n-r} = \frac{n!}{r!(n - r)!}\,p^r(1 - p)^{n-r} \qquad (r = 0,1,2\ldots). \quad [26]$$

This distribution can be used to determine whether the various levels of a membrane containing n channels are distributed binomially and thus whether the n channels indeed operate independently. For example, binomial independence has been verified for a bilayer containing four sarcoplasmic reticulum potassium channels (53) and an excised patch containing three Ca^{2+}-activated potassium channels (54). Another interesting application is the case of a chloride channel from *Torpedo* electroplax membrane which behaves as a dimer of two independent channel subunits in one phase of its kinetic cycle (55).

It can be shown that the mean of a binomial distribution is equal to np and its variance is equal to $np(1-p)$ (56). In one type of noise analysis, these relationships can be exploited to estimate unitary conductance from a homogeneous population of many channels by measuring the variance of the noise of a macroscopic current as a function of the mean current amplitude (57). The relationship of the binomial distribution to the Gaussian distribution can be intuitively understood by considering the measurement of a noisy macroscopic current resulting from a binomial distribution of many channels, each with an opening probability of 0.5. From moment to moment the number of open channels varies. When a channel opens it adds to the current. When a channel closes it reduces the current. This situation is equivalent to the probability distribution of a measurement that includes many independent errors, each of which have an equal probability of adding or not adding to the measurement. Thus, the binomial distribution approaches the Gaussian distribution in the limit of $p \rightarrow 0.5$ and $n \rightarrow \infty$.

Instead of the probability of the simultaneous occurrence of r openings of n channels, consider the probability of observing r openings from one channel in a unit of time, t. In this case we express the mean by a rate constant, λ, which defines the mean number of events per unit time. Thus, λt is the mean number of events observed in time, t. To treat this problem using the binomial distribution, t is divided by n small intervals and the probability, p, in Equation 26, is replaced with $\lambda t/n$, since this is the intrinsic probability of observing just λt events out of n 'attempts'. For this situation, the relevant distribution is the binomial distribution in the limit of $n \rightarrow \infty$, since the rate constant may be considered to be the mean frequency of successful events in an infinite number of trials, that is, $\lambda = np/t$. If this problem is solved analytically as stated, one finds that the binomial distribution reduces to the Poisson distribution (e.g. see ref. 56, p. 102):

$$P(r) = \frac{(\lambda t)^r}{r!} \exp\{-\lambda t\} \qquad (r = 0,1,2\ldots) \qquad [27]$$

where $P(r)$ gives the discrete probability of observing an integral number of events r in time t. The Poisson distribution is widely known to be applicable to processes such as radioisotope disintegration counting and the quantal release of acetylcholine at synapses. Although single-channel fluctuations may be analysed according to the Poisson distribution, the finite lifetime of opening and closing events makes it more convenient to consider the probability of finding no openings in time, t. According to Equation 27, this probability evaluated using $r = 0$ is equal to $P(0) = \exp(-\lambda t)$. If no openings occur in time, t, this means that the lifetime of a closing is greater than the time interval, t. Thus, if the time interval is varied, the probability that the observed lifetime of silent periods, τ, is greater than time, t is

$$P(\tau > t) = \exp(-\lambda t). \qquad [28]$$

Equation 28 is the cumulative form of a continuous distribution called the exponential distribution. Thus, we see that the Poisson distribution reduces to the exponential distribution by considering the length of silent periods or the waiting time between events. The Poisson and the exponential distributions are noted for their simple relationships between the mean and variance. For the Poisson distribution, the mean of the distribution equals the variance; for the exponential distribution, the mean is equal to the standard deviation and the reciprocal of the rate constant:

Poisson distribution: $\mu = \sigma^2 = \lambda t$ (units of events) [29A]

exponential distribution: $\mu = \sigma = \lambda^{-1}$ (units of time) [29B]

If a particular single-channel process is shown to conform to an exponential distribution by measuring a large population of events, a good estimate of the rate constant can be obtained from a small sample by simply measuring the mean of n events. According to Equation 29B and the properties of sampling distributions (56), this sample mean should have a standard deviation that varies according to $1/\sqrt{n}$. Thus a sample of 100 events from a population that has an exponential distribution should have 68% probability of being within ±10% of the population mean.

The particular form of Equation 28 is called a cumulative probability distribution function because it describes the summed probability of all events that are longer than a certain time, t. In practice, dwell-time distributions of single-channel events are usually compiled and analysed in the form of a probability density function or pdf, $f(t)$. For example, the probability density form of the Gaussian distribution is given in Equation 24. For a probability density function, the area under any segment of the curve corresponds directly to probability when the total area is normalized to 1. In exact terms, the area under $f(t)$ between any two times, t_1 and t_2, is the probability of finding events that are longer than t_1 and less than or equal to t_2. Thus, the density function and the cumulative function are related as derivative and integral. To find the pdf of the exponential distribution, Equation 28 is first converted to the cumulative probability of all events that are less than or equal to t and the derivative is taken as follows:

$$F(t) = P(\tau \leqslant t) = 1 - P(\tau > t) = 1 - \exp(-\lambda t) \quad [30A]$$

$$\frac{dF(t)}{dt} = f(t) = \lambda \exp(-\lambda t). \quad [30B]$$

Figure 7e shows an example of a pdf of an exponential distribution plotted using linear co-ordinate axes. A variety of other plotting methods are commonly utilized to enhance the detection of multiple exponential components. In particular, a logarithmically binned time axis with a linear or square root ordinate axis is a useful representation (see *Figure 8*) because each exponential component can be recognized as a distinct peak in the histogram with

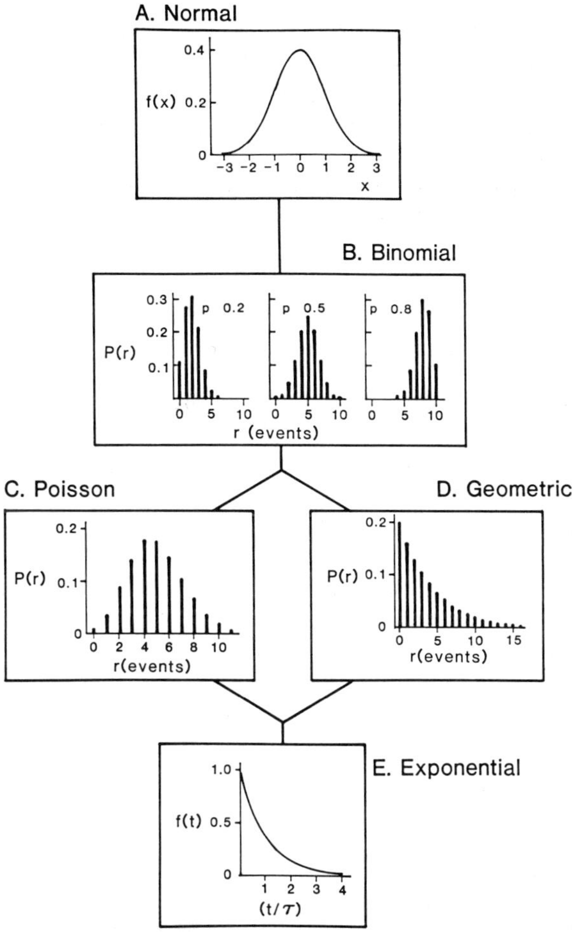

Figure 7. Various types of probability distributions encountered in the analysis of ion channels. (a) Normal or Gaussian distribution plotted according to Equation 24 using $\mu = 0$ and $\sigma = 1$. (b) Binomial distribution according to Equation 26 for a membrane patch containing $n = 10$ independently gating channels, where the intrinsic probability of channel opening is $p = 0.2$, 0.5, or 0.8. (c) Poisson distribution according to Equation 27 using $\lambda t = 5$. (d) Geometric distribution according to Equation 31 using $q = 0.2$. (e) Exponential distribution according to Equation 30B where the time constant, $\tau = 1/\lambda$.

the maximum corresponding to the mean or time constant, $\tau = 1/\lambda$, of that component (57).

The final type of distribution to be discussed is another discrete distribution called the geometric distribution. This distribution is encountered in the analysis of bursting behaviour of ion channels (48, 49). The geometric distribution has the form:

$$P(r) = q(1 - q)^r \qquad (r = 0,1,2\ldots) \qquad [31]$$

where q is the probability of a channel going from a short-lived closed state to a long-lived inter-burst state, and $1 - q$ is the probability of a channel going from a short-lived closed state to the bursting open state. $P(r)$ therefore gives the probability that r consecutive openings followed by one long inter-burst closing event will occur. The geometric probability distribution is named after the geometric series, $\sum_{n=0}^{\infty} p^n$, which converges to $(1 - p)^{-1}$ for values of $|p| < 1$. The cumulative probability given by the geometric distribution is thus a geometric series in $p = 1-q$. By comparison of the geometric distribution with Equation 26, it can be observed that the geometric distribution is a special case of the binomial distribution. Where the binomial distribution is concerned with r simultaneous openings and $n-r$ closings of n channels arranged ${}^nC_r$ ways, the geometric distribution considers only r rapid openings and one long closing of one channel, occurring only in that particular combination. The geometric distribution decreases by a constant factor with increasing r, a characteristic that is also true of the exponential distribution with respect to time. In fact, it can be shown that the geometric distribution approaches the exponential distribution for large values of r using the relationship

$$\lim_{r\to\infty}(1 - q)^r = \exp(-qr) \qquad [31A]$$

which holds for $q < 1$. *Figure 7* summarizes examples of five different probability distributions that are important in single-channel analysis and indicates the relationships that have been described between the various functions.

5. Analysis of channel gating according to the Q-matrix formulation

5.1 Channel gating can be described by a Markov process

The preceding section introduced the application of probability theory and statistics to analysis of single-channel activity. Since single-channel fluctuations generally appear to be due to transitions of the channel protein between discrete conformational states, a major goal of stochastic analysis is to elucidate steps that occur in the pathway of channel opening, closure, and block. The central assumption underlying this analysis is that each step in this pathway constitutes a transition between two kinetically distinct states of a 'Markov process' (59). Hence, it is sometimes referred to as Markov analysis. A time-homogeneous Markov process is one in which transitions from any given state do not depend on how or at what time the system reached that state; i.e. the system has no memory. To use a familiar example, in flipping a coin, one assumes that a coin which landed showing heads will, on the next toss, still have the same 50% chance of showing either heads or tails no matter how it landed on previous occasions, or at what time of day it is thrown.

Quantitative analysis of single-channel gating between many discrete states necessarily involves the solution of many simultaneous linear equations. The problem of solving such systems of equations is central to the field of linear algebra and is most conveniently handled by formulating the equations in matrix notation. Colquhoun and Hawkes (48–50) pioneered the application of this matrix approach to analysis of the gating behaviour of single channels. In keeping with mathematical precedent, they denoted the central matrix with the character 'Q' and hence established the 'Q-matrix' approach to channel gating. This approach has recently been extended by Crouzy and Sigworth (60) to include analysis of gating-current noise. The Q-matrix is elegant and versatile, making otherwise difficult calculations relatively straightforward. However, the Q-matrix method appears daunting to new students because of its mathematical content. With this situation in mind, the purpose of this section is two-fold. First, it provides an introductory level source for readers who wish to learn and access more advanced material. Secondly, for those whose primary interest is in using computer programs based on this technique, it is designed to demystify the underlying calculations.

For clarity, analysis in this section is restricted to the simplest possible kinetic schemes and the notation of Colquhoun and Hawkes is used wherever practical. Furthermore, the chosen examples are those for which results have already been derived directly (51) without the use of matrix notation. For readers who need a review of basic matrix manipulations, Appendix 2 at the end of this chapter provides a minimal summary. Those looking for further background should consult the introduction to matrix algebra given by Stephenson (61) and the texts by Strang (62) and Hollingsworth (63).

5.2 The probabilistic nature of ion channel gating

As illustrated in *Figures 1* and *9*, single-channel records consist of a consecutive display of random opening and closing events of varying duration. The length of each event (an opening, closure, or burst) is referred to as a dwell time. Important information concerning channel gating is contained in the *distribution* of dwell times. At a given instant when the channel is in the open state one cannot predict how long the channel will remain open. However, by examining the distribution of open times, the *probability* that an opening will last for a certain length of time can be determined. A typical problem in channel analysis is to identify kinetic schemes that predict observed probability distributions of open and closed dwell times. Such studies involve the statistical comparison of populations of measured dwell times compiled from single-channel records with theoretical dwell-time distributions of various kinetic schemes. The Q-matrix method simplifies this task by allowing a single computer program to be devised which will handle any kinetic scheme of interest.

In the following presentation, the initial goal is to derive closed and open dwell-time distributions for a simple two-state scheme. This is done by first obtaining equations that relate transition probabilities to rate constants. Next the Q-matrix is derived and probability density functions are calculated by manipulation of the Q-matrix. Although these steps are quite lengthy and the same distributions can be derived more directly by traditional methods, they are necessary to illustrate fully the application of the Q-matrix approach.

5.3 Conditional probabilities and transition rates

$$\underset{2}{\mathrm{C}} \underset{k_{12}}{\overset{k_{21}}{\rightleftharpoons}} \underset{1}{\mathrm{O}} \qquad \text{Scheme 2}$$

Scheme 2 is a simple two-state gating equilibrium with the states arbitrarily numbered as 1 for the open (O) state, and 2 for the closed (C) state. Using this notation, the rate of transition from state 1 to state 2 is k_{12} and the transition rate in the opposite direction is k_{21}. Intuitively, one might expect that the probability of a transition occurring from state 1 to state 2 during a period of time (Δt) is simply the product of the transition rate (or frequency) and the period of time. Thus, the probability of a transition from state 1 to state 2 during the period Δt would be $k_{12}\Delta t$. However, this is not quite exact. To give such rate constants a precise mathematical definition the idea of a conditional probability must be introduced. Conditional probabilities are written as $Prob(\mathrm{A}|\mathrm{B})$, which is read as the probability of event A given event B. (For example, one might describe the probability of eating chocolate cake given that there is some in the kitchen, or the probability of eating chocolate cake given that one has to go to the store to buy it. In these terms it is clear that the probability of event A may vary considerably with condition B!) To describe k_{12} and k_{21} in terms of conditional probabilities, suppose that a channel open at time, t, is observed for a short time-interval from t to ($t + \Delta t$). The conditional probability expression for k_{12} is then given by

$$Prob(\text{Closes between } t \text{ and } t + \Delta t | \text{Open at } t) = k_{12}\Delta t + o(\Delta t). \quad [32]$$

The left-hand term is the probability that a channel which is open at time t will close in the interval from t to ($t + \Delta t$). The right-hand term, $o(\Delta t)$, accounts for terms such as the probability of multiple transitions within the Δt (47). As Δt tends to zero so does the term $o(\Delta t)$. Furthermore, $o(\Delta t)$ approaches zero at a faster rate than Δt, so that the term $o(\Delta t)/\Delta t$ also approaches zero in this limit.

An expression similar to Equation 32 may be written for the probability that a channel does *not* close between t and ($t + \Delta t$):

$$Prob(\text{Does not close between } t \text{ and } t + \Delta t | \text{Open at } t) = 1 - [k_{12}\Delta t + o(\Delta t)]. \quad [33]$$

Since the channel must either close or remain open between t and $t = t + \Delta t$, the probabilities of Equations 32 and 33 must sum to unity. Similar expressions follow for k_{21} (i.e. the transition rate from closed to open):

$$Prob(\text{Opens between } t \text{ and } t + \Delta t|\text{Closed at } t) = k_{21}\Delta t + o(\Delta t) \quad [34]$$

$$Prob(\text{Does not open between } t \text{ and } t + \Delta t|\text{Closed at } t) = 1 - [k_{21}\Delta t + o(\Delta t)]. \quad [35]$$

To define the rate constant k_{12} in terms of a transition probability, divide both sides of Equation 32 by Δt and evaluate in the limit of $\Delta t \rightarrow 0$:

$$k_{12} = \lim_{\Delta t \rightarrow 0}[Prob(\text{Closes between } t \text{ and } t + \Delta t|\text{Open at } t)/\Delta t]$$

$$= \lim_{\Delta t \rightarrow 0}[Prob(\text{Closed at } t + \Delta t|\text{Open at } t)/\Delta t]. \quad [36]$$

In other words, rate constants k_{12} and k_{21} are defined as the probability of a transition between states per unit time. In the general case, rate constant q_{ij} represents the probability of a transition between any two states i and j per unit time and is defined as:

$$q_{ij} = \lim_{\Delta t \rightarrow 0}[Prob(\text{State } j \text{ at } t + \Delta t|\text{State } i \text{ at } t)/\Delta t]. \quad [37]$$

This equation is sometimes written in the more compact form

$$q_{ij} = \lim_{\Delta t \rightarrow 0}[P_{ij}(\Delta t)/\Delta t]. \quad [38]$$

Since this expression for $P_{ij}(\Delta t)$ is valid at any time, t, it could also be written as $Prob$ (State j at Δt | State i at 0). Hence, the general definition of transition probability P_{ij} at time t is:

$$P_{ij}(t) = Prob(\text{State } j \text{ at time } t|\text{State } i \text{ at time } 0). \quad [39]$$

Thus, as $\Delta t \rightarrow 0$, Equations 32–35 can be rewritten as follows:

$$P_{12}(\Delta t) = k_{12}\Delta t + o(\Delta t) \quad [40]$$

$$P_{11}(\Delta t) = 1 - [k_{12}\Delta t + o(\Delta t)] \quad [41]$$

$$P_{21}(\Delta t) = k_{21}\Delta t + o(\Delta t) \quad [42]$$

$$P_{22}(\Delta t) = 1 - [k_{21}\Delta t + o(\Delta t)]. \quad [43]$$

5.4 Derivation of the Q-matrix

Equations 40–43 relate the transition rates of Scheme 2 to conditional probabilities. The next step is to determine how these probabilities vary in time. For this purpose, consider a channel observed in the open state at both $t =$ 0 and $t = t + \Delta t$. The probability of such an event is

$$P_{11}(t + \Delta t) = Prob(\text{State 1 at } t + \Delta t|\text{State 1 at } 0). \quad [44]$$

Assume now that the channel's transition probabilities do not change with time regardless of how long the channel remains in its present state or how it arrived at that state. (As discussed earlier, this assumption is the basis of Markov analysis.) The probability expressed in Equation 44 can now be rewritten as the sum of two components:

$$\begin{aligned} P_{11}(t + \Delta t) = &\ Prob(\text{State 1 at } t + \Delta t | \text{State 1 at } t) \\ &\cdot Prob(\text{State 1 at } t | \text{State 1 at } t = 0) \\ &+ Prob(\text{State 1 at } t + \Delta t | \text{State 2 at } t) \\ &\cdot Prob(\text{State 2 at } t | \text{State 1 at } t = 0) \end{aligned} \quad [45]$$

The first term in the expression on the right of Equation 45 represents the probability that a channel open at $t = 0$ is also open (i.e. in state 1) at $t = t$ and $t = t + \Delta t$. This term is itself written as a product of two probabilities which are the probability of a channel being open at $t = t$ given that it was open at $t = 0$, and the probability of a channel being open at $t = t + \Delta t$ given that it was open at $t = t$. These probabilities are multiplied because the probability of both event A *and* event B is the product of their individual probabilities, $P_A P_B$. The second term on the right-hand side of Equation 45 represents the probability that a channel open at $t = 0$ is closed at t and finally open again at $t = t + \Delta t$. This latter term is written as a product analogous to the first term. The two product terms on the right of Equation 45 must be added since the probability of event A *or* event B occurring is $P_A + P_B$. Using the earlier relationships of Equation 41 and Equation 42, Equation 45 can be rewritten as

$$P_{11}(t + \Delta t) = \{1 - k_{12}\Delta t - o(\Delta t)\}P_{11}(t) + \{k_{21}\Delta t + o(\Delta t)\}P_{12}(t). \quad [46]$$

Rearranging and dividing by Δt:

$$\frac{P_{11}(t + \Delta t) - P_{11}(t)}{\Delta t} = -\left\{k_{12} + \frac{o(\Delta t)}{\Delta t}\right\} P_{11}(t) + \left\{k_{21} + \frac{o(\Delta t)}{\Delta t}\right\} P_{12}(t). \quad [47]$$

The expression on the left-hand side of Equation 47 may be recognized from differential calculus as the definition of the derivative of P_{11} in the limit of $\Delta t \rightarrow 0$. Thus, taking the limit as $\Delta t \rightarrow 0$:

$$\frac{dP_{11}(t)}{dt} = -k_{12}P_{11}(t) + k_{21}P_{12}(t). \quad [48]$$

Similar reasoning leads to expressions for the derivatives of the remaining transition probabilities, P_{12}, P_{22}, and P_{21}, of Scheme 2:

$$\frac{dP_{12}(t)}{dt} = k_{12}P_{11}(t) - k_{21}P_{12}(t) \quad [49]$$

$$\frac{dP_{22}(t)}{dt} = k_{12}P_{21}(t) - k_{21}P_{22}(t) \quad [50]$$

$$\frac{dP_{21}(t)}{dt} = -k_{12}P_{21}(t) + k_{21}P_{22}(t). \quad [51]$$

The four simultaneous equations 48–51 involve four unknowns $P_{11}(t)$, $P_{12}(t)$, $P_{22}(t)$, and $P_{21}(t)$. These equations can be written together (and solved) in matrix form:

$$\frac{\mathrm{d}}{\mathrm{d}t}\begin{bmatrix} P_{11} & P_{12} \\ P_{21} & P_{22} \end{bmatrix} = \begin{bmatrix} P_{11} & P_{12} \\ P_{21} & P_{22} \end{bmatrix} \begin{bmatrix} -k_{12} & k_{12} \\ k_{21} & -k_{21} \end{bmatrix}. \qquad [52]$$

Equation 52 may be verified by multiplying the two right-hand matrices together and comparing the result with Equations 48–51. In short form, Equation 52 is written as

$$\frac{\mathrm{d}}{\mathrm{d}t}\boldsymbol{P}(t) = \boldsymbol{P}(t)\boldsymbol{Q}. \qquad [53]$$

The last matrix on the left ($\boldsymbol{Q}$), containing the rate constants q_{ij}, is the Q-matrix. Notice from Equation 52 that the Q-matrix can be obtained directly by placing the transition rate q_{ij} in the ith row and jth column, and then filling in diagonal elements ($i = j$) with a number such that the sum of each row in the matrix is zero. This result is entirely general regardless of the complexity of the particular kinetic scheme. The index variables i and j assume values ranging from 1 to N, where N is the number of states in the scheme. Thus an N-state scheme always produces an $N \times N$ matrix.

As a simple differential equation, Equation 53 has a solution of the form $P(t) = P(0)\exp(\boldsymbol{Q}t)$, which is also the correct form of the matrix solution. In the matrix solution, $\boldsymbol{P}(0)$ is the unit matrix $\boldsymbol{I}$ (which has unit elements for $i = j$ and zero for all other elements). This follows from the definition of the conditional probabilities evaluated at $t = 0$. For example, $\boldsymbol{P}_{11}(0)$, the probability of being in state 1 at $t = 0$ given that the channel was in state 1 at $t = 0$ must equal one, etc. Since multiplying by the unit matrix is equivalent to multiplying by one, $\boldsymbol{P}(0)$ is usually omitted from the expression, giving the solution:

$$\boldsymbol{P}(t) = \mathrm{e}^{\boldsymbol{Q}t}. \qquad [54]$$

The term, $\exp(\boldsymbol{Q}t)$, a matrix exponential, may seem difficult to interpret. In fact, it is simply another matrix that can be found by standard manipulations. This can be shown by writing the exponential as a Maclaurin series expansion:

$$\mathrm{e}^{\boldsymbol{Q}_{AA}t} = \boldsymbol{I} + \boldsymbol{Q}_{AA}t + \frac{(\boldsymbol{Q}_{AA}t)^2}{2!} + \frac{(\boldsymbol{Q}_{AA}t)^3}{3!} + \ldots + \frac{(\boldsymbol{Q}_{AA}t)^n}{n!} \qquad [55]$$

where $\boldsymbol{I}$ is the unit matrix. Thus, $\exp(\boldsymbol{Q}t)$ can be obtained simply by adding and multiplying other matrices.

5.5 Relationship between conditional and macroscopic probability equations

It is worth noting that the probability expressions of Equations 45–51 are directly related to familiar equations of macroscopic behaviour that are usually derived on the basis of mass action. Consider that a channel can be in state 1 at time t by either of 'two' routes. It can start in state 1 at time $t = 0$ and be in state 1 at time $t = t$ or, alternatively, it can start in state 2 at $t = 0$ and be in state 1 by time $t = t$. Therefore, the probability of being in state 1 at time $t = t$ by either route is $P_1(t) = P_{11}(t) + P_{21}(t)$. Similarly, $P_2(t) = P_{22}(t) + P_{12}(t)$. Thus, Equations 48 and 51 can be combined to obtain a basic equation of chemical kinetics:

$$\frac{\mathrm{d}}{\mathrm{d}t}P_1(t) = -k_{12}P_1(t) + k_{21}P_2(t). \qquad [56]$$

For applications dealing with the macroscopic time course of a reaction following Scheme 2, Equation 56 is normally written directly according to the standard rules for deriving rate laws in chemical kinetics.

5.6 Eigenvalues and probability density functions of the Q-matrix

Earlier, in Section 4 of this chapter, the probability density function, $f(t)$, was introduced. The pdf for dwell times in state j is defined as

$$f_j(t) = \lim_{\Delta t \to 0}[Prob(\text{Dwell time in state } j \text{ is} > t \text{ and} \leq t + \Delta t)/\Delta t]. \qquad [57]$$

As described earlier, the area under $f(t)$ between any two times is the probability that a dwell time in state j is greater than t_1, but less than or equal to t_2. In practice, channel data are analysed by choosing an appropriate time-binning interval and measuring the number of dwell times that fall into consecutive bins. The number of observations in each bin can be compared to the number of events predicted by a theoretical pdf for a particular kinetic model. *Figure 8* shows an actual distribution of open-state dwell times compiled from a recording of a single Ca^{2+}-activated K^+ channel as plotted in two different ways and fit to a single exponential pdf.

The first step in finding $f(t)$ for any particular kinetic scheme is to partition the Q-matrix into submatrices containing the state(s) of interest. The example of Scheme 2 has only two types of states, open or closed. Scheme 2 also has only one open state but in general these will be A in number. The 'group' of A open states is therefore collectively designated with an $A \times A$ matrix. Similarly, the 'group' of B closed states is collectively represented with a $B \times B$ matrix, thus defining four submatrices $\boldsymbol{Q}_{AA}$, $\boldsymbol{Q}_{BB}$, $\boldsymbol{Q}_{BA}$, and $\boldsymbol{Q}_{BA}$. Equation 52 may thus be rewritten:

$$\frac{\mathrm{d}}{\mathrm{d}t}\begin{bmatrix} \boldsymbol{P}_{AA} & \boldsymbol{P}_{AB} \\ \boldsymbol{P}_{BA} & \boldsymbol{P}_{BB} \end{bmatrix} = \begin{bmatrix} \boldsymbol{P}_{AA} & \boldsymbol{P}_{AB} \\ \boldsymbol{P}_{BA} & \boldsymbol{P}_{BB} \end{bmatrix} \begin{bmatrix} \boldsymbol{Q}_{AA} & \boldsymbol{Q}_{AB} \\ \boldsymbol{Q}_{BA} & \boldsymbol{Q}_{BB} \end{bmatrix}. \qquad [58]$$

In partitioning the Q-matrix, $\boldsymbol{Q}_{AA}$ includes rate constants for transitions between A states and the related $i = j$ diagonal terms of $\boldsymbol{Q}$. Likewise $\boldsymbol{Q}_{BB}$ contains rate constants for transitions between B states and related diagonal

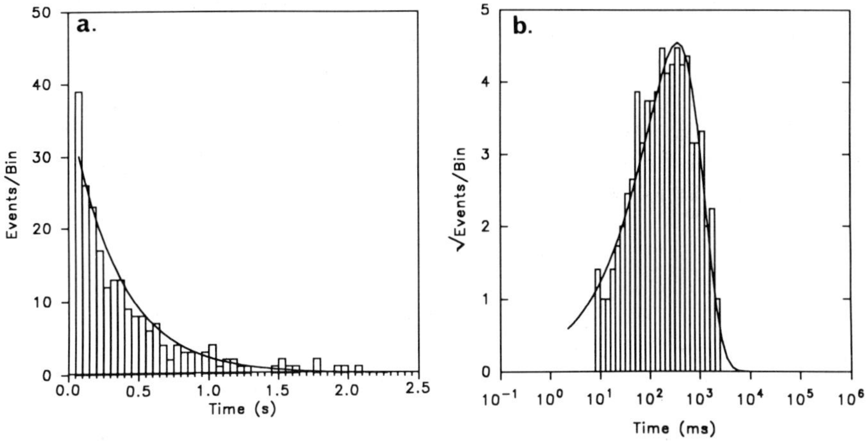

Figure 8. Comparison of single-exponential distributions of channel dwell-time events plotted in two different ways. Open channel dwell-times were accumulated from a bilayer recording of a single Ca^{2+}-activated K^+-channel. (a) Channel dwell-time events were compiled into a probability density histogram using a time bin of 50 msec. The histogram data are overlayed with a single-exponential pdf, Equations 30B or 69 in the text. The first bin is omitted since brief events are not fully resolved. (b) The same data in (a) are plotted according to Sigworth and Sine (58) using constant binwidth on a logarithmic time scale (abscissa) and a square-root axis for the ordinate. Using this transformation the probability density function is $g(z) = \sqrt{\exp(z - \exp(z))}$ where $z = \ln(10)[\log_{10}(t) - \log 10(\tau)]$ and τ is the reciprocal of the rate constant.

terms. $\boldsymbol{Q}_{AB}$ and $\boldsymbol{Q}_{BA}$ contain the rate constants for transitions between $A \rightarrow B$ and $B \rightarrow A$ states, respectively. Each $\boldsymbol{Q}$ submatrix in Equation 58 obeys relations such as

$$\boldsymbol{P}_{AA}(t) = e^{\boldsymbol{Q}_{AA}t}. \qquad [59]$$

In the present example of Scheme 1 there is only a single open state (state 1) in 'group' A, hence $A = 1$. Similarly, there is a single closed state (state 2) in the 'group' B, so $B = 1$. Thus, there are no actual transitions to consider within groups A or B, and only the diagonal terms are used: $\boldsymbol{Q}_{AA} = [-k_{12}]$ and $\boldsymbol{Q}_{BB} = [-k_{21}]$. The probability density functions for the open and closed states can now be calculated. The formula given by Colquhoun and Hawkes (51) and derived in ref. 49, when applied to the open state pdf of Scheme 2, is

$$f_A(t) = \Phi_A \boldsymbol{P}_{AA}(t)\,(-\boldsymbol{Q}_{AA})\boldsymbol{U}_A = \Phi_A e^{\boldsymbol{Q}_{AA}t}(-\boldsymbol{Q}_{AA})\boldsymbol{U}_A. \qquad [60]$$

The function $f_A(t)$ is the pdf of dwell times in state(s) A. In calculating probability density functions for other schemes, the same formula is used but the subscripts 'A' in Equation 60 are substituted by the particular states of interest. $\boldsymbol{U}_A$ is a vector matrix of unit elements and Φ_A is a row-vector containing the relative probabilities of an opening, starting in each of the open states within group A. Since there is only one open state in the present example, the probability that an opening will start in this state is unity and thus, $\Phi_A = [1]$. A more general formula for calculating Φ_A is given in ref. 49. The only term in Equation 60 that is difficult to calculate is the matrix, $\exp(\boldsymbol{Q}_{AA}t)$. One approach is to use the exponential series expansion (Equation 55). However, a more simple method is to write $\exp(\boldsymbol{Q}_{AA}t)$ as a sum according to the following equation of Colquhoun and Hawkes (48):

$$\boldsymbol{P}_{AA}(t) = \mathrm{e}^{\boldsymbol{Q}_{AA}t} = \sum_{i=1}^{i=A} \boldsymbol{A}_i \mathrm{e}^{\lambda_i t} \qquad [61]$$

where λ_i are called the eigenvalues of $\boldsymbol{Q}_{AA}$, and the $\boldsymbol{A}_i$, matrices are called the projectors or idempotents of matrix $\boldsymbol{Q}_{AA}$. The $\boldsymbol{A}_i$ matrices have the following properties:

$$\boldsymbol{A}_i\boldsymbol{A}_i = \boldsymbol{A}_i \qquad [62]$$

$$\boldsymbol{A}_i\boldsymbol{A}_j = 0 \qquad \text{for } i \neq j \qquad [63]$$

$$\sum_{i=1}^{i=A} \boldsymbol{A}_i = \boldsymbol{I}. \qquad [64]$$

The $\boldsymbol{A}_i$ matrices can be found once the eigenvalues are known by using the equation

$$\boldsymbol{A}_i = \prod_{j=1}^{j=A} \frac{(\boldsymbol{Q}_{AA} - \lambda_j \boldsymbol{I})}{(\lambda_i - \lambda_j)} \qquad (i \neq j). \qquad [65]$$

The symbol Π indicates the product of terms for all j. This method of calculating $\boldsymbol{A}_i$ is slightly different to that given by Colquhoun and Hawkes (48) as outlined in Appendix 2 of this chapter. Before proceeding to calculate the matrix, $\exp(\boldsymbol{Q}_{AA}t)$, a method for obtaining eigenvalues and eigenvectors must be introduced. The term 'eigen' means 'special' in German, and eigenvalues and eigenvectors are mathematically special because they enable the solutions of equations to be obtained. Eigenvectors $\boldsymbol{v}_i$ and eigenvalues λ_i of a matrix $\boldsymbol{Q}$ are defined by the relationship

$$\boldsymbol{Q}\boldsymbol{v}_i = \lambda_i \boldsymbol{v}_i. \qquad [66]$$

The product of matrix $\boldsymbol{Q}$ and its eigenvector, v_i is a column vector in the same direction as $\boldsymbol{v}_i$ but 'resized' by a scalar quantity called the eigenvalue. Equation 66 can be rearranged using the unit matrix:

$$[\boldsymbol{Q} - \lambda_i \boldsymbol{I}]\boldsymbol{v}_i = 0. \qquad [67]$$

A well-known result of linear algebra states that this equation has non-trivial solutions only when the determinant of matrix $[\boldsymbol{Q} - \lambda_i \boldsymbol{I}]$ is zero. This fact can be utilized to calculate the values of λ_i and consequently the vectors $\boldsymbol{v}_i$. In the present example the matrix of interest is $\boldsymbol{Q}_{AA}$. Since $\boldsymbol{Q}_{AA} = [-k_{12}]$, the condition for a zero determinant of the matrix $[\boldsymbol{Q}_{AA} - \lambda_i \boldsymbol{I}]$ leads to the equation

$$\mathrm{Det}[\boldsymbol{Q}_{AA} - \lambda_i \boldsymbol{I}] = -k_{12} - \lambda_i = 0 \qquad [68]$$

and thus, $\lambda_1 = -k_{12}$.

Since $A = 1$ for Scheme 2, there can only be one eigenvalue and one eigenvector for the 1×1 matrix $\boldsymbol{Q}_{AA}$. In general, there will be A eigenvalues and A eigenvectors for an $A \times A$ matrix. To calculate the matrices $\boldsymbol{A}_i$, Equation 64 can be used directly to determine that $\boldsymbol{A}_1 = [1]$, since in this example there is only one matrix $\boldsymbol{A}_i$. Thus, from Equation 61, $\exp(\boldsymbol{Q}_{AA}t) = [\exp(-k_{12}t)]$. Having used the eigenvalue to calculate the matrix $\exp(\boldsymbol{Q}_{AA}t)$, the solution for $f_A(t)$ can now be completed. Substituting Φ_A, $\boldsymbol{U}_A$, and $\exp(\boldsymbol{Q}_{AA}t)$ into Equation 60 gives

$$f_A(t) = k_{12}\mathrm{e}^{-k_{12}t}. \qquad [69]$$

Equation 69 reveals that the pdf for the open-state dwell time of Scheme 2 is a decaying exponential function, one of the probability distributions discussed in Section 4. This implies that the probability of finding dwell times in a time bin of short-duration events is always greater than the probability of finding dwell times in a time bin of long-duration events. This can be rationalized in terms of Eyring rate theory discussed previously in reference to ion permeation. At a microscopic level, thermal energy causes the chemical bonds of an open channel protein to stretch and vibrate at very high frequency ($k_BT/h \simeq 6 \times 10^{12}$/sec). Each vibration can be considered as an attempt to overcome the activation energy barrier that prevents channel closure. Therefore, if a channel is to remain open throughout the period from 0 to t, every attempt to close during this period must *fail*. Clearly, at longer times the number of attempts to surmount the barrier is greater and so is the chance of success. Since closure only requires one successful attempt, there is a higher probability of observing short rather than long dwell times.

The distribution of dwell times in the closed state can be calculated by the same method used for the open state and has the following pdf:

$$f_B(t) = k_{21}\mathrm{e}^{-k_{21}t}. \qquad [70]$$

Thus, Scheme 2, with a single open and a single closed state, predicts single-exponential dwell-time distributions characterized by a lifetime, τ, equal to the reciprocal of the rate constant for leaving each state. In general, if there are N closed states then the closed state pdf will be a sum of N exponential terms. The next section illustrates this rule by appling the Q-matrix method to the slightly more complicated example of Scheme 3 which includes two closed and one open state.

5.7 Q-matrix analysis of a three-state scheme

$$\underset{2}{C} \underset{k_{12}}{\overset{k_{21}}{\rightleftharpoons}} \underset{1}{O} \underset{k_{31}}{\overset{k_{13}}{\rightleftharpoons}} \underset{3}{C} \qquad \text{Scheme 3}$$

Using the arbitrarily assigned indices to each state in Scheme 3, the Q-matrix can be obtained directly using the rules outlined earlier. In position i, j of the Q-matrix where $i \neq j$, the rate constant q_{ij} is written. If there is no possible route for a particular transition, such as directly from state 2 to state 3, then this element of the matrix is set equal to 0. For the matrix diagonal positions, $i = j$, a term is written such that this row of the matrix sums to zero. Thus, the Q-matrix for Scheme 3 is

$$\begin{array}{cccc} i/j & j=1 & j=2 & j=3 \end{array}$$
$$\boldsymbol{Q} = \begin{array}{c} i=1 \\ i=2 \\ i=3 \end{array}\begin{bmatrix} -k_{12}-k_{13} & k_{12} & k_{13} \\ k_{21} & -k_{21} & 0 \\ k_{31} & 0 & -k_{31} \end{bmatrix}. \qquad [71]$$

Next, the Q-matrix is partitioned to define the open-state and closed-state submatrices $\boldsymbol{Q}_{AA}$ and $\boldsymbol{Q}_{BB}$, respectively:

$$\boldsymbol{Q}_{AA} = [-(k_{12} + k_{13})] \qquad [72]$$

$$\boldsymbol{Q}_{AB} = [k_{12} \quad k_{13}] \qquad [73]$$

$$\boldsymbol{Q}_{BA} = \begin{bmatrix} k_{21} \\ k_{31} \end{bmatrix} \qquad [74]$$

$$\boldsymbol{Q}_{BB} = \begin{bmatrix} -k_{21} & 0 \\ 0 & -k_{31} \end{bmatrix}. \qquad [75]$$

Since there is only one open state, the procedure outlined in Section 5.6 can be followed to derive the various terms of Equation 60. This gives the following pdf for the open state:

$$f_A(t) = (k_{12} + k_{13})e^{-(k_{12} + k_{13})t}. \qquad [76]$$

Since there is only a single open state in Scheme 3, the open state pdf is described by a single exponential. In this case, however, the distribution is governed by a rate constant which is the sum of both k_{12} and k_{13}. In general, the observed rate constant for a transition out of any one state is equal to the sum of all rate constants for leaving that state. Hence the time constant or the mean dwell time is simply the reciprocal of this sum.

The pdf for the closed state is expected to be a sum of two exponentials since there are two closed states. To compute the eigenvalues for this example, matrix $\boldsymbol{Q}_{BB}$ is substituted into Equation 67:

$$[\boldsymbol{Q}_{BB} - \lambda_i \boldsymbol{I}]\boldsymbol{v}_i = \begin{bmatrix} -k_{21} - \lambda_i & 0 \\ 0 & -k_{31} - \lambda_i \end{bmatrix} \boldsymbol{v}_i = 0. \qquad [77]$$

Setting the determinant of matrix $[\boldsymbol{Q}_{BB} - \lambda_i \boldsymbol{I}]$ to zero:

$$\mathrm{Det}[\boldsymbol{Q}_{BB} - \lambda_i \boldsymbol{I}] = (k_{21} + \lambda_i)\,(k_{31} + \lambda_i) = 0. \qquad [78]$$

The two solutions of Equation 78 are $\lambda_1 = -k_{21}$ and $\lambda_2 = -k_{31}$. Applying Equation 65 to find the matrices $\boldsymbol{A}_1$ and $\boldsymbol{A}_2$ gives

$$\boldsymbol{A}_1 = \frac{\begin{bmatrix} -k_{21} & 0 \\ 0 & -k_{31} \end{bmatrix} + \begin{bmatrix} k_{31} & 0 \\ 0 & k_{31} \end{bmatrix}}{k_{31} - k_{21}} = \begin{bmatrix} 1 & 0 \\ 0 & 0 \end{bmatrix} \qquad [79]$$

$$\boldsymbol{A}_2 = \frac{\begin{bmatrix} -k_{21} & 0 \\ 0 & -k_{31} \end{bmatrix} + \begin{bmatrix} k_{21} & 0 \\ 0 & k_{21} \end{bmatrix}}{k_{21} - k_{31}} = \begin{bmatrix} 0 & 0 \\ 0 & 1 \end{bmatrix} \qquad [80]$$

Using matrices $\mathbf{A}_1$ and $\boldsymbol{A}_2$ to calculate matrix $\exp(\boldsymbol{Q}_{BB}t)$ according to Equation 61 leads to

$$\mathrm{e}^{\boldsymbol{Q}_{BB}t} = \boldsymbol{A}_1 \mathrm{e}^{-k_{21}t} + \boldsymbol{A}_2 \mathrm{e}^{-k_{31}t} \qquad [81]$$

or

$$\mathrm{e}^{\boldsymbol{Q}_{BB}t} = \begin{bmatrix} 1 & 0 \\ 0 & 0 \end{bmatrix} \mathrm{e}^{-k_{21}t} + \begin{bmatrix} 0 & 0 \\ 0 & 1 \end{bmatrix} \mathrm{e}^{-k_{31}t} = \begin{bmatrix} \mathrm{e}^{-k_{21}t} & 0 \\ 0 & \mathrm{e}^{-k_{31}t} \end{bmatrix}. \qquad [82]$$

Next the vector Φ_B must be calculated. The first component of Φ_B is

Probability of a closure starting in state 2 =

$$\frac{\text{Rate of leaving state 1 for state 2}}{\text{Rate of leaving state 1 for state 2 or 3}} = \frac{k_{12}}{(k_{12} + k_{13})} \qquad [83]$$

Similarly, the probability of a channel closure starting in state 3 (the second term in Φ_B is $k_{13}/(k_{12} + k_{13})$. These are the two terms composing the vector Φ_B. Finally, the vector $\boldsymbol{U}_B$ will be a vector with two unit elements since B = 2. Upon substituting for $e^{\boldsymbol{Q}_{BB}t}$, $\boldsymbol{U}_B$, and Φ_B into Equation 60 we obtain

$$f_B(t) = \begin{bmatrix} \dfrac{k_{12}}{k_{12} + k_{13}} & \dfrac{k_{13}}{k_{12} + k_{13}} \end{bmatrix} \begin{bmatrix} \mathrm{e}^{-k_{21}t} & 0 \\ 0 & \mathrm{e}^{-k_{31}t} \end{bmatrix} \begin{bmatrix} k_{21} & 0 \\ 0 & k_{31} \end{bmatrix} \begin{bmatrix} 1 \\ 1 \end{bmatrix}. \qquad [84]$$

Finally, multiplying out these matrices yields the solution:

$$f_{\mathrm{B}}(t) = \left(\frac{k_{12}}{k_{12} + k_{13}}\right) k_{21}\mathrm{e}^{(-k_{21}t)} + \left(\frac{k_{13}}{k_{12} + k_{13}}\right) k_{31}\mathrm{e}^{(-k_{31}t)}. \qquad [85]$$

Note that in this case, unlike the open state distribution, the pdf of channel closures is a sum of two exponentials with one component corresponding to a pdf for dwell times in state 2 (k_{21}) while the other component corresponds to a pdf for dwell times in state 3 (k_{31}). Overall, the $f_B(t)$ function for Scheme 3 is a weighted average of pdf for dwell times in states 2 and 3. Although it is generally true that a sum of two exponentials describe the closed state

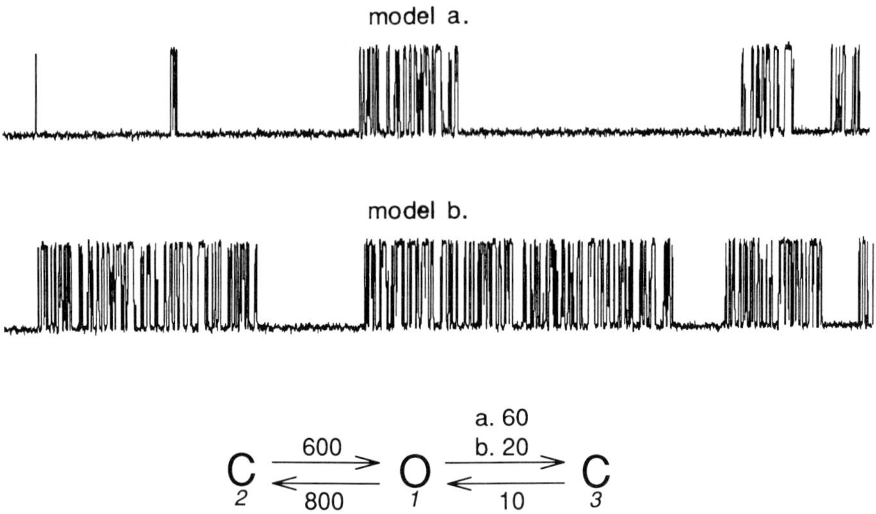

Figure 9. Two simulations of channel bursting behaviour using a three-state, closed–open–closed kinetic scheme. Bursting behaviour due to rapid transitions between states 1 (open) and 2 (closed) of Scheme 3 was simulated by a computer program. The lower scheme indicates the values of rate constants (in units of $\sec^{-1}$ used in the simulations of (a) and (b). Each record shows a one-second simulation. Gaussian noise was added to each simulation before filtering at 1 kHz. The comparison of (a) and (b) shows the effect of decreasing one rate constant (k_{13}) from 60 $\sec^{-1}$ to 20 $\sec^{-1}$. The average burst length is lengthened in (b) because bursts are terminated less frequently as k_{13} is decreased.

distribution whenever there are two closed states, the two exponential components do not always have such a simple physical interpretation.

The Q-matrix can also be used to analyse bursting behaviour. For example, in the case of Scheme 3, if transitions between states 1 and 2 are relatively more frequent than transitions between states 1 and 3, the single-channel record will have the appearance of 'bursts' of openings punctuated by long-lived closures (*Figure 9*). This behaviour is commonly observed when a blocking molecule with a brief residence time is responsible for the closures to state 2. As in previous examples, analysis of bursting behaviour is begun by dividing the Q-matrix into submatrices. In this case, one matrix contains the burst states 1 and 2 ($\boldsymbol{Q}_{12}$) and a second matrix contains the long-lived closed state 3 ($\boldsymbol{Q}_3$). The distribution of dwell times in the burst state will be a bi-exponential distribution, with lifetimes determined by the eigenvalues of the matrix $\boldsymbol{Q}_{12}$. Proceeding as before:

$$[\boldsymbol{Q}_{12} - \lambda_i \boldsymbol{I}]\boldsymbol{v}_i = \left\{ \begin{bmatrix} -(k_{13} + k_{12}) & k_{12} \\ k_{21} & -k_{21} \end{bmatrix} - \lambda_i \begin{bmatrix} 1 & 0 \\ 0 & 1 \end{bmatrix} \right\} \boldsymbol{v}_\mathrm{i}$$

$$= \begin{bmatrix} -(k_{12} + k_{13}) - \lambda_i & k_{12} \\ k_{21} & -k_{21} - \lambda_i \end{bmatrix} \boldsymbol{v}_i = 0. \qquad [86]$$

As before, the determinant of matrix $[\boldsymbol{Q}_{12} - \lambda_i \boldsymbol{I}]$ is set equal to zero and solved for the eigenvalues which provide the exponents in the probability distribution of the burst length. The eigenvalues are solutions to the quadratic equation:

$$\mathrm{Det}[\boldsymbol{Q}_{12} - \lambda_i \boldsymbol{I}] = \lambda_i^2 + \lambda_i(k_{21} + k_{12} + k_{13}) + k_{21}k_{13} = 0. \qquad [87]$$

The two λ_i eigenvalues of Equation 87 can be then used to find the matrices $\boldsymbol{A}_i$ and used to complete the solution of the pdf as outlined above. This is straightforward but laborious, and we will not follow the complete derivation but merely give the result:

$$f_{\mathrm{burst}}(t) = \frac{k_{13}}{\lambda_1 - \lambda_2}\{(\lambda_1 + k_{21})\mathrm{e}^{\lambda_1 t} - (\lambda_2 + k_{21})\mathrm{e}^{\lambda_2 t}\}. \qquad [88]$$

As the complexity of such solutions increases, the necessity of carrying out such calculations by computer becomes obvious. For example, a kinetic scheme with three closed states would produce a 3 × 3 matrix, $\boldsymbol{Q}_{BB}$. This would result in a cubic equation instead of the quadratic Equation 78. In general, a scheme composed of N closed states requires the solution of a polynomial of degree N. For this reason, the approach that we have illustrated must be enhanced by more more advanced techniques. Fortunately, a number of matrix methods exist for finding the eigenvalues of large matrices. One of the most popular approaches is the 'QR' method, outlined by Strang (62). Public domain computer code is available for a set of useful algorithms called 'Eispack algorithms' which use the 'QR' appproach to find eigenvalues and eigenvectors of large matrices. Further information concerning such methods may be found in the 'numerical recipes' series (64).

In summary, a Q-matrix composed of rate constants between discrete states can be written for any kinetic scheme, regardless of its complexity. The Q-matrix can then be used to predict probability density functions of dwell times in closed, open, or burst states. The Q-matrix approach can be adapted readily to other problems such as the modelling and prediction of current relaxation kinetics. These calculations generally involve a standard set of mathematical manipulations that can be implemented by a computer program. Thus, the Q-matrix approach pioneered by Colquhoun and Hawkes is a powerful and versatile tool for investigating and understanding single channel behaviour.

Appendix 1: Alternative calculation of $\boldsymbol{A}_i$

The method used by Colquhoun and Hawkes (48) to calculate the matrices $\boldsymbol{A}_i$ is illustrated below. In this particular example, Equations 79 and 80 are re-derived. Taking the eigenvalues from Equation 77 and proceeding to calculate the eigenvectors by substituting one eigenvalue at a time back into Equation 67 gives

$$\begin{bmatrix} 0 & 0 \\ 0 & -(k_{31} + k_{12}) \end{bmatrix} \begin{bmatrix} x_1 \\ y_1 \end{bmatrix} = \begin{bmatrix} 0 \\ 0 \end{bmatrix}$$

and

$$\begin{bmatrix} -(k_{12} + k_{13}) & 0 \\ 0 & 0 \end{bmatrix} \begin{bmatrix} x_2 \\ y_2 \end{bmatrix} = \begin{bmatrix} 0 \\ 0 \end{bmatrix}$$

where x_i and y_i are the unknown components of eigenvector $\boldsymbol{v}_i$. Thus, using the eigenvalue $\lambda_1 = -k_{12}$ gives the equations:

$$0x_1 + 0y_1 = 0$$

$$0x_1 - (k_{13} + k_{12})y_1 = 0.$$

A normalized eigenvector that satisfies this equation is

$$\boldsymbol{v}_1 = \begin{bmatrix} 1 \\ 0 \end{bmatrix}.$$

By a similar process the second vector can be obtained as:

$$\boldsymbol{v}_2 = \begin{bmatrix} 0 \\ 1 \end{bmatrix}.$$

These latter two vectors can be used to calculate the matrices $\boldsymbol{A}_i$. First, a new matrix $\boldsymbol{V}$ is introduced which is a matrix whose rows are the eigenvectors of matrix $\boldsymbol{Q}_{AA}$. The matrices $\boldsymbol{A}_i$ are then calculated as the product of the ith column of $\boldsymbol{V}$ multiplied by the ith row of $\boldsymbol{V}^{-1}$, where $\boldsymbol{V}^{-1}$ is the inverse matrix of matrix $\boldsymbol{V}$ (see Appendix 2). In the current example, matrix $\boldsymbol{V}$ is

$$\boldsymbol{V} = \begin{bmatrix} 1 & 0 \\ 0 & 1 \end{bmatrix}.$$

In this particular example $\boldsymbol{V}$ is a unit matrix. The unit matrix is a special case where $\boldsymbol{V} = \boldsymbol{V}^{-1}$ (since $\boldsymbol{V}\boldsymbol{V}^{-1} = \boldsymbol{I}$). Therefore the matrices $\boldsymbol{A}$ can be calculated according to the above rules as:

$$\boldsymbol{A}_1 = \begin{bmatrix} 1 \\ 0 \end{bmatrix} [1 \quad 0] = \begin{bmatrix} 1 & 0 \\ 0 & 0 \end{bmatrix}$$

$$\boldsymbol{A}_2 = \begin{bmatrix} 0 \\ 1 \end{bmatrix} [0 \quad 1] = \begin{bmatrix} 0 & 0 \\ 0 & 1 \end{bmatrix}.$$

This is the same result given in Equations 79 and 80 previously obtained by using the formula of Equation 65. For those following Colquhoun and Hawkes (51), note that they find eigenvalues of the matrix $-\boldsymbol{Q}$ (or $-\boldsymbol{Q}_{AA}$, etc.). Consequently, they write the exponential terms of matrices such as $\exp(\boldsymbol{Q}t)$ as $\exp(-\lambda_i t)$.

Appendix 2: A basic summary of matrix algebra

Matrices are a convenient mathematical tool for handling groups of related objects.

(1) A 2×3 matrix $\boldsymbol{A}$ is written as

$$\boldsymbol{A} = \begin{bmatrix} a_{11} & a_{12} & a_{13} \\ a_{21} & a_{22} & a_{23} \end{bmatrix}.$$

(2) A row vector is the name given to a single row matrix. A 3-element row vector $\boldsymbol{R}$ may be written

$$\boldsymbol{R} = [r_1 \quad r_2 \quad r_3].$$

(3) A column vector is the name given to a single column matrix. A 3-element column vector $\boldsymbol{C}$ is written

$$\boldsymbol{C} = \begin{bmatrix} c_1 \\ c_2 \\ c_3 \end{bmatrix}.$$

(4) Matrices are added or subtracted on an element by element basis. Thus the difference between two 2×2 matrices $\boldsymbol{A}$ and $\boldsymbol{B}$ is written:

$$\boldsymbol{A} - \boldsymbol{B} = \begin{bmatrix} a_{11} & a_{12} \\ a_{21} & a_{22} \end{bmatrix} - \begin{bmatrix} b_{11} & b_{12} \\ b_{21} & b_{22} \end{bmatrix} = \begin{bmatrix} a_{11} - b_{11} & a_{12} - b_{12} \\ a_{21} - b_{21} & a_{22} - b_{22} \end{bmatrix}.$$

(5) When a matrix is multiplied by a scalar (a number), the result is a new matrix, whose elements are simply the product of the original elements and the scalar. Thus a 2×2 matrix $\boldsymbol{A}$ multiplied by a scalar lambda is calculated as:

$$\lambda \boldsymbol{A} = \lambda \begin{bmatrix} a_{11} & a_{12} \\ a_{21} & a_{22} \end{bmatrix} = \begin{bmatrix} \lambda a_{11} & \lambda a_{12} \\ \lambda a_{21} & \lambda a_{22} \end{bmatrix},$$

(6) Matrices $\boldsymbol{A}$ and $\boldsymbol{B}$ are multiplied by taking each row in $\boldsymbol{A}$ and multiplying it by the corresponding column in $\boldsymbol{B}$. Hence a matrix multiplication such as $\boldsymbol{AB}$ can only be performed when the number of columns in matrix $\boldsymbol{A}$ equals the number of rows in matrix $\boldsymbol{B}$. If $\boldsymbol{A}$ is a matrix of order $(m \times p)$ with elements a_{ij} and $\boldsymbol{B}$ is a matrix of order $(p \times n)$ with elements b_{ij}, then the product matrix $\boldsymbol{C}$ has elements c_{ij} where $c_{ij} = \sum_{s=1}^{s=p} a_{is} b_{sj}$. An example of 2×2 matrix multiplication is

$$\boldsymbol{AB} = \begin{bmatrix} a_{11} & a_{12} \\ a_{21} & a_{22} \end{bmatrix} \begin{bmatrix} b_{11} & b_{12} \\ b_{21} & b_{22} \end{bmatrix} = \begin{bmatrix} a_{11}b_{11} + a_{12}b_{21} & a_{11}b_{12} + a_{12}b_{22} \\ a_{21}b_{11} + a_{22}b_{21} & a_{21}b_{12} + a_{22}b_{22} \end{bmatrix}.$$

(7) The determinant of a matrix is simply a single number that is only defined for a square $(n \times n)$ matrix and is denoted Det($\boldsymbol{M}$). The general pro-

cedure for calculating determinants is given by Stephenson (61). The results for a 2 × 2 matrix and a 3 × 3 matrix are:

$$\mathbf{A} = \begin{bmatrix} a_{11} & a_{12} \\ a_{21} & a_{22} \end{bmatrix}, \quad \mathrm{Det}(\mathbf{A}) = a_{11}a_{22} - a_{12}a_{21}$$

$$\mathbf{A} = \begin{bmatrix} a_{11} & a_{12} & a_{13} \\ a_{21} & a_{22} & a_{23} \\ a_{31} & a_{32} & a_{33} \end{bmatrix}, \quad \begin{aligned} \mathrm{Det}(\mathbf{A}) = {} & a_{11}a_{22}a_{33} - a_{11}a_{32}a_{23} + a_{21}a_{32}a_{13} \\ & - a_{21}a_{12}a_{33} + a_{31}a_{12}a_{23} - a_{31}a_{22}a_{13}. \end{aligned}$$

(8) The unit matrix $\boldsymbol{I}$ has unit elements on the diagonal and zero elements elsewhere. Thus the 3 × 3 unit matrix is written as

$$\boldsymbol{I} = \begin{bmatrix} 1 & 0 & 0 \\ 0 & 1 & 0 \\ 0 & 0 & 1 \end{bmatrix}.$$

(9) The adjoint matrix is written adj($\mathbf{A}$). Although the calculation of adj($\mathbf{A}$) is simple, it is lengthy, and so the result is quoted here for the 2 × 2 matrix solution:

$$\mathbf{A} = \begin{bmatrix} a_{11} & a_{12} \\ a_{21} & a_{22} \end{bmatrix}, \quad \mathrm{adj}(\mathbf{A}) = \begin{bmatrix} a_{22} & -a_{12} \\ -a_{21} & a_{11} \end{bmatrix}.$$

(10) The inverse matrix, of a matrix $\mathbf{A}$, is written $\mathbf{A}^{-1}$. It is related to the adjoint and determinant of $\mathbf{A}$ by the following relations:

$$\mathbf{A} = \begin{bmatrix} a_{11} & a_{12} \\ a_{21} & a_{22} \end{bmatrix}$$

$$\mathbf{A}^{-1} = \frac{\mathrm{adj}(\mathbf{A})}{\mathrm{Det}(\mathbf{A})}$$

$$\mathbf{A}^{-1} = \begin{bmatrix} \dfrac{a_{22}}{a_{11}a_{22} - a_{21}a_{12}} & \dfrac{-a_{12}}{a_{11}a_{22} - a_{21}a_{12}} \\ \dfrac{-a_{21}}{a_{11}a_{22} - a_{21}a_{22}} & \dfrac{a_{11}}{a_{11}a_{22} - a_{21}a_{12}} \end{bmatrix}.$$

Note the important property that $\mathbf{A}\mathbf{A}^{-1} = \boldsymbol{I}$.

References

1. Hamill, O. P., Marty, A., Neher, E., Sakmann, B., and Sigworth, F. J. (1981). *Pflügers Arch. Eur. J. Physiol.* **391**, 85.
2. Sakmann, B. and Neher, E. (ed.) (1983). *Single-channel recording*. Plenum Press, New York.
3. Miller, C. (ed.) (1986). *Ion channel reconstitution*. Plenum Press, New York.
4. Andersen, O. S. (1989). In *Methods in enzymology* (ed. S. Fleischer and B. Fleischer), Vol. 171, pp. 62–112. Academic Press, New York.

5. Wonderlin, W. F., French, R. J., and Arispe, N. J. (1990). In *Neuromethods* (ed. A. A. Boulton, G. B. Baker and C. M. Vanderwolf), Vol. 14, Neurophysiological techniques: basic methods and concepts, pp. 35–142. The Hamana Press Inc, Clifton, N.J.
6. Andersen, O. S. and Koeppe, R. E. (1992). *Physiol. Rev.* **72**, S89.
7. Hille, B. (1992). *Ionic channels of excitable membranes*, 2nd edn. Sinauer Associates, Sunderland, MA.
8. Rudy, B. and Iverson, L. E. (ed.) (1992). *Methods in enzymology*, Vol. 207, Ion channels. Academic Press, New York.
9. Jackson, M. B. (ed.) (1993). *Thermodynamics of membrane receptors and channels*. CRC Press, Boca Raton, FL.
10. Wonderlin, W. F., Finkel, A., and French, R. J. (1990). *Biophys. J.* **58**, 289.
11. Andersen, O. S. (1983). *Biophys. J.* **41**, 147.
12. Latorre, R., Labarca, P., and Naranjo, D. (1992). In *Methods in enzymology*, ed. B. Rudy and L. E. Iverson, Vol. 207, pp. 471–50. Academic Press, New York.
13. Green, W. N. and Andersen, O. S. (1991). *Ann. Rev. Physiol.* **53**, 341.
14. Heinemann, S. H., Terlau, H., Stühmer, W., Imoto, K., and Numa, S. (1992). *Nature* **356**, 441.
15. Miller, C. (1993). *Science* **261**, 1692.
16. Rosenberg, R. L. and Chen, X-H. (1991). *J. Gen. Physiol.* **97**, 1207.
17. Tsien, R. W., Hess, P., McCleskey, E. W., and Rosenberg, R. L. (1987). *Ann. Rev. Biophys. Biophys. Chem.* **16**, 265.
18. Läuger, P., Stephan, W., and Frehland, E. (1980). *Biochim. Biophys. Acta* **602**, 167.
19. Heinemann, S. H. and Sigworth, F. J. (1993). In *Thermodynamics of membrane receptors and channels* (ed. M. B. Jackson), pp. 407–22. CRC Press, Boca Raton, FL.
20. Neyton, J. and Miller, C. (1988). *J. Gen. Physiol.* **92**, 569.
21. Dani, J. A. and Levitt, D. G. (1990). *J. Theor. Biol.* **146**, 289.
22. Chiu, S.-W., Subramaniam, S., Jakobsson, E., and McCammon, J. A. (1989). *Biophys. J.* **56**, 253.
23. Levitt, D. G. (1986). *Ann. Rev. Biophys. Biophys. Chem.* **15**, 29.
24. Levitt, D. G. (1991). *Biophys. J.* **59**, 271.
25. Eisenman, G. and Horn, R. (1983). *J. Membr. Biol.* **76**, 197.
26. Hille, B. and Schwarz, W. (1978). *J. Gen. Physiol.* **72**, 409.
27. Woodhull, A. (1973). *J. Gen. Physiol.* **61**, 687.
28. Alvarez, O., Villarroel, A., and Eisenman, G. (1992). In *Methods in enzymology*, ed. B. Rudy and L. E. Iverson, Vol. 207, pp. 816–54. Academic Press, New York.
29. Hodgkin, A. L. and Katz, B. (1949). *J. Physiol.* **108**, 37.
30. Begenisich, T. B. and Cahalan, M. D. (1980). *J. Physiol.* **307**, 217.
31. Garber, S. S. (1988). *Biophys. J.* **54**, 767.
32. Fatt, P. and Ginsborg, B. L. (1958). *J. Physiol.* **142**, 516.
33. Hess, P., Lansman, J. B., and Tsien, R. W. (1986). *J. Gen. Physiol.* **88**, 293.
34. Fersht, A. (1985). *Enzyme structure and mechanism*. W. H. Freeman, New York.
35. Läuger, P. (1973). *Biochim. Biophys. Acta*, **311**, 423.
36. Coronado, R., Rosenberg, R. L., and Miller, C. (1980). *J. Gen. Physiol.* **76**, 425.
37. Hill, J. A., Coronado, R., and Strauss, H. C. (1989). *Biophys. J.* **55**, 35.
38. Olah, G. A., Huang, H. W., Liu, W., and Wu, Y. (1991). *J. Mol. Biol.* **218**, 847.
39. Finkelstein, A. and Andersen, O. S. (1981). *J. Membr. Biol.* **59**, 155.

40. Sandblom, J., Eisenman, G., and Neher, E. (1977). *J. Membr. Biol.* **31**, 383.
41. Urban, B. W., Hladky, S. B., and Haydon, D. A. (1980). *Biochim. Biophys. Acta* **602**, 331.
42. Neher, E. (1975). *Biochim. Biophys. Acta* **401**, 540.
43. Eisenman, G., Latorre, R., and Miller, C. (1986). *Biophys. J.* **50**, 1025.
44. Ravindran, A., Kwiecinski, H., Alvarez, O., Eisenman, G., and Moczydlowski, E. (1992). *Biophys. J.* **61**, 494.
45. Yellen, G. (1987). *Ann. Rev. Biophys. Biophys. Chem.* **16**, 227.
46. Begenisich, T. (1987). *Ann. Rev. Biophys. Biophys. Chem.* **16**, 247.
47. Colquhoun, D. (1971). *Lectures on biostatistics*. Clarendon Press, Oxford.
48. Colquhoun, D. and Hawkes, A. G. (1977). *Proc. R. Soc. Lond.* **199**, 231.
49. Colquhoun, D. and Hawkes, A. G. (1981). *Proc. R. Soc. Lond.* **211**, 205.
50. Colquhoun, D. and Hawkes, A. G. (1982). *Phil. Trans. R. Soc. Lond. B* **300**, 1.
51. Colquhoun, D. and Hawkes, A. G. (1983). In *Single-channel recording* (ed. B. Sakmann and E. Neher), pp. 135–75. Plenum Press, New York.
52. Yellen, G. (1984). *J. Gen. Physiol.* **84**, 157.
53. Labarca, P., Coronado, R., and Miller, C. (1980). *J. Gen. Physiol.* **76**, 397.
54. Barrett, J. N., Magleby, K. L., and Pallotta, B. S. (1982). *J. Physiol.* **331**, 211.
55. Miller, C. (1982). *Phil. Trans. R. Soc. Lond. B* **299**, 401.
56. Walpole, R. E. and Myers, R. H. (1978). *Probability and statistics for engineers and scientist*. MacMillan, New York.
57. Sigworth, F. J. (1980). *J. Physiol.* **307**, 97.
58. Sigworth, F. J. and Sine, S. M. (1987). *Biophys. J.* **52**, 1047.
59. Cox, D. R. and Miller, H. D. (1965). *The theory of stochastic processes*. John Wiley, New York.
60. Crouzy, S. C. and Sigworth, F. J. (1993). *Biophys. J.* **64**, 68.
61. Stephenson, G. (1973). *Mathematical methods for science students*. Longman, London.
62. Strang, G. (1976). *Linear algebra and its applications*. Academic Press, New York.
63. Hollingsworth, C. A. (1967). *Vectors, matrices, and group theory for scientists and engineers*. McGraw-Hill, New York.
64. Press, W. H., Flannery, B. P., Teukolsky, S. A., and Vetterling, W. T. (1986). *Numerical recipes, the art of scientific computing*. Cambridge University Press, Cambridge.

4

Assaying ion channels: design and practice

MICHIKI KASAI, TORU IDE, and WILLIAM S. AGNEW

1. Introduction

Isolation of the molecule in chemically pure form is an important approach in the characterization of ion channels. This enables determination of molecular composition and size, and may permit functional reconstitution in artificial membranes for biochemical or electrophysiological measurements. Purification requires that the protein be solubilized before separation from other proteins associated with the membrane sample. However, unlike other enzymes, channel molecules typically lack chemical catalytic activity, and quantification in detergent solution must either be performed with a specific biochemical marker, such as a ligand, toxin, or drug that binds stoichiometrically, or the protein must be returned to a membrane environment for measurement of ion-transporting function. Ligand binding assays constitute the single most important tool for channel and receptor quantification and purification, and the following discussion and examples are aimed at providing insights useful for designing new assay protocols. Some channels will be resistant to this approach, lacking any useful ligand, and we also describe how the progress of an ion channel purification protocol can be followed by bilayer reconstitution.

2. Ligand binding assays

2.1 Introduction

Ion channels, whether receptor activated, voltage-activated, or second-messenger activated, can be studied in binding assays when specific stoichiometric ligands are available in radiolabelled form. Such assays can be used to measure receptor abundance or surface density, can assist in purifications, and when analysed appropriately can reveal conformational mechanisms of regulation. The appropriate ligands can be selected from among physiological receptor agonists, selective neurotoxins, specific pharmacological agents, products of combinatorial libraries, or combinations of specific and non-specific markers (1–3). Binding assays provide an advantage over transport or enzymatic assays because they give a direct measure of the *number* of receptor molecules present. Combined with estimates of molecular weight, this can give rigorous estimates of protein purification (= $R_t\ M_r$/total membrane protein) or, combined with a knowledge of membrane surface area, estimates of channel density.

Receptor assays measure binding to a limited number of specific sites, either through non-equilibrium separation of bound from free ligand, or by using equilibrium measurements which do not require actual separation of receptor–ligand complexes. For the design of a new assay, several general rules may be applied in the selection of the protocol, tissue source, and marker ligand. For example, a ligand, aside from being 'specific' for the target receptor site, is typically most useful if the equilibrium dissociation constant is of the same order of magnitude as, or lower than, the receptor concentration in the assayed samples. The selection of equilibrium or non-equilibrium protocols depends on the quantitative stability of the receptor–ligand complex. Different types of information may be extracted depending on the properties of the ligand–receptor interaction and the kinds of analysis applied. Some practical insights emerge from even elementary consideration of ligand–receptor binding relationships.

2.2 Theoretical considerations

We can simply consider binding of a ligand (l) to a receptor site (r) according to

$$\mathrm{r} + \mathrm{l} \underset{k_2}{\overset{k_1}{\rightleftharpoons}} \mathrm{rl}$$

where the rate of appearance of complexes (rl) is given by

$$\mathrm{d[rl]/d}t = k_1[\mathrm{r}][\mathrm{l}] - k_2[\mathrm{rl}] \qquad [1]$$

which, at equilibrium, is set to d[rl]/dt = 0. Taking into account conservation of mass of total ligand (L_t = [l] + [rl]) and receptor (R_t = [r] + [rl]), the fractional occupancy (f.o.) is given by the Langmuir binding relationship

$$\mathrm{f.o.} = \frac{[\mathrm{rl}]}{R_t} = \frac{1}{1 + \dfrac{K_d}{[\mathrm{l}]}} \qquad [2]$$

where $K_d = k_2/k_1$ is the equilibrium dissociation constant, specifying the free energy of binding ($\Delta G = -RT \ln [K_d]$). Affinity is expressed as the inverse of the K_d, or as the free energy of binding. Relative affinity, or specificity, is given by ratios of K_ds; a relatively specific ligand binds to the target receptor with higher affinity (lower K_d) than to other components in the sample.

Similarly, the relative specificity of a receptor site for different ligands is reflected in the ratio of K_ds, either directly or in the difference in free energies of binding ($\triangle G_1 - \triangle G_2 = RT \ln (K_{d2}/K_{d1})$).

In Equation 2, if receptor is much lower in concentration than ligand ($R_t \ll L_t$), the free ligand concentration may be approximated by L_t. However, in most protocols, for reasons cited below, this is not the case. One approach is simply to compute free ligand as $[l] = L_t - [rl]$. However, it is instructive to examine the analytical solution to Equation 2 when conservation of mass for the ligand is explicitly taken into account by substituting $L_t - [l]$ for $[l]$ in Equation 2. The result is simply a quadratic equation,

$$[rl]^2 - (L_t + K_d + R_t)[rl] + R_tL_t = 0 \qquad [3]$$

which has the solution

$$[rl] = \frac{\beta \pm \sqrt{\beta^2 - 4\,R_tL_t}}{2} \qquad (\beta = R_t + L_t + K_d). \qquad [4]$$

This can be solved for the ligand concentration for half maximal saturation by simply setting $[rl] = [R]_t/2$. The resulting value, $L_{0.5}$, is

$$L_{0.5} = R_t/2 + K_d. \qquad [5]$$

Thus, rigorously, the half maximal saturation point always reflects *both* receptor abundance and the dissociation constant. This means that, when receptor abundance is high relative to the K_d, correction of total ligand for net binding cannot be ignored. It is worth noting that in considering ligand–receptor interactions within a small compartment, for example the interior of a cell or dendritic spine, small numbers of receptors create appreciable concentrations and the ligand concentration for half maximal response may reflect receptor abundance as much or more than K_d.

We may also evaluate Equation 4 for the limiting case where the dissociation constant is much lower than the receptor concentration ($K_d \ll R_t$). Under these conditions the equation has two real roots which have physical meaning:

$$[rl] = L_t \qquad (\text{for } L_t \leqslant R_t) \qquad [6]$$

$$[rl] = R_t \qquad (\text{for } L_t \geqslant R_t). \qquad [7]$$

This effectively describes saturation of a receptor with a very high affinity ligand (e.g. the nicotinic acetylcholine-receptor-binding α-bungarotoxin (4)). The receptor is essentially being *titrated* by the ligand and, as shown by equation (5), the mid-point for saturation reflects merely $R_t/2$. *Figure 1a* illustrates the saturation curve for the values of K_d relative to R_t.

The dissociation constant reflects only equilibrium binding, or the ratio of the unbinding and binding rate processes. In non-equilibrium assays, where unbound ligand is separated from bound ligand, if the time of isolation is slow

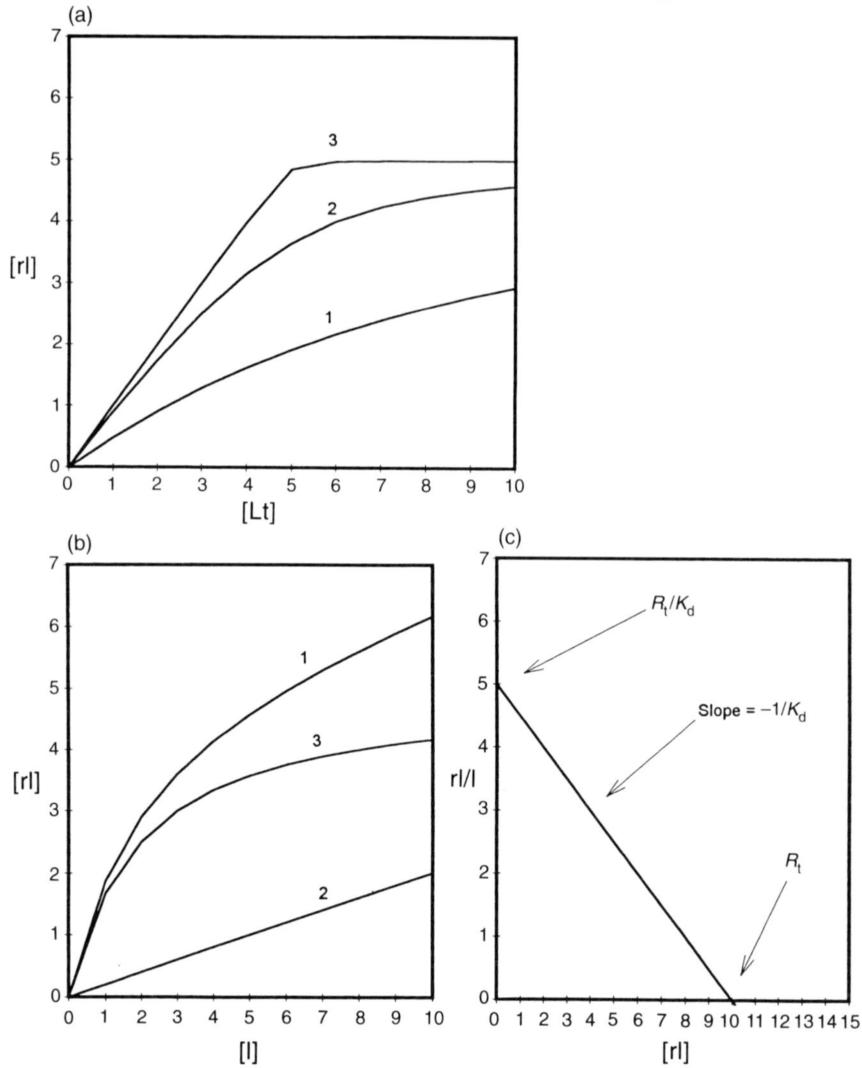

Figure 1. Saturation behaviour for single class of sites under different conditions of assay. Saturation curves computed in (a) when K_d is equal to R_t (1), $0.1R_t$ (2), and 0.001 R_t. As a limit (curve 3), values of [rl] are given by Equations 6 and 7, and the midpoint for saturation no longer reflects K_t, whereas in curve 1, the midpoint reflects both K_d and R_t. The 'plateau' appearance of curves 2 and 3 indicates that depletion due to binding must be appreciable. In (b), binding is shown according to Equation 9: curve 1 shows total binding (specific plus non-specific), curve 2 shows non-specific binding remaining in the presence of excess inhibitor, curve 3 depicts the difference, or the net specific binding obtained by subtracting values from curve 2 from curve 1. (c). Depicts the expected analytical parameters when the Langmuir equation is rearranged according to the Scatchard formula, e.g. $[rl]/[l] = R_t/K_d - [rl]/K_d$. Deviations from such linear behaviour may reflect positively or negatively interacting binding sites, or mixed populations of sites with markedly different binding affinities.

relative to the rate of dissociation, underestimates of the receptor–ligand complex will occur. The time frame involved can be estimated from Equation 1, setting [l] = 0, which becomes, when integrated,

$$[\mathrm{rl}]_t = [\mathrm{rl}]_0 \exp^{-k_2 t} = [\mathrm{rl}]_0 \exp(-0.69t/t^{1/2}) \qquad [8]$$

where $t^{1/2}$ is the half life of the complex and t is the time at which the concentration of l initially becomes 0: this is approximated by the separation time of the assay.

In practice, specific (saturable) binding must be distinguished from non-specific background signals (n.s.b.) which may reflect non-specific binding or imperfect separation of free and bound ligand. Specific binding, by definition, occurs at a limited number of high affinity sites and thus may be blocked with a competitive inhibitor, ideally the unlabelled form of the labelled marker. In the presence of inhibitor (i) and ligand:

$$\mathrm{r} + \mathrm{l} \underset{k_2}{\overset{k_1}{\rightleftharpoons}} \mathrm{rl} \qquad [9]$$

$$\mathrm{r} + \mathrm{i} \underset{k_4}{\overset{k_3}{\rightleftharpoons}} \mathrm{ri} \qquad [10]$$

predicts the corresponding Langmuir isotherm of simple competitive inhibitor, and the relationship for observed binding becomes

$$[\mathrm{rl}]_{\mathrm{obs}} = \frac{R_t}{1 + \frac{K_d}{[\mathrm{l}]}\left(1 + \frac{[\mathrm{i}]}{K_i}\right)} + k[\mathrm{l}] \qquad [11]$$

Figure 1b shows the pattern of binding seen for total, non-specific, and net specific binding as a function of [l]. Non-specific signals are estimated by performing the assays in the presence of excess inhibitor (i); the value of [rl] is obtained and subtracted from $[\mathrm{rl}]_{\mathrm{obs}}$ to obtain [rl]. To measure K_i for various inhibitors, net specific binding is measured at varying concentrations of i. The observed dissociation constant, $K_{d,obs} = K_d(1 + [\mathrm{i}]/K_i)$, can be replotted against [i] to obtain K_d (y intercept) and K_i (reciprocal of the slope). It should be mentioned here that the corresponding equation for competitive inhibition in the case where depletion of *both* inhibitor and ligand due to receptor binding have to be taken into account is quite complex:

$$\begin{aligned} 0 = {} & [\mathrm{rl}]^3\Delta \\ & + [\mathrm{rl}]_2(L_tK_i + I_tK_d - (K_d + B_{max})\,\Delta) \\ & + [\mathrm{rl}]\,[-L_tR_tK_i - L_t^2K_i - L_tI_tK_d - L_tK_iK_d + L_tR_t\,\Delta) \\ & + L_t^2R_tK_i \qquad \Delta = (K_d - K_i). \end{aligned} \qquad [12]$$

The reason for illustrating this equation is that it, together with its value under limiting conditions, can show oft-neglected limitations for certain types of assays used for receptors of very low abundance, mentioned below.

2.3 Practical guidelines for receptor assay protocols

With the foregoing in mind, some practical considerations in the choice of ligands, receptor preparations, and assay protocols can be summarized. Assays may be generally segregated into equilibrium and non-equilibrium methods. The latter seek to separate complexed from unbound marker before complex dissociation can take place. Such procedures include gel filtration assays, ion-exchange assays, and filter binding assays. To be reasonably reliable, that is for >90% recovery, the time of separation from Equation 8, must be faster than $t_s = 0.15\ t^{1/2}$.

A second rule of thumb is that for a ligand to be useful as a marker, aside from the obvious necessity of being specific for the channel or receptor, the K_d should be approximately equal to or less than the concentration of the receptor sites in samples to be assayed. This rule has its exceptions, but is nearly always true for different reasons in different assay protocols. For example, in most non-equilibrium assays (e.g. gel filtration or filter binding assays), the non-specific signal is 0.1–1% of total added ligand. To complete a saturation isotherm to within 90% of maximal binding, free ligand must be $[l] = 10\ K_d$ (cf. Equation 1). At the same time, to obtain a specific to non-specific signal of 10:1, it follows that $R_t \simeq 10$ (n.s.b) $= 10\ (0.01–0.001)L_t = 10\ (0.01–0.001)(10\ K_d)$, thus $R_t \simeq (0.1–1)K_d$. Thus, because receptor concentrations *are generally high* in such assays, the problem of ligand or inhibitor depletion due to binding cannot usually be ignored (cf. Equation 4).

For different reasons, similar conclusions apply to most equilibrium assay methods, such as supernatant depletion assays. In these procedures, membrane bound receptor is mixed with ligand and pelleted in the presence or absence of excess specific, unlabelled inhibitor. The ligand left in the supernatant, in the absence of inhibitor is equal to the concentration of free ligand [l] in equilibrium with the binding site. In the presence of excess inhibitor, the repairing ligand is equivalent to the (displaced) specifically bound label plus free ligand ([rl] + [l]). Subtracting the first of these values from the second leaves [rl]. For a reasonably accurate determination, specific binding should deplete the ligand in the supernatant by ~10% or more. Again, to attain 90% saturation, $[l] = 10\ K_d$ and must be equal to approximately $10\ R_t$, and $R_t \simeq K_d$ is seen as a desirable *upper* limit for K_d. These constraints can influence both the choice of ligand and the source of receptors which are to be studied.

In cases of rare receptors (e.g. the receptors for inositol-1,4,5-trisphosphate (5, 16) which are present at low abundance in many tissues), high specific activity ligand is required to obtain a measurable binding signal. Saturating concentrations of this labelled ligand, aside from being prohibitively expensive, result in high non-specific backgrounds. In such extreme cases, non-saturating levels of ligand are used; varying concentrations of unlabelled ligand are used to displace bound ligand competitively. If the unlabelled ligand is used as an inhibitor, the assay is equivalent of keeping total label constant, but changing the specific activity at every point. When *different* ligands are used as inhibitors, this is no longer the case. In many instances, the displacement midpoint is taken incautiously as approximately equal to the K_d of the displacing ligand. For same-molecule displacement, Equation 12 applies, because depletion due to binding of both inhibitor and labelled ligand *must* be taken into account. However, as $K_d = K_i$, the equation simplifies to a quadratic, with the solution

$$[\mathrm{rl}] = \left[\frac{L_t}{L_t + I_t}\right]\frac{\beta \pm \sqrt{\beta^2 - 4\,(L_t + I_t)R_t}}{2} \qquad [13]$$

where $\beta = (K_d + R_t + L_t + I_t)$ and the inhibitor concentration for 50% inhibition is given by the relationship:

$$I_{0.5} = \frac{K_d + R_t + L_t - (3/2)b}{1 - (b/2L_t)} \qquad [14]$$

where b is the level of binding in the absence of inhibitor.

This is an important relationship, because if an estimated K_d and R_t, generated from a supernant displacement curve of a relatively high concentration of ligand (L_t) do not satisfy the conditions of Equation 14, very large errors can result. Thus, if from Equation 13 binding observed in absence of inhibitor ([rl]), the highest level of observed binding ($\sim R_t$), and the level of L_t used are small compared to the midpoint concentration for displacement ($I_{0.5}$), the midpoint approximates to K_d. Finally, it should be noted, for competitive displacement assays where the dissociation constant of the inhibitor is markedly different from that of the marker ligand, Equation 1 does not simplify, and $I_{0.5}$ cannot yield a reliable estimate of K_i or K_d.

3. Receptor assays

There are few limitations to the actual methods that can be applied in receptor assays, other than those dictated by the rate and equilibrium interactions of ligand and protein. Examples are given of three assays which have proven to be useful in the characterization of ligand-activated and voltage-activated ion channels.

3.1 Assay for inositol 1,4,5-trisphosphate receptors

The intracellular receptors for inositol-1,4,5-trisphosphate are intracellular, ligand-activated, highly regulated Ca^{2+}-release channels (5–7, 16). These receptors are typically found at low abundance in virtually all cells, and are obtained at higher levels in membranes derived either from mammalian brain, particularly cerebellum, or in cell lines transfected with cloned cDNAs. The syringe assay described here (*Protocol 1*) is a non-equilibrium procedure (8), permitting separation of bound from free ligand in ~ 1 sec, with a high signal-to-background ratio, and may be used with membrane fragments or detergent-solubilized receptor preparations.

Protocol 1. Ion-exchange assay for inositol 1,4,5-trisphosphate receptors

Equipment and reagents

- [^{3}H]-inositol 1,4,5-trisphosphate ([^{3}H]InsP$_3$) (e.g. New England Nuclear)
- High-purity unlabelled InsP$_3$ (e.g. Boehringer-Mannheim) for diluting specific activity
- Sonicator for membrane dispersal (e.g. Laboratory Supplies)
- Buffer A: 20 mM EPPS pH 8.5, 20 mM NaCl, 100 mM KCl, 1 mM EDTA
- Anion exchange resin: Dowex 1-X2 (Cl^{-1} in 50 mM Tris-OH pH 8.5, 1 mM EDTA
- Bovine serum albumin (15 mg/ml)
- 3 ml tuberculin syringes; nylon mesh, 10 μm pore size (Spectramesh) and Whatman No 1 filter paper to fit barrel of syringe

Method

1. Homogenize 15–20 fresh rat cerebella in 10 volumes of ice-cold buffer A and sediment at 100 000 *g* for 30 min (9). Resuspend pellets in half the original volume of buffer A and resediment. Resuspend final pellets in 4 ml buffer A per g wet weight of tissue (protein yield ~20 mg/ml). Aliquots may be stored indefinitely at −70°C.
2. For assay as sonicated membrane fragments, dilute 1 ml thawed membranes five-fold in buffer A, homogenize with seven strokes in a Potter Elvehjem glass homogenizer, transfer to a 30 ml Corex test tube and sonicate for 15 sec in a bath sonicator. This reveals cryptic sites and reduces the size of the fragments.
3. Solubilized extracts of receptor are prepared by diluting the membrane suspension five-fold in buffer A containing 2% (w/v) Lubrol-PX and homogenizing on ice with 15 strokes of a Potter Elvehjem homogenizer. Incubate the suspension with rotation for 20 minutes at 4°C and sediment at 100 000 *g* for 40 min (supernatants should contain 1.2–1.6 mg protein/ml, with 20–30 pmol receptor sites/mg protein).
4. Prepare a slurry of Dowex 1-X2 (Cl$^-$) anion-exchange resin in 50 mM Tris-OH (pH 8.5), 1 mM EDTA, and equilibrate with 10 volumes of 15 mg/ml BSA: wash resin with BSA-free buffer. Remove plungers from 3 ml tuberculin syringes; insert a nylon disc (Spectramesh, 10 μm pore size), followed by a paper disc (Whatman No. 1) into the barrel. Pour settled columns of 0.6 ml resin. Mix aliquots of solubilized or membrane-bound receptor with varying amounts of radiolabelled ligand in a final volume of 0.2 ml and incubate for 10 minutes in the

cold. Apply equilibrated samples to the syringe wall, insert and rapidly depress the plunger, forcing it smoothly down the barrel of the syringe to extrude receptor-bound complexes into a scintillation vial for counting. Non-specific binding is measured in the presence of 12 μM unlabelled $InsP_3$.

As illustrated in *Figures 2* and *3*, saturation binding of approximately 30 pmol/mg protein is obtained in soluble extract assays; given the molecular mass of the receptor subunit of 315 000, this corresponds to nearly 1% of the protein in the extract. Scatchard analysis indicates more than one component of binding and may be described by two classes of sites, a high-affinity population (8–10 pmol/mg protein, $K_d \simeq 5$ nM) and a low-affinity population (15–20 pmol/mg protein, $K_d \simeq 120$ nM). This is not an artefact of the assay protocol: when the same procedure is performed at 20°C (*Figure 3a,b*) a single class of high-affinity sites remains active, while the low-affinity population appears to be (reversibly) inhibited. Dissociation-rate experiments reveal two components of ligand dissociation, a slowly dissociating component with a dissociation rate of approximately $k_2 = 0.003\ \text{sec}^{-1}$, and a more rapid process with $k_2 = 0.098\ \text{sec}^{-1}$: both processes are relatively slow compared with the ~ 1 sec separation time of the assay.

3.2 Assay for voltage-sensitive Na^+ channels using labelled tetrodotoxin

Voltage-sensitive Na^+ channels are readily assayed by the specific binding of

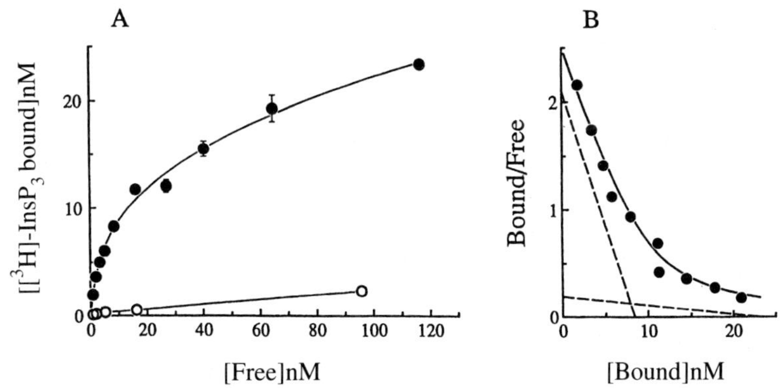

Figure 2. Binding measured on solubilized cerebellar $InsP_3$ receptor. Panel A shows total (closed symbols) and non-specific (open symbols) binding. Panel B reveals the re-plot according to Scatchard, indicating more than one population of receptor sites, as described in the text. Our studies have indicated the presence of at least two populations, a high affinity population (R_t = 8–12 pmol/mg protein, K_d = 5 nM) and a low affinity population (R_t = 18–25 pmol/mg protein, K_d = 100–120 nM).

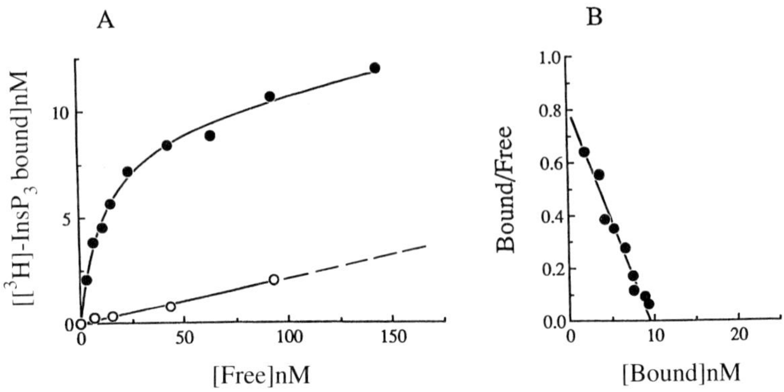

Figure 3. Saturation binding of [^{3}H] InsP$_3$ to its receptor at 20°C. Part A: Equilibrium binding isotherm, non-specific (open symbols) and total (closed symbols) binding. Part B: Scatchard analysis, showing only one class of binding site.

labelled forms of the neuropoisons tetrodotoxin (TTX) and saxitoxin (STX) in membrane-bound and detergent-solubilized form (10–12).

The main electric organ of *Electrophorus electricus* offers a tissue with a particularly high abundance of a single isoform of Na^+ channel: these tissues are exceptionally dense in connective tissue (13, 14). Soluble receptor is readily prepared from membranes derived by differential sedimentation of homogenates. Membrane vesicles suitable for receptor assays require additional manipulations. Spun-column gel filtration assays, first applied to the Na^+ channel from the eel electroplax, can be conducted as described in *Protocol 2* (10).

Protocol 2. Spun-column assay for binding of [^{3}H]TTX to voltage-sensitive Na^+ channels

Equipment and reagents

- [^{3}H]tetrodotoxin (TTX) labelled and purified according to ref. 12; 50–60 μCi/μmol, radiochemical purity 60–70%, **OR**
- High specific activity, exchange-labelled [^{3}H]saxitoxin (STX) (New England Nuclear (15))
- Electric eels (World Wide Scientific Supply)
- Polytron homogenizer
- Homogenization buffers: 50 mM Sodium phosphate pH 6.8, 5 mM EDTA containing protease inhibitors (AEBSF, aprotinin A); 50 mM sodium phosphate pH 7.2
- Beckman VTI-50 vertical gradient tubes
- Sephadex G-50 fine grade gel and 50 mM sodium phosphate pH 7.4 to equilibrate
- Cheesecloth for filtering homogenate

Method

1. Sacrifice animals by hypothermia and decapitation, remove the main organs and trim of connective tissue. Freeze at −80°C. Dice partially frozen tissue into 1–2 cm^3 cubes, add to four volumes of homogeniza-

tion buffer, homogenize in a Polytron tissue homogenizer, and filter through two layers of cheesecloth. Centrifuge at 30 000 *g* for 20 min, discard supernatants, and resuspend pellets in five volumes of the same buffer. Resediment at 100 000 *g* for 40 min. Discard supernatants and resuspend the packed pellets in 2 ml of buffer/g wet weight.

2. Prepare collagenase-treated membranes by diluting approximately 35 g of crude membranes into 80 ml of 50 mM sodium phosphate, pH 7.2. Mix with light homogenization and add sufficient $CaCl_2$ to give 0.2 mM free Ca^{2+} together with 150 units/ml of collagenase (Worthington CLS III). Mix suspension and incubate at 30°C with occasional stirring, then add 2 ml of 0.2 M EDTA. Layer membranes on to eight sucrose gradients (15 ml per tube) in Beckman VTI-50 vertical gradient tubes. Gradients comprise 12 ml of 18% (w/v) sucrose on top of 12 ml of 40% sucrose in the buffer chosen for the application (for binding assays, 50 mM sodium phosphate pH 7.2, 0.1 mM AEBSF). Centrifuge at 45 000 r.p.m. for 2.5 h and collect the flocculent material at the sharp boundary between the 18% and 40% interface (typically about 10 ml/tube). Dilute to 360 ml in buffer, homogenize lightly, and centrifuge at 100 000 *g* for 60 min. Resuspend final pellets (representing approximately 1 g per 12 g of starting pellet) in 2 ml buffer per g, and store frozen at −80°C. (These membranes contain approximately 30 pmol $[^3H]$TTX binding sites/mg protein, corresponding to approximately 1% Na^+ channel by weight. The high recovery of binding sites confirms that about 80% or more of the initial pellet is connective-tissue protein.)
3. Swell and equilibrate Sephadex G-50 (fine grade) gel with 50 mM sodium phosphate buffer (pH 7.4). Pour small gel-filtration columns into the barrels of 3 ml disposable plastic syringes fitted with a disc of Whatman No 1 filter paper. Allow gel to settle by gravity to a final volume of 1.4 ml, and suspend syringes by their finger tabs in plastic culture tubes (17 × 100 mm). To prepare soluble extracts, thaw membrane suspensions, make 0.1% (w/v) with Lubrol-PX (mix without homogenization), and centrifuge at 100 000 *g* for 30 min. Remove supernatant and store on ice.
4. Incubate varying amounts of $[^3H]$TTX (0.1–10 × K_d) alone or in combination with excess (10 μM) unlabelled TTX with the extract for 10 minutes on ice. Layer 0.2 ml aliquots on to gel columns which are placed immediately in a bench-top centrifuge and accelerated rapidly to 1300 *g* (see ref. 10 for a discussion). Bring centrifuge to a halt after 45 sec and remove the solution eluted into the culture tube for counting.

Figure 4 illustrates results from binding measurements performed on suspensions of collagenase-treated membranes solubilized in Lubrol immediately

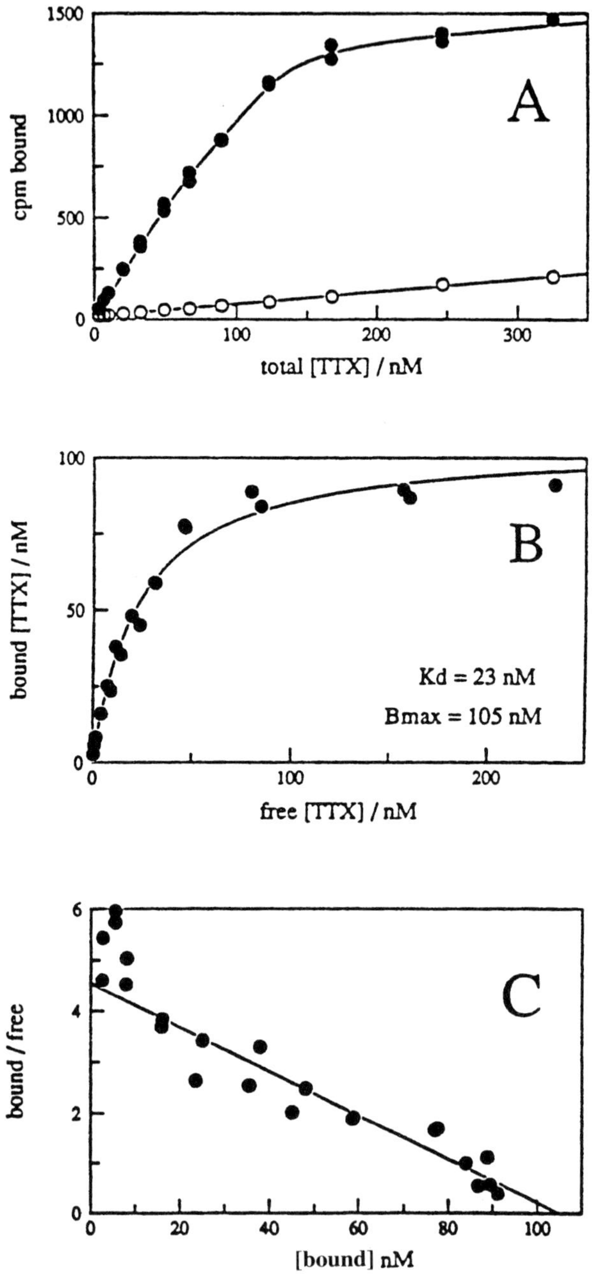

Figure 4. Spun-column assay of [^{3}H]TTX binding to electroplax membranes. Part A: Specific (closed symbols) and non-specific (open symbols) binding; Part B: least-squares fitting to show binding parameters. Note imperfect fit. Part C: As A and B, but presented as a Scatchard analysis.

before assay. *Figure 4*, which shows both specific and non-specific binding, illustrates the 'plateau-effect' predicted by Equation 4 and discussed above. Despite being relatively slow (0.25–1 min) this assay provides high recovery of toxin–channel complexes because of a comparatively slow dissociation rate for TTX from the electroplax channel (half-life ~28 min). In contrast, dissociation of [^{3}H]STX is comparatively more rapid, and more rapid elution procedures (11) enhance recovery.

3.3 Assay for voltage-sensitive Na^+ channels using radiolabelled STX

Although binding measurements for [^{3}H]TTX to electroplax sodium channels are readily performed with the spun-column assay, binding of [^{3}H]STX is not. Despite the fact that STX binds with slightly higher affinity than TTX and exhibits competitive displacement interactions with TTX, the rates of binding and unbinding may be between 100 and 1000 times faster. Because STX dissociation is so rapid, an equilibrium procedure such as supernatant depletion is desirable. Also, because receptor concentrations in assayed samples can be high (e.g. 100–200 nM) compared with the dissociation constant (K_d = 1–5 nM), this preparation lends itself to supernatant depletion protocols (even with nearly saturating levels of free [^{3}H]STX, a substantial percentage of total ligand is bound and thus depleted by pelleting the membranes from suspension). In addition, non-specific binding is low. In contrast, our experience for total cerebellar IP_3 receptor assay suggests that despite the fact that receptor abundance in assayed samples is high (~50–100 nM), the low affinity ($K_d \simeq 120$ nM) of a substantial subpopulation of the receptors renders the supernatant depletion assay problematic: complete saturation isotherms are difficult to prepare. In addition, because there is evidence for multiple components of binding, the competition displacement assays cited above are very difficult to interpret, except under highly restricted experimental conditions.

Protocol 3. Suspension–depletion assay for binding of [^{3}H]STX (or [^{3}H]TTX) to electroplax membranes

Equipment and reagents

- See *Protocol 2*
- Beckman airfuge or equivalent

Method

1. Suspend collagenased electroplax membranes in assay buffer (50 mM Na phosphate, pH 7.2) to a final concentration of approximately 3 mg protein/ml, typically corresponding to 75–100 nM [^{3}H]STX binding sites. Equilibrate duplicate or triplicate aliquots (250 μl) of receptor suspension with varying concentrations of [^{3}H]STX (250 μCi/μmol) (1–300 nM) in the absence or presence of 10 μM unlabelled STX or TTX.

Protocol 3. *Continued*

2. Place 0.2 ml aliquots in 220 μl cellulose proprionate centrifuge tubes for the A 100/30 rotor of the Beckman airfuge. Take 25 μl aliquots from the remaining sample for scintillation counting. Centrifuge samples for 5 min at maximum speed and withdraw 50 μl aliquots for counting (the total concentration of ligand is determined from the radioactivity of the uncentrifuged samples).
3. Ligand left in the supernatant of samples containing excess unlabelled TTX or STX constitutes the concentration of free [^{3}H]STX in equilibrium with the binding site and the (displaced) specifically bound label: non-specific counts are depleted by association with the pellet.
4. Ligand remaining in the samples incubated in the absence of cold toxin reflect free ligand concentrations alone. Subtracting these values from those of the competition samples yields specifically bound ligand, to be analysed as a function of free ligand concentration.

This constitutes a relatively standard procedure which may be used to assess receptor abundance in samples in which ligand dissociation rates and receptor abundance are unknown. This enables examination of the efficiency of receptor solubilization procedures and non-equilibrium assays. However, many ionic channels have no specific marker. Thus, for purification of such ion channel proteins, new techniques to assay the protein are required. These are described in the next section.

4. Functional assay by bilayer reconstitution

The planar bilayer technique (see Chapter 2) has been used in this way for sarcoplasmic reticulum (SR) channel proteins (16, 17), with four main steps: membrane solubilization with detergent, separation by chromatography, reconstitution into proteoliposomes, and incorporation into planar bilayers. The main steps are summarized in the *Protocols 4* and *5*.

Protocol 4. Solubilization and reconstitution of SR channel proteins into planar lipid bilayers

Equipment and reagents

- CHAPS
- Protease inhibitors: aprotinin, leupeptin, pepstatin A, and antipain
- Asolectin (type IIS)
- Cellulose dialysis tubing
- Solubilization solution: 300 mM KCl, 2 mM EGTA, 20 mM Tris-HCl, pH 7.0
- Dialysis buffer: 100 mM KCl, 2 mM EGTA, 20 mM Tris–HCl, pH 7.0
- Storage solution: 300 mM sucrose, 100 mM KCl, 20 mM Tris–HCl, pH 7.0

Method

1. Prepare light SR (LSR) vesicles from rabbit skeletal muscle (e.g. as in ref. 16) and resuspend final pellets in 0.1 M KCl, 0.3 M sucrose, 20 mM Tris–HCl, pH 7.0 at 15–20 mg/ml. The suspension may be frozen and stored at −80°C after snap freezing in liquid nitrogen.
2. Suspend membranes at room temperature in solubilization solution to give a final protein concentration of 1–2 mg/ml. Add a mixture of protease inhibitors (0.5 μg/ml aprotinin, 1 μg/ml leupeptin, 1 μg/ml pepstatin, 1 μg/ml antipain; each given as final concentration) just before use. Solubilization is initiated by adding 1% (w/v) CHAPS, and the sample is stirred for 1 h at 4°C, then centrifuged at 100 000 *g* for 30 min to remove insoluble material.
3. Prepare proteoliposomes by supplementing proteins solubilized in CHAPS with asolectin solubilized in 1% (w/v) CHAPS, 0.1 M KCl, 20 mM Tris–HCl, pH 7.0 to give a final concentration of 1–5 mg/ml. Stir for 30 min, then transfer to dialysis tubing and dialyse against 1000 volumes of dialysis buffer at 4°C, with an equal volume of polystyrene resin (Bio-Beads SM-2) added to the buffer. Continue dialysis for 72 h and finally centrifuge turbid sample at 100 000 *g* for 30 min. The pellet may then be suspended in storage solution and stored at −80°C.
4. Fuse proteoliposomes with planar lipid bilayers (see Chapter 2 and ref. 18). Fusion is facilitated by

- presence of negative lipids in either the planar bilayer or liposome membrane;
- millimolar concentrations of divalent cations in the *cis*-solution;
- hyperosmolarity of the vesicular interior with respect to the *cis* chamber;
- hyperosmolarity of the *cis* chamber with respect to the *trans* chamber.

The detection of specific single-channel currents can be used to follow the progress of a purification protocol (see *Protocol 5*). A simple purification protocol for SR membrane channel proteins can require as few as three steps: anion exchange, gel filtration, and concanavalin-A-affinity chromatography.

Protocol 5. SR on channel purification—a three-step procedure

Equipment and reagents

- Cosmogel DEAE column
- TSK-G4000SW column
- Concanavalin-A sepharose column
- Elution buffer A: 0.3 M KCl, 1% CHAPS, 2 mM EGTA, 20 mM Tris–HCl, pH 7.0
- Elution buffer B: 0.6 M KCl, 1% CHAPS, 2 mM EGTA, 20 mM Tris–HCl, pH 7.0
- Elution buffer C: 0.1 M KCl, 0.8% CHAPS, 2 mM EGTA, 20 mM Tris–HCl, pH 7.0
- Elution buffer D: 0.5 M KCl, 0.8% CHAPS, 1 mM $CaCl_2$, 1 mM $MnCl_2$, 20 mM Tris–HCl, pH 7.0

Protocol 5. *Continued*

Method

1. Apply solubilized membranes from *Protocol 4* to an anion-exchange column (Cosmogel DEAE) equilibrated in elution buffer A containing the mixture of inhibitors described in *Protocol 4.* Elute with a continuous gradient of KCl.
2. Reconstitute fractions into proteoliposomes, and assay for channel activity by the planar bilayer method (see *Protocol 4* and Chapter 2).
3. Inject the fraction with maximal activity into a size-exclusion column, e.g. TSK-G4000SW, and elute with buffer C at a flow rate of 3 ml/min. (Equilibrate the column in the same buffer). Assay fractions by planar bilayer reconstitution, as before.
4. Apply gel filtration fractions which have channel activity to a con-A–sepharose column (10 mm × 7 mm) equilibrated in buffer D, and elute with 1 M α-methyl-D-mannoside. (Remove EGTA in gel filtration fractions by a short gel filtration column, e.g. PD-10 from Pharmacia.) Reconstitute glycoproteins into bilayers after incorporating into liposomes as before.

Channel proteins may denature when solubilized with detergent and it is possible that channel activity is changed during the purification procedure. In addition, some detergents may form pores in lipid bilayers and produce artefactual current records resembling ion-channel currents. Application of the method therefore requires that the properties of native channels be known in as much detail as possible, in order not to confuse artefactual current with the desired channel current. *Figure 5* shows the elution profile from DEAE chromatography and accompanying appearances of fractions following SDS polyacrylamide gel electrophoresis while *Figure 6* shows reconstituted SR anion channel activity after all three chromatography steps.

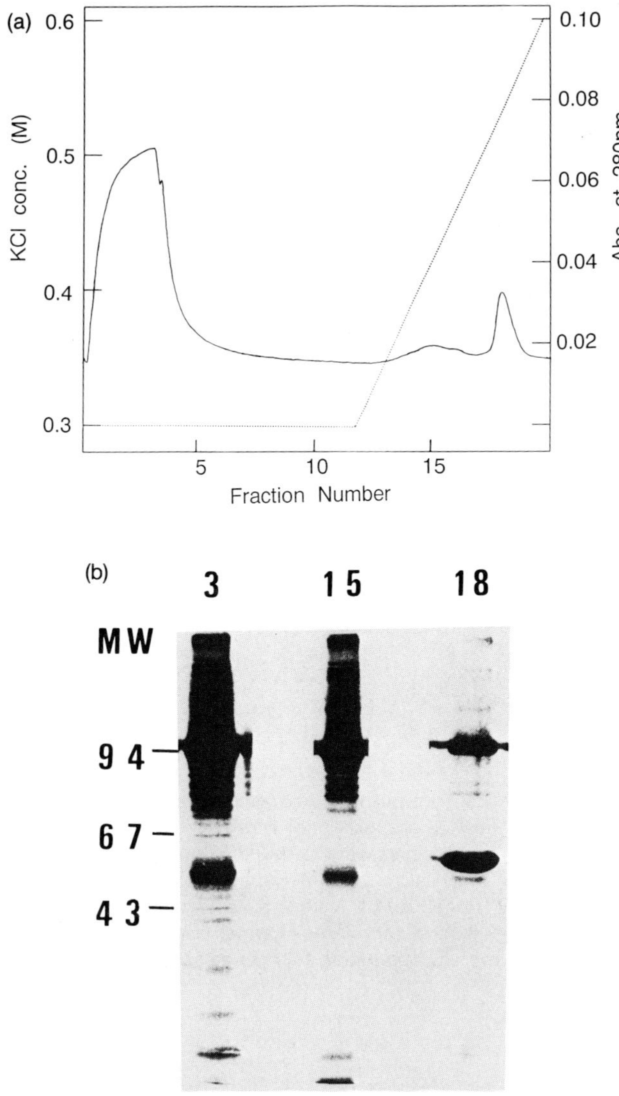

Figure 5. (a) DEAE column elution profile with a continuous KCl gradient. LSR proteins solubilized in 1% (w/v) CHAPS, 2 mM EGTA, and 20 mM Tris–HCl, pH 7.0 in the presence of a mixture of protease inhibitors applied to the DEAE column (7.5 × 75 mm), and eluted with a continuous gradient of KCl from 0.3 to 0.6 M (dotted line). OD_{280} and KCl concentration are shown. (b) SDS-PAGE of fractions from the DEAE column.

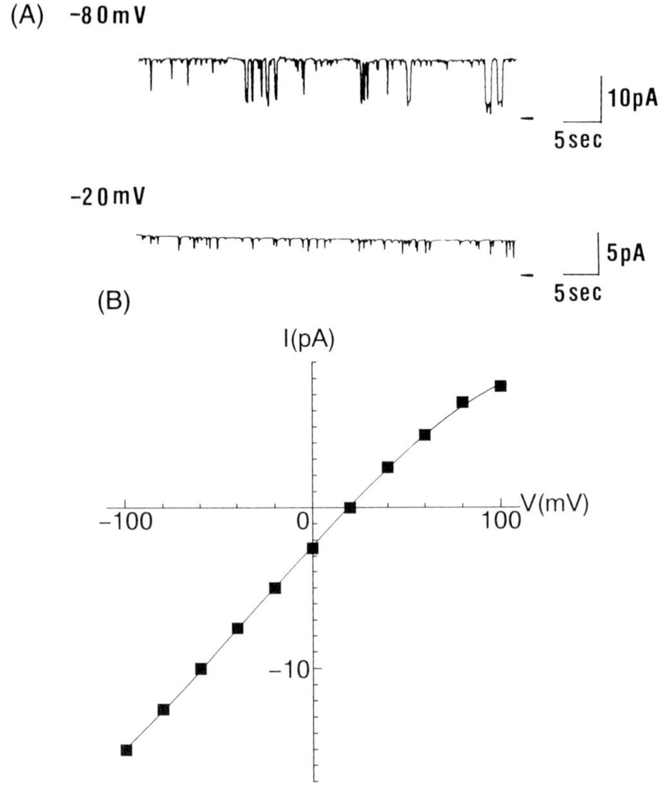

Figure 6. (A) Current records from purified anion-selective channels incorporated into a lipid bilayer. A concanavalin-A absorption pool was added with asolectin to give a final concentration of 2 mg/ml, and vesicles were formed by dialysis. Current fluctuations were recorded in 0.3 M KCl, 1 mM $CaCl_2$, 20 mM Hepes-Tris, pH 7.2 (*cis*), and 0.1 M KCl, 20 mM Hepes-Tris, pH 7.2 (*trans*). (B) Cl^- versus K^+ selectivity of purified anion channel from SR. The slope of the line is the single-channel conductance, g_{Cl} = 115 pS. The reversal potential was 20 mV corresponding to a permeability ratio, P_{Cl}/P_K of 7.

References

1. Agnew, W. S., Cooper, E. C., Henkel, S., Correa, A. M., James, W. M., Ukomadu, C., and Tomiko, S. A. (1991). *Ann. N.Y. Acad. Sci.* **625**, 200.
2. Morad, M., Naylor, W., Kazda, S., and Shramm, M. (ed.) (1988). *The calcium channel: structure, function and implications*. Springer, Berlin.
3. Tometsko, M. and Richards, F. M. (ed.) (1980) The applications of photochemistry in probing biochemical targets. *N.Y. Acad. Sci.*, New York.
4. Schmidt, J. and Raftery, M. A. (1973). *Anal. Biochem.* **52**, 349.
5. Irvine, R. F. (1992). *Advances in second messenger and phosphoprotein research* (ed. J. W. Putney), Vol. 26, pp. 161–85. Raven Press, New York.
6. Furuichi, T., Yoshikawa, S., Miyawaki, A., Wada, K., Maeda, N., and Mikoshiba, K. (1989). *Nature* **342**, 32.

7. Supattapone, S., Worley, P. F., Baraban, J. M., and Snyder, S. H. (1988). *J. Biol. Chem.* **263**, 1530.
8. Hingorani, S. R. and Agnew, W. S. (1991). *Anal. Biochem.* **194**, 204.
9. Hingorani, S. R. and Agnew, W. S. (1994). *J. Biol. Chem.* (In press.)
10. Levinson, S. R., Curatalo, C. J., Reed, J., and Raftery, M. A. (1979). *Anal. Biochem.* **99**, 72.
11. Moczydlowski, E., Olivera, B. M., Gray, W. R., and Strichartz, G. R. (1986). *Proc. Natl Acad. Sci. USA* **83**, 5321.
12. Agnew, W. S., Levinson, S. R., Brabson, J. S., and Raftery, M. A. (1978). *Proc. Natl Acad. Sci. USA* **75**, 2606.
13. Ritchie, J. M., Rogart, R. B., and Strichartz, G. (1976). *J. Physiol.* **261**, 477.
14. Miller, J. A., Agnew, W. S., and Levinson, S. R. (1983). *Biochemistry* **22**, 462.
15. James, W. M., Emerick, M. C., and Agnew, W. S. (1989). *Biochemistry* **28**, 6001.
16. Ide, T., Morita, T., Kawasaki, T., Taguchi, T., and Kasai, M. (1991). *Biochim. Biophys. Acta* **10067**, 213.
17. Ide, T., Sakamoto, H., Morita, T., Taguchi, T., and Kasai, M. (1991). *Biochem. Biophys. Res. Commun.* **176**, 38.
18. Tanifuji, M., Sokabe, M., and Kasai, M. (1987). *J. Membr. Biol.* **99**, 103.

5

Channel cloning, mutagenesis, and expression

PAUL J. WHITING, KEITH A. WAFFORD, IRIS PRIBILLA, and THOMAS PETRI

1. Introduction

The application of the techniques of molecular biology to the understanding of the structure and function of ion channels has resulted in a quantum leap in our knowledge which 15 years ago would have seemed inconceivable. The cloning of a cDNA encoding a channel protein is now considered only the first step, opening up many areas for detailed study:

(a) Computer analysis of the deduced amino acid sequence to predict putative transmembrane topology, secondary structure, and post-translational modification.

(b) Expression *in vitro* of the recombinant channel for both pharmacological and functional studies, and also for studies of channel structure, assembly, and regulation.

(c) Site-directed mutagenesis for structure/function studies.

(d) Generation of site-directed antibodies as biochemical probes.

In this chapter we focus on approaches to clone cDNAs encoding the ion channel of interest, methods to express the cloned channel *in vitro*, and methods for site-directed mutagenesis of the cloned channel.

It is important to note that every channel will present its own individual problems, and that a successful approach in one case may not work in another. For instance, expression cloning is really only suitable for channels which function as homo-oligomers. Similarly, the type of recombinant expression system used will depend to some degree upon the questions one is trying to answer. For site-directed mutagenesis and analysis of structure and function a transient expression system is required, whereas for detailed pharmacological and functional studies a permanent expression system would be more efficient, reproducible, and robust.

2. Isolation of cDNAs by hybridization screening of cDNA libraries

2.1 General principles

cDNA libraries are screened with cDNA probes, the sequences of which are homologous to the cDNA sequence the researcher is attempting to isolate. This approach can be used to isolate species homologues (e.g. the human version of a rat cDNA) and to isolate cDNAs encoding related members of a gene family.

The choice of cDNA library is often critical to the outcome. It is obviously important to choose a cDNA library prepared from a source where the target cDNA(s) is likely to be enriched. In the case of channels/receptors one would use knowledge of target distribution derived from functional or radioligand binding studies. This could mean using a cDNA library prepared from a particular region of the brain, or from a cell line. It is not always necessary to construct a cDNA library. A number of companies (Stratagene, Clontech, Invitrogen) now market a wide range of cDNA libraries prepared from different sources, and these have proved successful in a number of cases for isolating channel protein cDNAs. However, if a particular cDNA library is not commercially available, or if the target cDNA is likely to be particularly long (i.e. greater than 3000 base pairs, where successful cloning would be dependent upon having a good quality library with long inserts), it would be necessary to construct the cDNA library oneself. Approaches to constructing cDNA libraries are beyond the scope of this chapter, but there are numerous texts giving detailed methodologies (1, 2), and kits are available from a number of companies for this purpose (e.g. Stratagene, Invitrogen, Pharmacia, Gibco-BRL). The most convenient vectors are those which allow automatic excision of the cDNA insert-plasmid from the λ sequences. Such vectors include λZAP (Stratagene) and λDR2 (Clontech).

2.2 DNA probes for screening cDNA libraries

2.2.1 cDNAs as probes

The easiest starting point for obtaining a cDNA probe is to 'clone by phone', i.e. obtain the cDNA from the laboratory that originally cloned it. If the cDNA is to be used as a probe to clone a species homologue then the entire cDNA sequence would be satisfactory, i.e. it would not be necessary to remove any 5′ or 3′ untranslated region. For isolating related cDNAs, i.e. other members of a gene family, the probe would be used under quite low-stringency hybridization conditions and therefore it would be desirable to remove most of the untranslated region since these sequences will not be conserved. Additionally, it may even be desirable to use as a probe a shorter region of the cDNA which contains only highly conserved sequences, e.g. in the case of the *N*-methyl-D-aspartate (NMDA)-type glutamate receptors, the sequences shown in *Figure 1*. Where possible, these short DNA fragments can be isolated by restriction enzyme digestion. Alternatively, it is possible to generate them by PCR using oligonucleotide primers derived from the cDNA sequence, and the cDNA as the template.

2.2.2 cDNA probes derived by PCR using redundant primers

Knowledge of the sequences of cDNAs encoding members of a gene family can be utilized in PCR (the polymerase chain reaction) to isolate related cDNA sequences. These can be used subsequently as probes to isolate the full length cDNA from a cDNA library. An example demonstrating this approach is shown in *Figure 1*. The deduced amino acid sequences of the conserved region of four rat NMDA receptor subunits are shown aligned according to their consensus sequences. There are clearly regions of complete conservation of amino acid sequence. Two such sequences (1 and 2) are overlined. These particular sequences are identified because they contain a high proportion of amino acids which have only one or two possible codons, and avoid residues such as leucine and serine which have six possible codons. In *Figure 1b* the two sequences are shown with the cognate nucleotide sequence of each subunit aligned below. From these aligned sequences it is possible to design oligonucleotide primers for use in PCR. The identification of conserved sequences which do not contain amino acids such as leucine and serine will result in consensus primers which are less redundant. In turn, the lower the redundancy of the primer, the more specific the priming will be during the PCR. The template for PCR would be 1st strand cDNA prepared from mRNA isolated from the target tissue. There are standard protocols for mRNA preparation (1, 2) and kits are available commercially from a number of sources (e.g. Stratagene, Invitrogen, Pharmacia, and Promega) which enable the rapid preparation of mRNA. Similarly, 1st strand cDNA synthesis can be performed using standard protocols or using commercially available kits. The 1st strand cDNA can be used directly in the PCR. Conditions for the PCR have, to some degree, to be determined empirically. For PCR protocols refer to refs 2, 3. Probably the most important factor which should be varied in the 'consensus primer' PCR is the annealing temperature. Since the purpose of this approach is to look for related sequences, use of a low annealing temperature, e.g. 40°C, would give the greatest possibility of success.

PCR products can be subcloned for subsequent analysis by two main approaches.

1. Incorporation of a restriction enzyme site on the 5′ end of each oligonucleotide primer, e.g. in *Figure 1* PCR primer 1 could be designed to have an EcoRI site (GAATTC), giving the primer sequence 5′ CGAATT-CGTCGC. . . . The PCR product can then be subcloned by digestion with the relevant restriction enzyme, purification on an agarose gel, and ligation

(a)

```
                      1
RAT NMDA2C 564 YSPAVWVMMFVMCLTVVAiTVFmFEYfSPVsYNqnLtkGKkPGGpsFTIGKSvWLLWALVF
RAT NMDA2D 578 YSPAVWVMMFVMCLTVVAvTVFiFEYlSPVGYNRsLAtGKrPGGstFTIGKSIWLLWALVF
RAT NMDA2A 553 FSAsVWVMMFVMLLIVSAiAVFVFEYFSPVGYNRnLAkGKaPtGllFTIGKAIWLLWGLVF
RAT NMDA2B 554 FSAdVWVMMFVMLLIVSAvAVFVFEYFSPVGYNRcLAdGrePgGpsFTIGKAIWLLWGLVF

               NNSVPiENPRGTTSKIMVLVWAFFAVIFLASYTANLAAFMIQEqYiDTVSGLSDkKFQRPQ
               NNSVPVENPRGTTSKIMVLVWAFFAVIFLASYTANLAAFMIQEEYVDTVSGLSDrKFQRPQ
               NNSVPVQNPKGTTSKIMVSVWAFFAVIFLASYTANLAAFMIQEEfVDQVtGLSDKKFQRPh
               NNSVPVQNPKGTTSKIMVSVWAFFAVIFLASYTANLAAFMIQEEyVDQVsGLSDKKFQRPn

               dQYPPfrFGTVPNGSTErNIRSNYrDMHthMVkfNQrsVEdALTsLKmGKLDAFIYDAAVL
               eQYPPlkFGTVPNGSTEkNIRSNYPDMHsYMVRyNQprVEeALTqLKaGKLDAFIYDAAVL
               DySPPFRFGTVPNGSTERNIRNNYPyMHqYMtRFNQRGVEDALvSLKTGKLDAFIYDAAVL
               DfSPPFRFGTVPNGSTERNIRNNYaeMHaYMgkFNQRGVdDALlSLKTGKLDAFIYDAAVL

               NYMAgKDEGCKLVTIGSGKVFATTGYGIAmqKdShWKRaIDLALLQlLGDgEtqkLEtvWL
               NYMArKDEGCKLVTIGSGKVFATTGYGIALhKGSrWKRpIDLALLQFLGDdEiEmLErLWL
               NYkAGRDEGCKLVTIGSGyiFASTGYGIALQKGSpWKRQIDLALLQFvGDGEMEELEtLWL
               NYmAGRDEGCKLVTIGSGkvFASTGYGIAiQKdSgWKRQvDLAiLQlfGDGEMEELEaLWL

                                    2
               SGICqNeKnEVMSSKLDIDNMAGVFYMLLVAMGLaLLVFAWEHLVYWkLRHsvpnss
               SGICHNdKiEVMSSKLDIDNMAGVFYMLLVAMGLSLLVFAWEHLVYWrLRHClgpth
               TGICHNEKNEVMSSQLDIDNMAGVFYMLaAAMALSLITFIWEHLFYWkLRfCFtGVC
               TGICHNEKNEVMSSQLDIDNMAGVFYMLgAAMALSLITFIcEHLFYWqfRhCFmGVC
```

(b)

	Sequence 1								Sequence 2								
	V	W	V	M	M	F	V	M	D	N	M	A	G	V	F	Y	M
NMDA2C	gtg	tgg	gtg	atg	atg	ttc	gtg	atg	gac	aac	atg	gcg	ggg	gtc	ttc	tac	atg
NMDA2D	gtc	tgg	gtg	atg	atg	ttc	gtc	atg	gac	aac	atg	gca	ggt	gtc	ttc	tac	atg
NMDA2A	gtc	ttg	gtg	atg	atg	ttt	gtg	atg	gat	aac	atg	gtg	ggc	gtg	ttc	tac	atg
NMDA2B	gtg	tgg	gtg	atg	atg	ttt	gtg	atg	gac	aat	atg	gca	ggt	gtc	ttc	tat	atg

PCR Primer 1 (sense): 5' gt(gc)tgggtgatgatgt(tc)gt(gc)atg 3'

PCR Primer 2 (antisense): 5' cat(ga)tagaa(tcg)ac(gac)cc(ct)gccat(ga)tt(ga)tc 3'

Figure 1. Design of redundant PCR primers for isolation of cDNAs encoding members of a gene family. (a) Primary amino acid sequences of rat NMDA receptor NR2A, B, C, and D subunits (32) are aligned. Note that only part of the full amino acid sequence of each subunit (around residues 550–850) is shown. The overlined sequences, 1 and 2, indicate the regions used for design of the PCR primers. (b) Alignment of the amino acid sequence and the cognate nucleotide sequences of each subunit used to design the PCR primers, the sequences of which are shown below.

into a cloning vector. This approach tends to be a very efficient way of subcloning PCR products but has the potential disadvantage that if the product has an internal restriction enzyme site the whole process will be complicated and one may inadvertently lose the sequences for which one is looking.

2. Subcloning of the PCR products directly into a cloning vector using commercially available kits (available from Amersham Invitrogen and Stratagene) which have been specifically designed for this purpose.

Sequence analysis of subcloned PCR products (e.g. using the GenBank database) will determine whether they represent novel sequences.

2.3 Screening of cDNA libraries

For general methodologies on screening of cDNA libraries the reader should consult refs 1 and 2. The most common and well-tested methods for screening libraries rely upon ^{32}P-labelled probes. These should be radiolabelled to high specific activity (greater than 10^9 c.p.m./μg DNA). This can be readily achieved using random priming, for which standard protocols are available, or again suitable kits can be purchased (e.g. Stratagene, Pharmacia, Amersham, DuPont-NEN). More recently, a number of companies (e.g. Stratagene, Promega, United States Biochemicals, Amersham, New England BioLabs) have introduced non-radioactive detection systems which may also be suitable. For the latter, the protocols will differ depending upon the non-radioactive system chosen, and will not be considered further here.

The stringency of hybridization that is chosen depends upon whether the aim is to isolate a cDNA which is identical to the probe sequence (e.g. the full length cDNA of a novel sequence obtained by PCR with redundant primers), or homologous to it (e.g. a species homologue, or a related member of a gene family). High stringency conditions should be used for the former and low stringency conditions for the latter.

Protocol 1. Hybridization of cDNA libraries

Equipment and reagents

- 20 × SSPE: 175 g NaCl, 28 g NaH_2PO_4, 7.4 g EDTA. Dissolve in 900 ml of H_2O, adjust pH to 7.4, make up volume to 1 litre with H_2O
- Denatured salmon testes DNA (Sigma)
- Sodium dodecyl sulphate (SDS)
- 50 × Denhardt's solution: 1 g polyvinylpyrrolidone; 1 g bovine serum albumin (BSA) fraction V, Sigma; 1 g Ficoll, Sigma; H_2O to 100 ml
- Formamide

Method

1. For high-stringency hybridization use the following solution: 5 × Denhardt's, 5 × SSPE, 0.1mg/ml salmon testes DNA, 0.5% SDS, 50% formamide.

Protocol 1. *Continued*

2. For low-stringency hybridization use the same solution as above, but without formamide.
3. Hybridization should be carried out in a volume of about 3 ml per 137 mm diameter nylon/nitrocellulose filter, ideally in a container of slightly greater diameter than the filter, to minimize the volume of hybridization solution required. The hybridization is performed at 42°C, in a slowly-shaking water bath. It is necessary to prehybridize (block non-specific sites) for about 4 h before hybridization. Use fresh solution for hybridization, containing at least 10^5 c.p.m. of boiled probe per ml of solution. Hybridize for at least 18 h.
4. Filter washes:
   (a) For high-stringency hybridization, washing should be carried out at 65°C in a gently shaking water bath: wash twice for 15 min each in 500 ml 5 × SSPE/0.5% SDS, followed by two 15 min washes in 500 ml 0.3 × SSPE/0.5%SDS.
   (b) For low-stringency hybridization, washing should be carried out at 42°C in a gently shaking water bath: initial washing conditions should be three times, 15 min each in 500 ml of 5 × SSPE/0.5% SDS.[a]
5. Expose filters to film (e.g. Kodak XAR) using intensifying screens (e.g. DuPont Cronex screens); exposure should be done at −70°C for an appropriate time period (1–7 days) depending upon the radioactivity of the filters.

[a] It is important to note that the conditions used for low-stringency hybridization have to be determined, to some degree, empirically. The conditions given are about the lowest stringency conditions worth attempting. If, under these conditions, the filters have a very high background then they should be rewashed at higher stringency (i.e. increasing the temperature and/or decreasing the SSPE concentration) until the background reaches an acceptable level.

2.4 Obtaining full length cDNAs

Very often the cDNAs isolated from libraries will not be full length, i.e. will lack part of the coding region, normally the 5′ end. This is dependent upon the quality of the cDNA library, and to some degree the luck of the researcher. Obviously the longer the coding region of the target cDNA the more likely that the cDNAs obtained will be incomplete. Since channel proteins tend to be quite large, this is likely to be a problem. Approaches to obtain the 5′ end of truncated cDNAs include the following.

1. Rescreening the cDNA library with a DNA probe generated from the very 5′ end of the truncated cDNA you have isolated, or if you are trying to isolate a species homologue, rescreening with a cDNA generated from the very 5′ end of the known sequence.

2. Use 'rapid amplification of cDNA ends' (RACE), a PCR method, to isolate the missing sequences. Protocols are available for this approach (2, 3).
3. Use anchored PCR. This is a simplified version of RACE which has proven successful for isolating the missing ends (both 5′ and 3′) of a number of cDNAs. Sequence-specific oligonucleotide primers are derived from the end of the known sequence, situated ideally so that the PCR product will contain an 'endogenous' restriction enzyme site already present in the cDNA, thus allowing the PCR product to be engineered directly on to the truncated cDNA. Alternatively, the sequence-specific primers can have a restriction enzyme site incorporated on to the 5′ end for ease of subcloning of the PCR product (see Section 2.2.2), or the blunt-ended PCR product can be subcloned directly (see Section 2.2.2). The other primer is derived from vector sequences just 5′ and 3′ to the insert; for instance, if the cDNA library is in λZAP (Stratagene) then primers (about 25mers) derived from the T3 and T7 sequences on either side of the polylinker are ideal. The PCR reaction is performed under 'standard' conditions (94°C, 45 sec; 55°C, 1 min; 72°C, 1 min) with *Taq* polymerase (for PCR protocols see refs 2, 3) using the cDNA library as the template (1 μl of a 10^9–10^{10} p.f.u./ml titred library in a 50 μl PCR reaction). PCR products can then be subcloned and characterized by DNA sequencing. Where possible, the PCR product can be engineered directly on to the truncated cDNA to yield a full length cDNA. If this is not possible then the sequence data derived from the anchored PCR, e.g. the 5′ untranslated sequence of the cDNA, can be used to engineer a full length cDNA, again using PCR.

3. Site-directed mutagenesis

3.1 General principles

As discussed above, the cloning of the cDNA encoding a protein is now considered just a first step, albeit a crucial one. In the case of channel proteins the second step is usually the functional expression of the recombinant channel *in vitro*. This is considered later in the chapter. Demonstration of functional expression opens the way to structure/function analysis using site-directed mutagenesis (SDM). Since channel proteins are often very large (e.g. in the case of the sodium channel, greater than 1500 amino acids long) the scope for mutagenesis is obviously enormous. There are several ways of identifying amino acids as good targets for mutagenesis:

1. Identifying amino acids which are consensus sites, e.g. for phosphorylation.
2. Identifying amino acids which, in homologous proteins, other members of the gene family have been shown to play a functional role.

3. Being guided by any biochemical studies on the channel (e.g. affinity labelling) which have identified individual amino acids as being important.
4. Intuition!

When using SDM to identify amino acids involved in function, pharmacology, or assembly of channels it has proved useful in a number of cases to make chimeric channels between homologous polypeptides (e.g. see ref. 4, where residues involved in binding of benzodiazepines to the GABA-A receptor were identified). This approach allows the researcher to delineate a region of the polypeptide within which individual amino acids can be subjected to SDM.

Interpreting results from SDM experiments can be fraught with difficulties. Changes which remove all functional responses are hard to interpret. Has the mutation actually had an effect on the channel function, or has it simply prevented the channel from folding or assembling correctly? One would have to use biochemical techniques, e.g. anti-channel antibodies, sedimentation analysis, etc. to assess whether the mutated channel was indeed being assembled, transported, and expressed on the cell surface. There is certainly a school of thought which says that it is sensible to make relatively conservative amino acid changes initially (e.g. removing a charge, as in aspartic acid to asparagine) and subsequently make more drastic changes (e.g. changing charge, as in aspartic acid to arginine).

There are numerous texts describing in detail different approaches to SDM (e.g. ref. 2). The two methods that we have found useful for SDM of channel proteins are discussed in the following sections.

3.2 Oligonucleotide-directed mutagenesis

This is a tried, tested, and robust method developed by Thomas Kunkel, and uses an approach which is able to select against the non-mutated strand of the double-stranded DNA. Precise experimental details are given in ref. 2. The method depends upon making single-stranded DNA from the double-stranded plasmid. As such, it is most convenient to have the cDNA encoding the channel polypeptide in a eukaryotic expression vector which contains the M13 bacteriophage origin of replication. The one that we have found most useful is pcDNAIAmp (available from Invitrogen). By using this vector it is possible to perform the SDM and the expression studies without having to subclone the cDNA between different vectors. If the mutagenesis is performed using pcDNAIAmp, which is a relatively large vector, it is important to use T7 DNA polymerase rather than T4 DNA polymerase, since the former is more efficient at extending longer sequences. The buffer and conditions of the polymerase step are the same for both enzymes.

The overall efficiency of performing the mutagenesis process can be increased by incorporating into the mutagenizing oligonucleotide a further base change which will result in the incorporation of a new restriction enzyme site or removal of an existing site. In this way, potential mutants can be screened simply by restriction enzyme digest rather than by having to sequence a number of clones. It is obviously important to ensure that the base change which results in the generation of the restriction site is a silent one, i.e. does not result in an amino acid change. There are computer programs available (e.g. Prosite program of PC/gene, Intelligenetics Inc.) which will search a DNA sequence and identify single base changes which will result in the generation of a restriction site.

It is possible to use SDM to change a single amino acid, or by using longer mutagenic oligonucleotides (up to 40mers) it is possible to mutate several amino acids at the same time. Using this approach, once a stock of single-stranded DNA has been prepared, it is possible to go from deciding which mutation to make, to having a stock of the mutated DNA ready for expression, in less than a week.

3.3 PCR mutagenesis

PCR has an ever increasing number of applications, one of which is SDM. PCR mutagenesis is certainly quick, although once the researcher has a stock of single stranded DNA it is not in fact any quicker than oligonucleotide directed mutagenesis (Section 3.2). Additionally, it has the disadvantage of requiring a minimum of two oligonucleotide primers per mutagenesis reaction, while oligonucleotide-directed mutagenesis requires only one. Since oligonucleotides are the major cost in a mutagenesis reaction, this is obviously an important consideration. The situation where we have found PCR most useful in SDM is shown diagrammatically in *Figure 2*. It relies upon having a unique restriction enzyme site in the flanking region of the vector (this will normally be a site in the polylinker), and a second unique restriction enzyme site within 20 bases of the site of mutation. Although the latter will not always occur, given the large number of restriction enzymes now available it is surprising how many times there is a site conveniently close to the mutation site. The mutagenic primer is designed to contain both the base change leading to the mutation and the unique restriction enzyme site. PCR is performed with a universal primer and the mutagenic primer using the cDNA clone as template. The PCR product is digested with the restriction enzymes and subcloned into the equivalently digested cDNA construct to yield a mutated cDNA.

This same approach can also be used to introduce a restriction enzyme site into the cDNA. This is particularly useful when making chimeric constructs between homologous polypeptides (i.e. members of a gene family), using a restriction enzyme site found in one of the two sequences. This is shown in *Figure 3*. Obviously when making chimeric constructs there will not often be a conveniently situated restriction enzyme site which is conserved on both cDNAs. In this case one would have to use SDM or PCR mutagenesis to

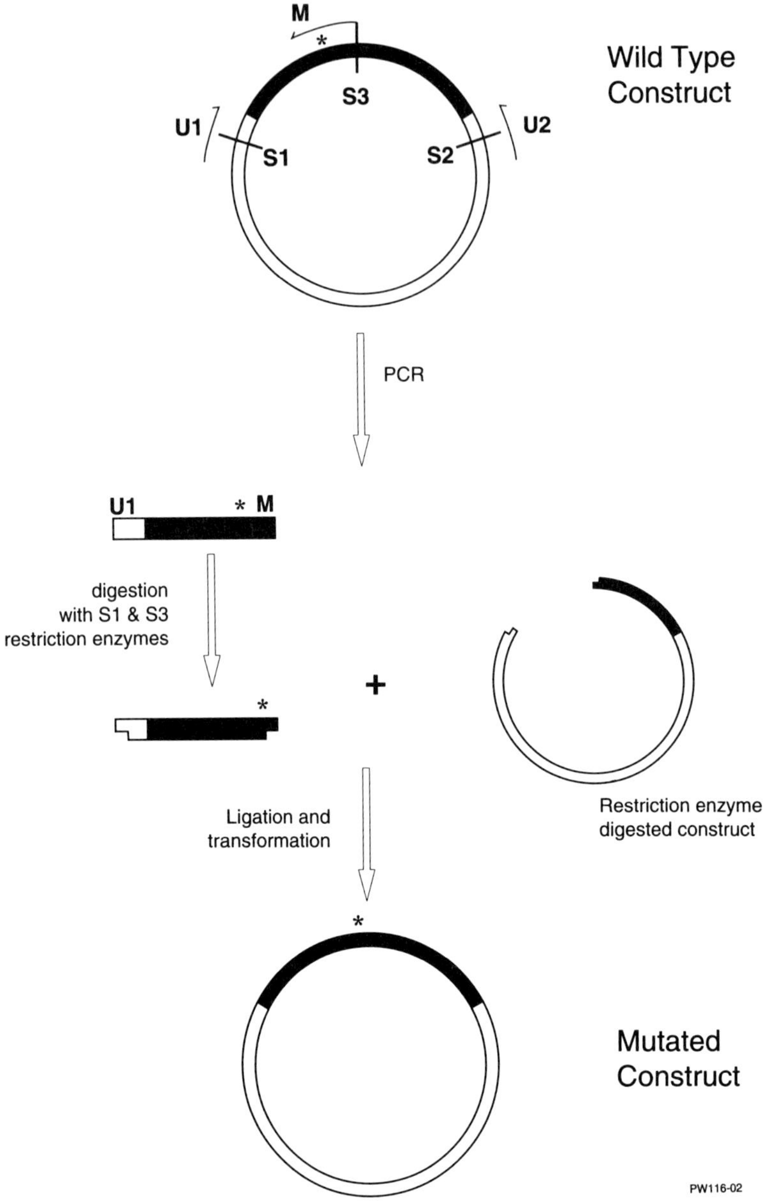

Figure 2. Site-directed mutagenesis using PCR. The DNA construct is shown at the top of the figure, the filled-in region representing the heterologous cDNA, and the open region representing the plasmid sequences. The asterisk represents the site of mutation, introduced by the mutagenic primer M. U1 and U2 are universal PCR primers located either side of the polylinker sequences. S1 and S2 are unique restriction enzyme sites also located in flanking regions, e.g. in the poly-linker. S3 is a unique restriction site in the cDNA within 20 bases of the site of mutation. PCR is performed using, in this case, primers U1 and M. The PCR product is cut with restriction enzymes S1 and S3 and then ligated into S1 and S3 cut vector to yield the mutated construct.

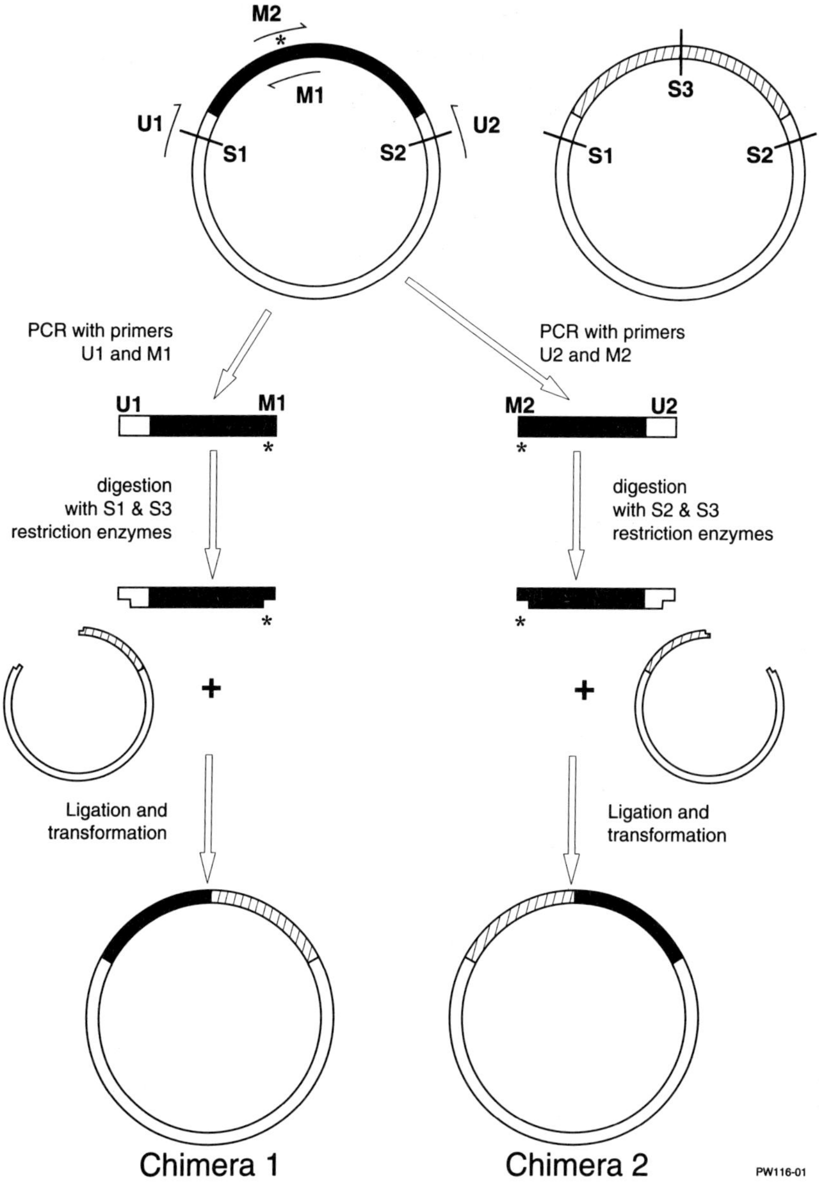

Figure 3. Construction of chimeric cDNAs using PCR. The two DNA constructs are shown at the top of the figure. The filled-in and hatched regions of the constructs represent two cDNAs encoding related members of a gene family, and the open region represents plasmid sequences. U1 and U2 are universal PCR primers located on either side of the polylinker sequences. S1 and S2 are unique restriction enzyme sites also located in the flanking regions, e.g. in the poly-linker. M1 and M2 are mutagenic primers introducing base change(s) (indicated by the asterisk) which will result in the introduction of restriction enzyme site S3 which is already present in one of the cDNAs. PCR is performed using primers U1 + M1, and U2 + M2. PCR products are cut with the appropriate restriction enzymes and then ligated into similarly cut vector to yield chimeras 1 and 2.

introduce a restriction enzyme site in both cDNAs, and then use these to construct the chimera.

When using PCR to manipulate cDNAs it is important to be aware that the polymerase may introduce unwanted additional mutations into the cDNA. This possibility may be reduced by using thermostable polymerases which are reported to have a higher fidelity (e.g. Vent polymerase, available from New England BioLabs) in place of Taq polymerase. Nevertheless it is prudent to sequence the amplified portion of DNA.

4. Choice of expression system

4.1 Overview

One of the primary aims of cloning the cDNA encoding a channel is to be able to subsequently express the recombinant channel and study its structure, function, etc. There are four principle systems for expressing recombinant channels: expression in *Xenopus* oocytes, transient or stable expression in mammalian somatic cells, and baculovirus-driven expression in insect cells. Their advantages and disadvantages are discussed below.

4.2 *Xenopus* oocyte expression

This is a transient expression system. RNA or cDNA is injected into the oocyte and the protein is expressed 1–2 days later. The system is relatively robust once set up under optimal conditions, although occasionally a 'bad batch of toads' can lead to a decrease or loss of expression. Additionally, it is very quick: nuclear injection of the cDNA construct (see Section 5.2) can give expression of the recombinant channel the next day. Oocyte expression is well suited to the electrophysiological study of voltage gated channels and, since it is a transient expression system, it is useful for mutagenesis studies and for studies of multimeric proteins formed from co-assembly of different subunits where there is a need to express a number of different subunit combinations. It is perhaps less ideal for functional studies of ligand-gated ion channels due to the large size of the cell and consequently the relatively slow application time of the drug. It is also a small-scale expression system and so not really suitable for radioligand binding studies, and is not convenient for biochemical studies on recombinant proteins, although some detailed studies of this sort have indeed been performed (5). One fact that should always be borne in mind is that the *Xenopus* oocyte is not a mammalian cell, and thus may not process and assemble membrane proteins in exactly the same way. Additionally, the lipid environment of the membrane will be different from that of a mammalian cell, and this could potentially affect the functional properties of the expressed channel.

4.3 Transiently transfected cells

As the name indicates, this is a transient expression system. Tissue culture cells are transfected with cDNAs and maximum levels of expression of the protein occur about 48 h later. As with *Xenopus* oocyte expression it is a relatively robust system once established within the laboratory. It is, however, quite dependent upon the condition of the cells and the quality of the DNA. If these are not optimal then only low levels of expression will be achieved. Such factors can obviously lead to variations both within a lab and between labs. Also, transient transfection is generally more labour-intensive than oocyte expression—tissue culture cells need to be maintained, and DNA stocks prepared. Also, while transient transfection can be done on a small scale, e.g. on coverslips for electrophysiology, or on a medium scale, e.g. in tissue culture dishes for biochemical experiments, or even on a large scale, e.g. on tissue culture plates/trays for radioligand binding studies. The latter can be very labour intensive. However, as with *Xenopus* oocytes, transient expression is ideal for mutagenesis studies and for co-expressing the subunits of a channel family in different combinations.

4.4 Stably transfected cells

In stable transfection the DNA encoding the protein is stably integrated and maintained in the genome of the transfected cell. To establish a stable, clonal cell line expressing a channel protein takes a minimum of 2–3 months. As such, stable transfection is unsuitable where a quick answer is required such as for rapid analysis of mutants. Stable expression is, however, probably the most suitable system for detailed, long term studies, or where relatively large amounts of expressed channel are required. Stable expression in mammalian cells is, in the long run, much less labour intensive than transient transfection, and is also more robust and reproducible.

4.5 Baculovirus expression

In this system the cDNA of interest is cloned and packaged into a baculovirus particle which is then used to infect tissue culture cells. It can best be defined as transient expression, since expression only occurs for a few days after infection. As such, many of the points addressed in Section 4.2 apply here. The infection efficiency (in terms of the proportion of cells which will express the recombinant protein) is, however, greater than achieved through transient transfection, leading to the possibility of large-scale expression. However, there are some potential disadvantages of the baculovirus system.

1. As with *Xenopus* oocytes, it is important to remember that insect cells are not mammalian cells, and thus could synthesize, modify, and assemble channel proteins in a different way. For instance, insect cells cannot synthesize complex oligosaccharide side chains.

2. The second potential disadvantage is specific to the expression of multimeric channels, the subunits of which are engineered into individual virus particles. The subunit stoichiometry of the recombinant channel can depend on the ratio of the virus particles used to infect the cells (6). Obviously this could lead to experimental variability. Alternatively, it could be used positively as an experimental tool to help understand channel structure and assembly.

Despite these potential problems, mammalian channel proteins such as the *Shaker* potassium channel (7), the AMPA-type glutamate receptor (8), and the benzodiazepine-sensitive GABA-A receptor (6) have been functionally expressed in insect cells. Indeed, expression of the latter suggests that even some complex channels comprised of different subunits can be synthesized and assembled in this system.

5. *Xenopus* oocyte expression system

5.1 Overview

The oocytes of the aquatic toad *Xenopus laevis* offer a unique tool for the study of exogenously expressed proteins, especially those expressed on the cell surface such as channels. The oocyte is well equipped with the cellular machinery needed to transcribe DNA rapidly and manufacture protein whether it be native or foreign to the cell. The ability of *Xenopus* oocytes to express foreign proteins was demonstrated by injecting mRNA extracted from rabbit reticulocytes and detecting radioactively labelled globin (9). Messenger RNA from other sources soon followed and in 1982, nicotinic acetylcholine receptors from mRNA extracted from *Torpedo* electric organ were expressed and shown to form ion channels in the oocyte membrane. The large size of the oocyte (1–1.5 mm diameter) makes the cell readily accessible for electrophysiological recording studies—although their large size can also be a problem—see Section 5.4. In the field of neuroscience, many different receptors found in the brain can be expressed in *Xenopus* oocytes, including G-protein linked receptors such as members of the 5-HT receptor family, voltage-gated ion channels such as potassium channels, and ligand-gated ion channels such as GABA and glutamate receptors. The *Xenopus* oocyte can also be used as a tool for identifying cDNAs which code for novel channels—expression cloning. Several techniques have been described which utilize the oocyte as a screen to gradually narrow down pools of mRNA which express receptor, to obtain an individual clone. Two methods which have been used successfully to obtain cDNA clones of neurotransmitter receptors are described here.

5.2 Preparation and injection of oocytes

This section describes the techniques used for preparation of oocytes and their injection with cDNA or cRNA. You should first become fully appraised of the legislation governing work of this sort involving live animals.

Protocol 2. Surgical removal of oocytes

Equipment and reagents

- Ethyl *m*-aminobenzoate
- Scalpel, scissors, forceps (sterilized)
- Sutures (dissolvable)
- Petri dish

Frog anaesthesia

1. Make approximately 1 litre of a solution of 0.4% ethyl *m*-aminobenzoate (also called Tricaine or MS222) (Sigma) using water taken from the frogs' tank in an ice bucket.
2. Place the frog in the anaesthetic solution for approximately 15–20 min, checking at intervals to ensure the frog is becoming flaccid. Remove the frog and place on a clean surface for surgery. The animal should appear limp but have clear pupils and breathing can be checked by observing the nostrils.
3. During the anaesthesia, prepare surgical instruments and sterile conditions for the ovary removal procedure.

Removal of ovary tissue

1. Place the anaesthetized frog on its back on a flat surface and make a small incision with the scalpel (1–2 cm) through both the skin and body wall musculature in the posterior ventral side.
2. Use forceps to locate ovary lobes and carefully pull out one lobe, excise lobe with scissors, and transfer to a Petri dish containing oocyte isolation medium (see *Table 1*). Repeat lobe removal as required. Each lobe can contain up to 2000 oocytes, several hundred of which will be stage V and VI which are usable for injection.
3. Suture the incision with 2–3 stitches in the body wall and 2–4 separate stitches in the skin using dissolvable sutures.
4. Allow the frog to recover in approximately 4 cm water with the head above water to avoid drowning. Make a small, tissue pad on which to rest the frog's head while recovering, and keep checking to make sure the head is above water level. Allow 2–4 h for recovery before replacing in aquarium. The frog should maintain clear pupils and check that the frog is still breathing by watching the nostrils.
5. The wound will heal and the procedure can be repeated after a 3–4 week period, the number of times depending on current animal procedures legislation.

Protocol 2. *Continued*

Alternatively, if the frog is to be killed

1. Kill frog by injection of barbiturate (e.g. 'Euthetal') or anaesthesia as above, followed by decapitation and destruction of the brain.
2. Make abdominal incision and remove complete ovary.
3. Transfer the ovary to oocyte isolation medium. One transfer into fresh medium may be neccessary to remove contaminating blood and yolk from broken oocytes.

The freshly isolated oocytes can be individually removed and defolliculated using a binocular dissecting microscope (see *Protocol 3*). Select the largest oocytes for isolation. These should be 1.0 to 1.3 mm in diameter and have a well defined animal (black or dark brown half) and vegetal (yellow half) pole. These are the fully grown but still immature oocytes described developmentally as stage V and VI oocytes. While developing in the ovary the oocyte is surrounded by a number of membranes and layers of cells. For injection and electrophysiological recording, these layers should be removed down to the vitelline membrane. This can be achieved in two ways: (i) manually, using fine watchmaker's forceps, or (ii) using collagenase or other proteolytic enzymes. Manual dissection is recommended because although difficult at first it is a much faster way of obtaining clean, healthy oocytes ready for injection.

Table 1. Media for use with *Xenopus* oocytes

	Isolation medium	**Modified Barth's (MBS)**
NaCl	108 mM	88 mM
KCl	2 mM	1 mM
EDTA	0.1 mM	–
Hepes	1 mM	10 mM
$MgSO_4$	–	0.82 mM
$Ca(NO_3)_2$	–	0.33 mM
CaCl	–	0.91 mM
$NaHCO_3$	–	2.4 mM
NaOH	pH 7.9	pH 7.5

Incubation medium

- 1 litre of regular modified Barth's medium
- 1 ml of penicillin/streptomycin (stock solution = 10 000 units/litre penicillin and 10 mg/litre streptomycin)
- 1 ml of gentamycin sulphate solution (stock solution = 50 mg/ml)

Once manually isolated, the only remaining layer to be removed is formed by the follicle cells which surround the vitelline membrane. A mild collagenase treatment is sufficient to remove these cells in order to facilitate handling the oocytes while injecting them. An alternative approach to injecting mRNA is to inject cDNA directly into the oocyte nucleus. For this purpose the cDNA should be subcloned into a eukaryotic expression vector (Section 6.1.2); we have found that vectors with a CMV promoter work very well. The oocyte nucleus is very large and located in the animal pole of the oocyte; it can be injected with cDNA using a very similar protocol to mRNA injection. Nuclear injected cells will actively transcribe the DNA and translate it into protein. This technique has several advantages:

1. DNA is less labile than mRNA and easier to handle;
2. a single plasmid preparation of the cDNA construct produces sufficient for several years of experiments;
3. cDNA injection results in the rapid appearance (within 24 h) of relatively large currents.

Procedures for the injection of mRNA or cDNA are described in *Protocol 4*.

Protocol 3. Manual dissection—removal of folliculated oocytes

1. Place the lobes of ovary into a large Petri dish containing oocyte isolation medium. This solution is calcium free and hypertonic and shrinks the follicle-enclosed oocyte making the manual defolliculation easier.
2. Gently pull open the lobes using fine forceps and swirl the ovary a little to wash the exposed oocytes. If many of the oocytes have broken and the medium is cloudy, or contaminated with blood, transfer the lobes of ovary to fresh isolation medium.
3. Incubate the oocytes in isolation medium for at least 15 min before starting manual defolliculation.
4. Using two pairs of fine watchmaker's forceps (size No 5 is ideal), which work better if they are not too sharp (it is easy to rupture the oocyte), remove the theca and outer layers as follows.
5. Select an oocyte for isolation and bring the right forceps underneath the oocyte as shown in *Figure 4a*. With the left forceps, hold the ovarian wall as close to the oocyte as possible for support, and gently squeeze the right forceps together to grip the outer membrane without damaging the oocyte itself. Carefully pull with the right forceps in a direction parallel to the ovarian wall and the outer membrane will tear; the oocyte will bulge out from the tear (*Figure 4b*) and the oocyte should pop out of the membrane.

Protocol 3. *Continued*

6. Do not try to remove the membrane if the oocyte becomes detached from the ovary, and regularly discard follicle-enclosed and any damaged oocytes. Follicle-enclosed oocyte can be identified by blood vessels on the surface of the oocyte.
7. Once removed, transfer the defolliculated oocytes into regular MBS with a transfer pipette or a wide-mouthed Pasteur pipette.
8. Place the freshly isolated oocytes into a Petri dish containing type 1A collagenase (Sigma) at 0.5 mg/ml diluted into regular MBS for 8 min, to remove follicular cells adhering to the cell membrane. Swirl the oocytes intermittently to facilitate removal of the follicle cells.
9. Transfer the oocytes back into regular MBS and swirl again to wash away any remaning collagenase. The oocytes are now ready to be injected.

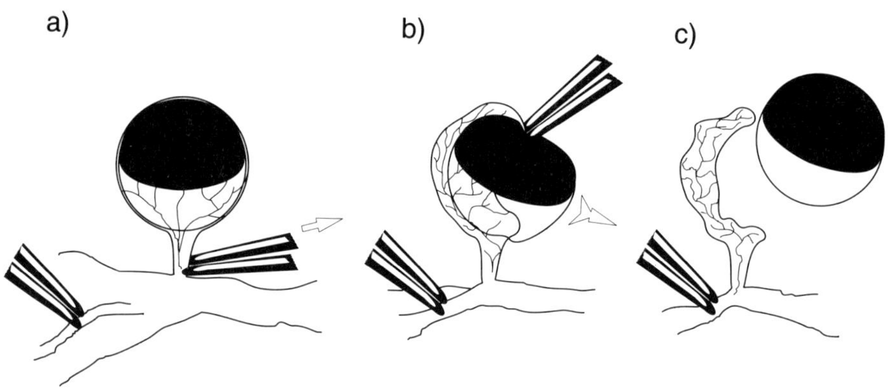

Figure 4. Illustration of the technique for manual removal of oocytes from the ovary-attached follicle using watchmaker forceps. With the left-hand forceps hold a section of ovary wall for support, while grasping and pulling back the outer membrane surrounding the oocyte, (a). The oocyte will protrude from the membrane sac (b) and can be gently teased out (c), ready for collagenase treatment and injection (for details see *Protocols 3* and *4*).

Protocol 4. Injection of mRNA or cDNA into *Xenopus* oocytes

Equipment and reagents

- Stereomicroscope with eyepiece graticule
- Fibre-optic light source if none available on microscope
- Microelectrode puller (e.g. Flaming/Brown, Model P-87, Sutter Instrument Co.)
- Micromanipulator
- Forceps (blunt)
- Drummond 10 μl pipette and glass capillaries, 2–10 μl bore (both available from many sources, e.g. Lazer Instruments Ltd, UK) mounted onto micromanipulator

A. *Preparing microinjection pipette*

1. Load the glass capillary into a microelectrode puller with the heating element as close to one end as possible and pull with appropriate settings to produce a tip of approximately 1 cm in length with a fairly even shank.

2. Using a graticule in the stereomicroscope, examine the tip and break it down to between 20 and 30 μm using a clean pair of fine scissors or forceps (for cDNA injection a finer tip of 10 μm is required). Try to bevel the tip when breaking it down as this does less damage to the oocyte when injecting.

3. When injecting mRNA it is important that micropipettes are sterilized to remove RNase (>1 h at 200°C). Calibration of the tip is not required.

B. *Loading mRNA or cDNA*

1. Back fill the pipette up to 1 cm with mineral oil or high grade Paraffin, and load the pipette on to the plunger making sure the barrel has been pushed through the rubber washer in the holder and up into the main part of the syringe. Tighten the screw holding the pipette and wind down the plunger until oil is being pushed out of the tip and the micropipette is completely filled with oil. The oil does not affect the mRNA/cDNA and visualization of the aqueous/oil interface allows the mRNA/cDNA loading to be observed clearly. If required a fat-soluble dye (e.g. Fast Red) can be included in the oil to ensure easy observation of the interface.

2. Place 1–3 μl of mRNA/cDNA solution on a small strip of Nescofilm or Parafilm using a 2 μl or 10 μl automatic pipette (use RNAse-free sterile tips). *Beware—larger tip sizes tend to cause problems when pipetting such small volumes, especially on to a flat surface.*

3. Make sure that the oil is ejected from the tip when the plunger is wound down. Set the pipette to a middle position (5–7 μl). Place the tip of the micropipette into the mRNA/cDNA solution[a] and slowly wind up the plunger, being careful not to draw up any air. The RNA should be drawn into the pipette and the interface will move up the pipette as more is drawn in. Take up an excess of RNA/DNA (3–4 μl of mRNA is enough for 60–70 oocytes, cDNA will go further as less is injected).

4. Place the oocytes in a Petri dish of MBS and swirl into a cluster roughly in the centre of the dish. Examine oocytes under microscope and remove any that are injured or dead. The animal pole should be dark and unblemished.

5. Supporting the oocyte carefully with a blunt pair of forceps, insert the micropipette tip into the oocyte making sure to penetrate both the vitelline membrane and the cell membrane.

Protocol 4. *Continued*

6. Inject 50–70 nl of mRNA (or 10–20 nl of cDNA) by slowly winding down the plunger (clockwise) 5–7 small divisions (one small division represents 10 nl). **Important—Inject mRNA into the interface of the animal and vegetal pole of the oocyte, but inject cDNA into the top of the animal pole or dark region.**
7. Remove pipette tip carefully, using forceps to hold the oocyte. A small amount of leakage from the cytoplasm may occur but the site will seal up within a few minutes.

[a] The amount of RNA/cDNA to be injected will vary depending upon how well that particular channel expresses in oocytes. A good starting point is a concentration of 10–20 ng/µl of cRNA, and 1–5 ng/µl of cDNA.

5.3 Endogenous receptors and ion channels

Many ion channels and receptors have been expressed in *Xenopus* oocytes, but the cell does possess a number of endogenous channels and receptors which it is important to be aware of when using the oocyte for expression studies. The presence of muscarinic acetylcholine receptors has been demonstrated by electrophysiological techniques. Responses to acetylcholine have been well characterized (10) and on receptor activation intracellular calcium is mobilized via the inositol trisphosphate pathway. This rise in calcium then activates Ca^{2+}-dependent chloride channels in the membrane, evoking the characteristic oscillatory response from G-protein linked receptors expressed in oocytes (11). Responses to adenosine, vasoactive intestinal peptide, some prostaglandins, catecholamines, and gonadotrophins have also been reported. However, these have been shown to be due to receptors present on the follicle cells surrounding the oocyte, which are electrotonically coupled to the oocyte membrane (12). When the follicle cells are removed with collagenase these responses are abolished. Some divalent cations also elicit calcium-dependent chloride responses when applied to oocytes (13). Defolliculated oocytes also contain several types of voltage-gated ion channel, the most predominant of which is the Ca^{2+}-dependent chloride channel (14) which can be activated by events which trigger a rise in intracellular calcium. Other ion channels identified in oocytes include voltage-dependent calcium channels and certain types of K^+ and Na^+ channels (15). The oocyte membrane also contains stretch-activated channels which can be blocked by gadolinium and calcium (16).

Oocytes contain the machinery necessary for the post-translational modification of proteins and the construction of large multisubunit proteins which are transported to the oocyte cell membrane. They also express endogenous G-proteins which are able to link to foreign G-protein-coupled receptors and produce appropriate responses via second messenger pathways (17). Oocytes are also able to glycosylate and phosphorylate residues on foreign proteins, and these processes have been shown to be critical for the functioning of a number of receptors.

5.4 Electrophysiological recording from *Xenopus* oocytes

The large diameter of *Xenopus* oocytes makes them amenable to electrophysiological recording. The most commonly used technique is that of two-electrode voltage clamp. Oocytes are placed in a small perspex chamber and held in place using either small pins or a nylon grid. The chamber is continuously perfused with MBS, either by gravity or a perfusion pump, and the set-up is grounded via a silver chloride pellet in contact with the perfusion medium. Current recordings can be made using a voltage-clamp amplifier and several commercially available amplifiers specifically designed for oocyte recording are now available, e.g. Warner Instruments, Oocyte clamp (Model OC-725B); Dagan, Clampator One (CA-1); Axon Instruments, Geneclamp 500. These amplifiers are designed to have a high voltage output compliance in order to clamp the large currents generated by expression in oocytes. The two electrodes containing 2M KCl are inserted into the oocyte and the cell can be voltage clamped at a suitable potential. Drugs or voltage-steps can now be applied to the cell to activate the receptors or channels expressed and currents can be measured directly.

In some cases it may not be necessary to voltage clamp the oocyte, for example when rapid throughput is required to see if a particular receptor is being expressed. This technique is particularly useful when expression screening—a strategy for identifying pools of mRNA which contain the receptor of interest (*Protocols 5* and *6*). For these purposes it necessary only to insert a single electrode, obtain a stable baseline, and apply the drug to activate the receptor. A depolarization or hyperpolarization indicates that the receptor is being expressed. In this way a large number of oocytes can be screened for expression in rapid succession. As voltage clamping is not required for this method a cheaper, simple amplifier will suffice for recording.

Oocytes can also be patch clamped, but this technique is a little more difficult than voltage clamping. The vitelline membrane surrounding the occyte will obstruct the formation of a seal on to the oocyte membrane and must be removed prior to clamping. To do this the oocyte must be immersed in a hypertonic solution (200 mM potassium aspartate, 20 mM KCl, 1 mM $MgCl_2$, 10 mM EGTA, 10 mM Hepes, pH 7.4) to shrink the oocyte from the vitelline membrane, which can then be removed with fine forceps. It should be stressed that the de-vitellinized oocyte is extremely fragile and should be handled very carefully. To avoid interference from stretch-activated channels it is best to remove a patch of membrane and measure single channels in the detached patch.

5.5 Expression cloning

The ability of *Xenopus* oocytes to express exogenous mRNA has been used as a tool to obtain single clones which code for receptor proteins, enabling the isolation and sequencing of receptors which have proved to be unobtainable using biochemical isolation techniques. Several procedures have been developed for carrying out expression screening, and these have been successfully used in obtaining sequences for many neurotransmitter receptors including the 5-HT_{1c} amd 5-HT_3 receptor (17) and glutamate receptors (18) The basic steps involved in two methods are outlined in *Protocols 5* and *6*. Both methods involve injecting fractionated batches of mRNA and screening for a functional response. The batches are subdivided and rescreened until a single clone is isolated. The two methods used successfully so far have taken slightly different approaches, each with their own advantages and disadvantages. The first method (*Protocol 5*) involves fractionating the cDNA clones and repeatedly subdividing until a single clone encoding the channel can be identified. The disadvantage with this method is that many ion channels and/or receptors are multi-subunit proteins, and often single subunits may not form a functional receptor/channel. This means that a functional receptor/ion channel may not be expressed unless more than one subunit is present, and once the pools containing each contributing subunit are separated, expression is lost. Despite this, the NMDA receptor NR1 subunit, which expresses very poorly without associated subunits, was isolated by this method.

The second method (*Protocol 6*) was used successfully to clone the 5-HT_{1c} receptor (17) and involves hybridizing pools of cDNA with total mRNA from the source tissue and looking for an inhibition of the agonist response. In a similar manner to the previous protocol, the pools of cDNA which cause a depletion of the response can be subdivided until a single clone is obtained. The individual clone can then be sequenced and expressed alone. This method avoids the problem of a single subunit not expressing as the number of clones which are expressed are not reduced in each subdivision. However, there are other problems associated with this method. It may not yield full length clones, and large pools of RNA are required for the hybrid-depletion assay, but when compared with the problems associated with the purification and isolation of receptors using biochemical techniques, expression cloning has proved in many cases to be a rapid and efficient way of cloning receptors and ion channels.

Protocol 5. Expression cloning

1. Isolate brain (or channel rich tissue) poly(A)$^+$ RNA using a standard RNA isolation procedure.
2. Size fractionate the RNA on a sucrose density gradient and divide into several pools.

3. Inject mRNA from each individual fraction into oocytes and screen for the presence of the receptor by measuring agonist-induced depolarizations.
4. Construct a cDNA library from the channel-rich fraction using a bacteriophage vector (Section 2.1).
5. Divide the library into about 20 individually amplified libraries (each library should contain about 50 000 clones).
6. Perform *in vitro* synthesis of mRNA from each sub-library in turn and again test each pool of transcripts in oocytes, for expression of the channel.
7. Take the pool which expressed the channel and subdivide this into 100 pools (each containing about 500 clones).
8. Inject RNA from each pool into oocytes and identify the pool which results in expression of the functional channel.
9. Plate phages from the positive pools at low density to allow picking of individual phage clones.
10. Inject cRNA *in vitro* transcribed from individual positive phage clones to identify a single clone encoding the channel.

Protocol 6. Hybrid depletion

1. Isolate poly(A)$^+$ mRNA from channel-rich tissue.
2. Size fractionate the poly(A)$^+$ mRNA as in *Protocol 5* and subdivide into several pools of mRNA.
3. Inject each pool into *Xenopus* oocytes and screen for those which contain the channel.
4. Make a cDNA library from this channel-enriched pool of mRNA using a standard procedure (1, 2).
5. Subdivide the cDNA library into several pools and hybridize each with total mRNA extracted from the target tissue.
6. Remove the unhybridized RNA from the RNA–DNA hybrids on a nitrocellulose column.

4. Express the remaining unhybridized RNA in oocytes and screen for a functional response.
5. Isolate those pools of cDNA which caused a depletion of the response, as these contain the required cDNA.
6. Subdivide the positive pools and repeat the hybridization procedure.
7. Repeat the above steps until a single cDNA is isolated which depletes the response of total mRNA when injected into oocytes.

Protocol 6. *Continued*

8. Make cRNA from the cDNA clone which caused depletion, or alternatively (if cDNA is in a eukaryotic expression vector) inject the cDNA into oocytes and screen for functional expression of the clone.
9. Sequence the cDNA obtained.

6. Transient and stable expression in mammalian cells

6.1 Transient expression

DNA can be introduced into mammalian tissue culture cells by transfection, or by using eukaryotic viruses as vectors. A number of viral vectors are either being developed or are available, including *vaccinia*, *Herpes simplex*, and adenovirus (19). However, they have not yet been shown to be of general use. Additionally, their manipulation requires a degree of proficiency in virology and is beyond the scope of this chapter.

6.1.1 Transient transfection of mammalian cells

In transient transfection the DNA is taken up by the cell and, by an unknown mechanism, some of it will enter the nucleus where it persists episomally for a few days, allowing transcription to occur. There are a number of different methodologies available:

(a) Calcium phosphate-mediated precipitation of DNA. This is the most widely used and best characterized method and is described below (Section 6.1.3).
(b) Electroporation. This requires specialized equipment and is less suitable for adherent cells and for larger scale transfections. It will not be considered further.
(c) Liposome-mediated transfection. In this approach the DNA is mixed with a lipid reagent resulting in the formation of DNA-containing liposomes which mediate uptake of the DNA into the cell. This approach is reportedly very efficient and can be adapted to transfect different cell types. Lipid reagents are available commercially from several companies (e.g. Boehringer-Mannheim, Gibco-BRL, Promega) and are supplied with experimental protocols. However, where transfection studies are performed on a larger scale, this approach could prove prohibitively expensive. Calcium phosphate-mediated transfection is in contrast very cheap, and is adequate for most requirements.

6.1.2 Vectors and cells for transient transfection

There is now a range of commercially available (e.g. Stratagene, Invitrogen, Pharmacia, Clontech) eukaryotic expression vectors. They all have the ColE1 origin of replication and ampicillin-resistance gene or SupF sequences (for replication and maintenance in *E. coli*), and a multiple cloning site. However, they do differ in a number of features, including:

(a) The promoter type (SV40 early promoter, SV40 late promoter, CMV enhancer/promoter, RSV long terminal repeat). The CMV promoter gives high levels of transcription, especially when transfected into HEK 293 cells, which constitutively express two proteins which trans-activate the promoter.
(b) The type of polyadenylation/transcription termination sequences (SV40 signal sequences or bovine growth hormone sequences). This is probably not a critical factor.
(c) The presence or absence of the M13 origin of replication, to allow preparation of single stranded DNA for SDM.
(d) The presence or absence of the SV40 origin of replication to allow episomal replication in cells expressing SV40 large T antigen (e.g. COS cells), resulting in higher transcription levels.

One vector which we have found useful is pcDNAIAmp (from Invitrogen). This has a number of features, including the CMV promoter for high levels of transcription in a number of different cell lines (especially HEK 293), and the M13 origin of replication for generation of single stranded DNA and SDM. Although it is possible to perform transient transfection with a wide range of cells, including primary cultures, the highest efficiencies and highest levels of expression are generally achieved in a limited number of well characterized cell lines. These include:

- HEK 293 (ATCC CRL 1573)
- CHO-K1 (ATCC CCL 61)
- CV 1 (ATCC CCL 70)
- COS 1 (ATCC CRL 1650)
- COS 7 (ATCC CRL 1651)
- L (ATCC CCL 1).

These cell lines have proved suitable for the expression of recombinant channels and subsequent functional, pharmacological, and biochemical studies. It is important to be aware, for currently unknown reasons, that a particular channel may be expressed better in one cell line than another. An example is the rat brain sodium channel, which could be expressed in CHO cells, but not in L tk-, COS, or Y 1 cells (20).

6.1.3 Calcium phosphate-mediated transfection

The calcium phosphate-mediated gene-transfer protocol which has been developed by Chen and Okayama is relatively robust and efficient and has proved useful for expression of channel proteins (e.g. 21). A detailed methodology which also provides useful tips has been published (22), and the reader should refer to this. When expressing a multimeric channel the cells will have to be co-transfected with plasmid constructs containing the different subunits. It is important to note that most efficient expression may not necessarily be achieved by transfecting with an equal amount of each subunit DNA; altering the ratio may give improved levels of expression of the multimeric channel.

6.1.4 Identification of 'positive' transiently transfected cells

In transient transfection, only a proportion of the cells (<1%–50%, depending upon a range of factors) will take up the DNA and express the protein. This is a drawback for electrophysiological experiments where single cells are studied. A method of identifying 'positive' cells (23) uses co-transfection of the DNAs with an expression construct containing a reporter gene such as β-galactosidase (e.g. pSV-β-galactosidase from Promega). Using this approach we have found that at least 80% of cells which are positive for the marker gene are also expressing the channel.

Protocol 7. Identification of 'positive' transiently transfected cells

1. Cells to be transfected are plated on poly-L-lysine coated coverslips at a low density to obtain single cells.
2. Transient transfection is performed (Section 6.1.3) with the DNAs, including the reporter gene construct.
3. Analyse 48 h later as follows:
   (a) Wash-cells with phosphate-buffered saline (PBS).
   (b) Incubate for 5 min, 37°C in 1 ml PBS.
   (c) Aspirate PBS and replace with 0.5 ml of 1 mM fluorescein di-β-D-galactopyranoside (FDG, from Molecular Probes Inc.) in 0.5 × PBS, pre-equilibrated to 37°C.
   (d) Incubate for 1 min at 37°C, then 5 min on ice.
   (e) View cells under a fluorescence microscope fitted with a fluorescein filter and identify the 'positive' fluoresceing cells in the field, which are selected for subsequent electrophysiological recording.

6.2 Stable expression

In stable expression the DNA is covalently integrated into the DNA of the cell, and its incorporation is maintained as the cell divides. This results in the creation of a transfected cell line which permanently expresses the recombinant protein.

6.2.1 Choice of vectors and cells

The choice of eukaryotic expression vector for use in stable expression is largely determined by the two following factors.

i. The type of selection system used

Generally, when stably transfecting cells, the eukaryotic expression vector in which the channel protein cDNA has been cloned should contain a selectable marker. This is not an absolute requirement, since it is possible to co-transfect the expression construct containing the channel cDNA with a separate expression vector containing the selection marker. A number of selection markers have been used, but genes conferring resistance to antibiotics such as neomycin or hygromycin have proved the most popular. Cells which integrate the transfected DNA into their genome will be resistant to the antibiotic, while those cells which do not will die. We have found that neomycin resistance is a robust selection system, and will describe its use in more detail below. A number of eukaryotic expression vectors containing the neomycin-resistance gene are available commercially (e.g. pXT1, Stratagene; pRc/CMV, pRc/RSV, and pcDNAIneo, Invitrogen; pMAMneo, Clontech). Additionally, some labs construct their own 'designer' vectors which they make available to other researchers.

ii. The type of expression: constitutive or inducible

If expression of the recombinant channel in the transfected cell is not likely to be toxic to that cell then constitutive expression is probably adequate. Constitutive expression of the channel protein can be driven by the eukaryotic promoters described in Section 6.1.2. If expression of the channel could be toxic to the transfected cell then an inducible promoter, such as the metallothionein promoter or the mouse mammary tumour virus (MMTV) long terminal repeat, would be preferable. The expression vector ZEMRVSP6 reported by Catterall and co-workers (20) has the metallothionein promoter and expression of the heterologous protein is induced by addition of zinc to the culture medium. The expression vectors pMAMneo (Clontech) or pMSGneo (constructed in our laboratory, ref. 24) contain the MMTV promoter and expression is induced by addition of dexamethasone to the culture medium. We have found the latter system suitable for stable expression of a number of channel proteins, including the nicotinic acetylcholine receptor (24) and the GABA-A receptor (25).

Regarding the choice of cells for stable expression, the comments made in Section 6.1.2 are equally applicable.

Protocol 8. Generation of stable cell lines expressing channel proteins

Equipment and reagents

- Channel cDNA, cloned into the chosen expression vector
- Geneticin (G418 sulphate, from Gibco-BRL). Make a 10 mg/ml stock in serum-free culture medium; filter, sterilize, and store at 4°C or −20°C.
- Cells to be transfected, in logarithmic growth, and low passage number
- Dexamethasone (tissue-culture grade, Sigma), 1 mM stock in ethanol, filtered and stored at −20°C

Method

1. Transfect cells (Section 6.1.3 and ref. 22). If the expression vector does not contain the neomycin-resistance gene then it is possible to co-transfect with a second vector containing the neomycin gene.
2. 72 h after transfection split the cells 1:4 and culture in medium containing 1 mg/ml Geneticin. At this stage there will be a great deal of cell death as non-transfected cells die.
3. After a further week, increase the concentration of Geneticin in the medium to 2 mg/ml.
4. When clones of cells reach 1–2 mm in diameter they should be subcloned into wells of a 48-well culture plate. This is done using cloning cylinders (6 mm diameter; Bellco Biotechnology, USA) as follows:
   (a) Prepare cloning cylinders: apply a thin layer of vacuum grease to the bottom of a glass Petri dish, place cloning cylinders end-up in grease, autoclave.
   (b) Remove culture medium from the plate.
   (c) Place cloning cylinder over a clone of cells using the vacuum grease to seal around the bottom of the cylinder.
   (d) Add a drop of trypsin-EDTA into the cylinder to release the cells, and transfer to the wells of a 48-well culture plate.
5. Expand cells for subsequent assay to detect expression of channel: for electrophysiological analysis, seed clones on to a cover slip; for radioligand binding, a confluent 10 cm culture dish of cells is required.[a]
6. If the expression is under control of the MMTV promoter, add dexamethasone to the culture medium at a final concentration of 1 μM. Since the cells are going to be harvested, there is no need to have Geneticin in the medium. Harvest cells for assay 2–7 days after induction with dexamethasone.[b]
7. Those clones expressing the recombinant channel which are to be kept should be recloned from single cells, and stocks of frozen cells prepared.

8. The stably transfected cell line should be maintained in medium containing 0.5 mg/ml Geneticin.

[a] To obtain the clone expressing the highest levels of recombinant channel it would be prudent to assay at least 50 clones. The proportion of clones expressing the recombinant channel varies in our hands from 10 to 50%.
[b] If expression of the recombinant channel proves to be toxic to the cells then addition of channel antagonists to the culture medium may be helpful.

7. Baculovirus expression

7.1 Introduction

Since Smith *et al.* (26) reported in 1983 the first expression of a heterologous protein in baculovirus-infected insect cells, a large number of foreign genes have been expressed in this system because of its ease of use and efficiency (27). The type of baculovirus used most widely is the *Autographa californica* nuclear polyhedrosis virus (AcNPV). During AcNPV replication two types of virus are produced: soon after infection (from approximately 12 h post infection (p.i.)) so-called non-occluded virus particles are released by budding from the cell membrane. They are responsible for transmission from cell to cell in infected animals and are also used for the transmission in cultured cells in the laboratory. Later in the replication process (from about 20 h p.i.), mainly occluded virus particles are formed in the nucleus of the cell. These particles are embedded in protein crystals consisting mainly of the virus-encoded polyhedrose protein. This protein can make up to 30–70% of the total protein in infected cells. The occluded virus particles assure the persistence and spreading of the virus in the field but are not essential for the propagation of the virus in cell culture (28).

In the baculovirus expression system the gene encoding the polyhedrose protein is replaced by the foreign cDNA. This results in the very strong polyhedrin promoter driving transcription of the heterologous gene.

7.2 Construction of recombinant baculoviruses

7.2.1 Principle

Because of its large size, direct insertion of a gene into the virus genome using restriction enzymes and ligase is impossible. Recombinant baculoviruses are therefore constructed by homologous recombination between regions of a transfer vector bearing the heterologous gene to be expressed, under the control of the desired virus promoter, and flanked by corresponding DNA regions of the viral genome. Recombination takes place in susceptible cells after transfection with virus and vector DNA. In earlier studies, where wild-type viral DNA was used for recombination, isolation of the rare recombinant viruses from the background of wild-type virus proved difficult. This problem was solved, however, by the development of a baculovirus mutant (29), from which a small but essential DNA region adjacent to the polyhedrin gene can be removed. Using such truncated and linearized viral DNA for co-transfection, only those recombinant viruses in which the missing DNA region is supplemented with the homologous region from the transfer vector can replicate.

7.2.2 Baculovirus transfer vectors

A number of different transfer vectors have been described (30, 31). pVL1392 and pVL1393 are used routinely in our laboratory because of their convenient polylinker placed behind the polyhedrin promoter. These and other vectors are commercially available (e.g. from Clontech, Invitrogen, or PharMingen). Insertion of the heterologous gene into the vector is achieved using standard molecular biology techniques. For transfection, the vector construct must then be purified by CsCl-gradient centrifugation or the Quiagen (Diagen) method.

7.2.3 Sf9 cell culture

The Sf9 cell line, derived from the moth *Spodoptera frugiperda*, is the most frequently used host cell line for the replication of AcNPV. The cells are easy to grow as attached cells and in suspension culture (in spinner or shaker flasks, or in fermenters) without any adaptation between phases. Because of their smooth surface they are better suited for electrophysiological patch-clamp studies than other insect cell lines.

Sf9 cells are grown in supplemented Grace's insect cell medium (TNM-FH, Gibco) with 10% fetal calf serum (FCS). No CO_2 application is required, because they grow best at pH 6.2 which can be maintained by buffer in the culture medium. For suspension cultures, 0.1% Pluronic F 68 (BASF) is added to the medium, to protect cells from shear forces. Good aeration is needed for optimal growth. Shaker flasks should therefore only be filled to one quarter of their nominal capacity. At 27°C the cells duplicate approximately daily, but temperatures above 29°C stop cell division. Care must therefore be taken to avoid temperatures higher than 27°C. Incubators with a cooling device and shaker platforms with low heat output (from Infors or Braun) are recommended. Furthermore, the use of an incubator without internal fan lowers the risk of evaporation and contamination (Heraeus, type BK 6160).

In suspension culture, shaken at approx. 100 r.p.m., cells are seeded at a density of 0.5×10^6 cells/ml. The maximal density of 6×10^6 cells/ml is achieved after 3–4 days. In static culture flasks, Sf9 cells adhere only loosely to the surface and can be detached easily by knocking the flask lightly (with the flat of the hand). 0.6×10^6 cells are seeded in 3 ml medium on a 10 cm^2 surface and split every 3–4 days.

7.2.4 Transfection of linearized defective virus and vector DNA

Two companies supply the linearized defective baculovirus DNA: PharMingen and Clontech laboratories. Both companies also supply very useful handbooks with their reagents, but recommend the use of relatively high amounts of their viral DNA. We found, however, that 1/10 of the indicated virus DNA amount will produce sufficient recombinant particles and allow one transfection kit to last 10 times longer. With the PharMingen kit transfection is performed by $CaPO_4$ precipitation, while the Clontech method uses lipofection. *Protocol 1* refers to the PharMingen method. Care should be taken to avoid breaking the linear virus DNA by shear forces (e.g. by repetitive pipetting).

Protocol 9. Cotransfection of virus and transfer vector DNA

All manipulations in this and the following protocols must be done under sterile conditions and virus-contaminated material should be inactivated before being discarded!

Method

1. Seed 2×10^5 logarithmically growing Sf9 cells in 0.5 ml of growth medium into one well of a 24-well plate.
2. Allow the cells to attach for about 1 h at 27°C.
3. Mix 0.05 μg linearized virus DNA and 0.2 μg transfer vector DNA in a sterile 1.5 ml Eppendorf tube.
4. Incubate the DNA for 5 min at room temperature.
5. Add 0.1 ml 125 mM Hepes, pH 7.1, 125 mM $CaCl_2$, 125 mM NaCl (PharMingen Transfection buffer B) and mix well.
6. Aspirate the medium from the cells and add 0.1 ml of fresh growth medium.
7. Add the DNA mixture dropwise to the cells while shaking the 24-well plate gently (a fine precipitate will form).
8. Incubate the cells for 4 h at 27°C.
9. Aspirate the transfection supernatant and add 0.5 ml of growth medium to the cells.
10. Incubate for 4–5 days at 27°C.
11. Harvest the supernatant containing the recombinant viruses and store at 4°C.

7.2.5 Plaque assay

The plaque assay is performed to determine the titre of infectious virus particles in a suspension as well as to isolate individual virus clones. The plaques correspond to virus-induced lysis of Sf9 cells and are visualized by staining of intact cells, which metabolize a vital dye (MTT, see below) into a coloured product. It is important to have Sf9 cells growing logarithmically both before and during the plaque assay, because baculoviruses multiply only in growing cells. To obtain clearly visible plaques, there should be no deviation from the indicated cell numbers. For transfection supernatants, titres between 10^2 and 10^4 plaque-forming units (p.f.u.) per ml can be expected. Virus stock titres rank usually from 2×10^7 to 5×10^8 p.f.u./ml. For the isolation of single recombinant virus clones, it is important to confirm the plaques by microscopical detection of lysed cells. Six to eight plaques are isolated routinely and amplified to give the 'primary virus stock'.

Protocol 10. Plaque assay

Equipment and reagents

- TNM-FH medium with 10% FCS
- Autoclaved 3% Seaplaque low melting temperature agarose (FMC Marine Colloids) in dist. H_2O
- Sterile plugged Pasteur pipettes (short form) with rubber teat
- 5 mg/ml MTT (3-[4,5-dimethylthiazol-2-yl]-2,5-diphenyltetrazolium bromide, Sigma) in phosphate-buffered saline (PBS), filtered and sterilized

Method

1. Seed 1.5×10^6 cells per well in 2 ml growth medium on a six-well plate.
2. Allow the cells to attach for 1 h.
3. Prepare serial ten-fold dilutions of the virus in growth medium.
4. Remove the growth medium from the cells and add, dropwise, 0.1 ml of the virus dilutions to the centre of the well.
5. Incubate the plate for 1 h at room temperature with occasional gentle shaking.
6. Mix one part of molten 3% Seaplaque agarose and two parts of growth medium, both pre-warmed to 37°C.
7. Aspirate the inoculum and carefully overlay the cells with 2 ml of the agarose mixture.
8. After the agarose has solidified, overlay it with 1 ml growth medium.
9. Incubate plate at 27°C for 5–6 days.
10. Add 0.2 ml of MTT-solution per well and incubate in the dark at 27°C for 0.5 to 24 h until the plaques are clearly visible.
11. Aspirate the liquid phase completely.
12. Count plaques at appropriate dilutions (i.e. 20–50 plaques per well) and calculate the original virus titre.

Protocol 11. Isolation of recombinant virus clones

Method

Proceed as in *Protocol 10,* then

1. Select a plate with well separated plaques.
2. Circle the plaques on the bottom of the plate with a fine marker pen.
3. Check and confirm these plaques by inspection under the microscope at $\times$ 100 magnification.
4. Pick the selected plaque by sucking a plug of agarose from above the plaque into a Pasteur pipette.
5. Transfer this plug into 0.5 ml of growth medium in one well of a 24-well plate.
6. Homogenize the agarose by pipetting it up and down.
7. Add 1 $\times$ 10^5 Sf9 cells in 0.5 ml medium.
8. Incubate for 4–5 days at 27 °C.
9. Harvest the medium from the well.
10. Remove cells and cellular debris by centrifugation at 3000 *g* for 10 min.
11. Store the supernatant at 4 °C as the primary virus stock.

7.2.6 Characterization of recombinant virus clones by PCR

Although only recombinant viruses are obtained when using the defective linearized virus DNA for co-transfection, correct insertion of the heterologous cDNA is not guaranteed. Therefore a quick check for the presence and size of the foreign cDNA in the viral genome is recommended. Viral genomic DNA is purified from an aliquot of the primary virus stock and the cDNA insert is amplified by PCR with primers corresponding to the recombination regions (supplied by Invitrogen and Clontech). As a control, genomic DNA from the wild-type virus is used for PCR. The presence and correct size of the inserted cDNA is then verified by agarose gel electrophoresis of the PCR product.

Protocol 12. Extraction of viral DNA and its characterization by PCR

Equipment and reagents

- TE buffer: 10 mM Tris–HCl, pH 7.8, 1 mM EDTA
- 10% SDS in dist. H_2O
- Water-saturated phenol/chloroform (90:10)
- 3 M sodium acetate, pH 5.6

Protocol 12. *Continued*

Method

1. Mix 100 μl of the primary virus stock with 350 μl TE buffer and 50 μl 10% SDS.
2. Extract the viral DNA with an equal volume of phenol/chloroform.
3. Precipitate the DNA with $^1/_{10}$ volume of 3 M sodium acetate and two volumes of 100% ethanol for 2 h at −20°C or 30 min at −70°C.
4. Pellet the DNA by centrifugation for 10 min at a maximum of 13 000 *g*.
5. Wash once with 70% ethanol and air dry.
6. Dissolve the DNA in 50 μl TE buffer.
7. Use 10 μl as template in the PCR according to the instructions of the primer supplier.
8. Analyse $^1/_5$ of the PCR reaction on a 0.8% agarose gel.

7.2.7 Secondary and working virus stocks

PCR positive virus clones must be amplified to 'secondary virus stocks' before testing for recombinant protein expression. Therefore, an Sf9 suspension culture is infected with a low virus-to-cell ratio (multiplicity of infection (m.o.i.) of approx. 0.1) and incubated for 4–5 days to allow several cycles of virus replication and cell infection. Because growing cells are needed during the whole incubation time to promote viral replication, infection is done at a rather low cell density of only 0.5×10^6 cells/ml. This procedure largely increases the virus quantity and avoids the generation of defective interfering virus particles.

For the production of recombinant protein, infections are done at a high m.o.i. of 3–10, to ensure that most cells are infected immediately. This condition synchronizes the production of the recombinant protein, which is started by activation of the polyhedrin promoter approximately 20 h post infection. As maximal production of recombinant protein is achieved about 48 h p.i., infections can be started at a cell density of 2×10^6. The presence of the recombinant protein can be assessed by immunohistochemistry or functionally by electrophysiological recordings in the case of an ion channel, or by binding assays for receptor proteins (see below). The secondary stocks of virus clones which induce good protein expression are stored in aliquots at −70°C. For large-scale protein production 'working stocks' are produced by infecting larger Sf9 cell cultures at low m.o.i. The titre of the working stock is determined by a plaque assay (see *Protocol 10*) to ensure comparable m.o.i. for different protein productions.

Protocol 13. Preparation of secondary and working virus stocks

Method

1. Add 100 μl of the primary virus stock to 20 ml of a logarithmically growing Sf9 cell culture (0.5–1 × 10^6 cells/ml) in a 100 ml shaker flask.
2. Incubate for 4–5 days at 27°C, shaking at 100 r.p.m.
3. Remove cellular debris by centrifugation (10 min, 3000 *g*).
4. Store the supernatant at 4°C as 'secondary virus stock'; it can be expected to contain 2 × 10^7–1 × 10^8 p.f.u/ml.
5. Infect another 20 ml of a logarithmically growing Sf9 cell culture (2 × 10^6 cells/ml) with 2 ml of this secondary stock (= high m.o.i.).
6. Incubate for 44–48 h at 27°C, shaking at 100 r.p.m.
7. Harvest cells and test for expression of the recombinant protein.
8. When expression has been demonstrated, store the secondary stock of a selected clone in 500 μl portions at −70°C.
9. Prepare 'working stocks' by infection of 250 ml Sf9 cultures (0.5 × 10^6 cells/ml in a 1 litre shaker flask) with 100–200 μl of the secondary stock.
10. Incubate for 4–5 days at 27°C, shaking at 100 r.p.m.
11. Remove cellular debris by centrifugation (10 min, maximum 3000 *g*).
12. Determine the titre of the working stock by plaque assay.
13. Store the working stock at 4°C for up to three months or at −70°C for longer periods.

7.3 Protein production

Protein production in quantitative amounts is performed in principle by scaling up the conditions described in *Protocol 13*, steps 5–7. The expression rates depend strongly on the protein to be expressed and the culture conditions. Under the conditions described here, soluble cytosolic proteins may yield up to 500 mg/litre Sf9 culture, while soluble secreted proteins may give 10–100 mg/litre. The yield of complex membrane proteins like the GABA-A receptor that comprises several subunits ranges only from 0.1–1 mg functional protein per litre. However, it may well be that only a small proportion of the synthesized subunit polypeptides assemble into the functional receptor complex. Some details of GABA receptor expression in Sf9 cells are described below.

To guarantee co-infection of a high percentage of Sf9 cells, m.o.i. of three or more are used of *each* subunit coding virus. Under this condition >90%

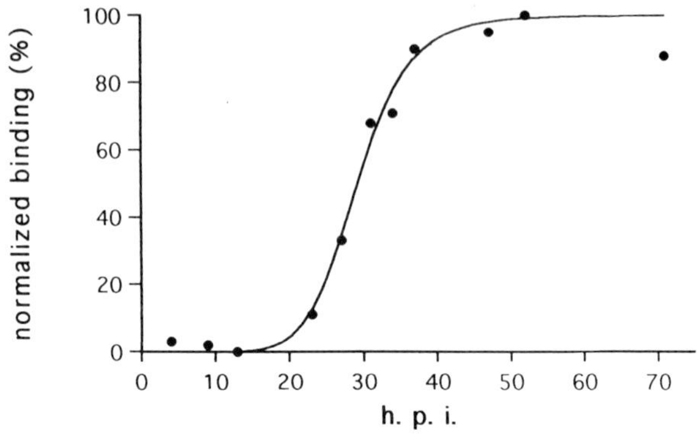

Figure 5. Time course of benzodiazepine binding site expression in Sf9 cells. Cells were co-infected with viruses harbouring cDNAs encoding α5, β2, and γ2 subunits of the GABA-A receptor. Specific binding was determined for membranes prepared from cells which were harvested at the indicated time post-infection (hours post-infection). The number of binding sites is expressed as a percentage of the maximum specific binding, found at about 52 h.p.i.

of the cells express the desired recombinant receptor subunits as judged by immunostaining at 48 h p.i. The time course of benzodiazepine binding-site expression, indicating the functional assembly of α-, β-, and γ-subunits of the GABA-A receptor, is illustrated in *Figure 5*. For electrophysiological recordings, infected cells are transferred 2 h p.i. to plates containing sterile glass cover slips. The cells attach rapidly to the cover slips and can be recorded 18–56 h p.i. with increasing numbers of cells responding to GABA. The optimal m.o.i. for infection, harvest and recording times may vary with the protein to be (co)expressed, promoters used for driving protein expression, and culture conditions. They should therefore be carefully monitored and adapted to give optimal results.

Acknowledgements

We would like to thank Dr Ruth McKernan (Merck, Sharpe and Dohme Research Laboratories, Harlow, UK) for reading this manuscript, and Dr J. Turner (Schering AG, Berlin) for kindly providing us with immunological data on GABA-A receptor expression in baculovirus-infected cells and fruitful discussion of the manuscript.

References

1. Sambrook, K., Fritsch, E. F., and Maniatis, T. (ed.) (1989). *Molecular cloning, a laboratory manual* (2nd edn). Cold Spring Harbor Press, Cold Spring Harbor, New York.

2. Ausubel, F. M., Brent, R., Kingston, R. E., *et al.* (ed.) (1987). *Current protocols in molecular biology*. Greene Publishing Associates and Wiley-Interscience, John Wiley, New York.
3. Innis, M. A., Gelfand, D. H., Sninsky, J. J., and White, T. J. (ed.) (1990). *PCR protocols*. Academic Press, San Diego, CA.
4. Pritchett, D. B. and Seeburg, P. H. (1991). *Proc. Natl Acad. Sci. USA* **88**, 1421.
5. Anand, R. A., Conroy, W. G., Schoepfer, R., Whiting, P., and Lindstrom, J. (1991). *J. Biol. Chem.* **266**, 11192.
6. Carter, D. B., Thomsen, D. R., Im, W. B., *et al.* (1992). *Bio/Technology* **10**, 679.
7. Klaiber, K., Williams, N., Roberts, T., Papazian, D. M., Jan, L. Y., and Miller, C. (1990). *Neuron* **5**, 221.
8. Kawamoto, S., Onishi, H., Hattori, S., Miyagi, Y., Mishina, M., and Okuda, K. (1991). *Biochem. Biophys. Res. Commun.* **181**, 763.
9. Gurdon, J. B., Lane, C. D., Woodland, H. R., and Marbaix, G. (1971). *Nature* **233**, 177.
10. Kusano, K., Miledi, R., and Stinnakre, J. (1977). *Nature* **270**, 739.
11. Parker, I., Sumikawa, K., and Miledi, R. (1987). *Proc. Roy. Soc. Lond. B* **231**, 37.
12. Miledi, R. and Woodward, R. M. (1989). *J. Physiol.* **416**, 601.
13. Miledi, R., Parker, I., and Woodward, R. M. (1988). *J. Physiol.* **400**, 33P.
14. Miledi, R. (1982). *Proc. Roy. Soc. B* **215**, 491.
15. Dascal, N. (1987). *CRC Crit. Rev. Biochem.* **22**, 317.
16. Yang, X. and Sachs, F. (1989). *Science* **243**, 1068.
17. Lubbert, H., Snutch, T. P., Dascal, N., Lester, H. A., and Davidson, N. (1987). *J. Neurosci.* **7**, 1159.
18. Seeburg, P. (1993). *Trends Neurosci.* **16**, 359.
19. Neve, R. L. (1993). *Trends Neurosci.* **16**, 251.
20. West, J. W., Scheuer, T. L., Maechler, L., and Catterall, W. A. (1992). *Neuron* **8**, 59.
21. Hadingham, K. L., Wingrove, P. B., Le Bourdelles, B., Palmer, K. J., Ragan, C. I., and Whiting, P. J. (1993). *Mol. Pharm.* **43**, 970.
22. Chen, C. A. and Okayama, H. (1988). *BioTechniques* **6**, 632.
23. Angelotti, T. P., Uhler, M. D., and Macdonald, R. L. (1993). *J. Neurosci.* **13**, 1418.
24. Whiting, P., Schoepfer, R., Lindstrom, J., and Priestley, T. (1991). *Mol. Pharm.* **40**, 463.
25. Hadingham, K. L., Harkness, P. C., McKernan, R. M., *et al.* (1992). *Proc. Natl Acad. Sci. USA* **89**, 6378.
26. Smith G. E., Fraser, M. J., and Sunners, M. J. (1983). *J. Virol.* **46**, 584.
27. Miller, L. K. (1988). Baculoviruses as gene expression vectors. *Ann. Rev. Microbiol.* **42**, 177.
28. Blissard, G. W. and Rohrmann, G. F. (1991). Baculovirus diversity and molecular biology. *Ann. Rev. Entomol.* **53**, 127.
29. Kitts, P. A. and Possee, R. D. (1992). *BioTechniques* **14**, 810.
30. O'Reilly, D. R., Miller, L. K., and Luckow, V. A. (1992). *Baculovirus expression vectors: a laboratory manual*. W. H. Freeman and Co., New York.
31. King, L. A. and Possee R. D. (1992). *The baculovirus expression system. A laboratory guide*. Chapman and Hall, London.
32. Ishii, T., Moriyoshi, K., Sugihara, H., *et al.* (1993). *J. Biol. Chem.* **268**, 2836.

6

Antibodies as tools for the study of the structure and function of channel proteins

F. ANNE STEPHENSON

1. Introduction

The exquisite specificity of antibody molecules combined with the ease with which they can be labelled makes them valuable in the study of the structure and function of proteins. Historically, the generation of an antibody against a defined protein required the purification to homogeneity of high concentrations of the antigen followed by animal immunization and the production of polyclonal antibodies. The specificity of these antibodies was determined by the major antigenic regions of the immunogen. In general, channel proteins are not easily isolated in the amounts required for polyclonal antibody production. However, the advent of monoclonal antibody technology has circumvented the need for a pure antigen. There was a requirement instead for a rapid, selective screening method for the production of antibodies against those proteins expressed at low abundance. The specificity of these antibodies was again dictated, for the most part, by the antigenic regions of the protein of interest. Today, the availability of cDNAs encoding channel polypeptides means that the natural proteins themselves are no longer required for antibody production, since recombinant or synthetic antigens can be employed instead. In many cases, the specificities of the antibodies thus produced are predetermined, which means that they are even more powerful research tools. In this chapter, I will describe recent advances in the production and characterization of these antibodies, together with selected examples of their uses. It is an update of a previous chapter from the *Practical approach* series (1). The use of the antibodies will focus on work from my group on the study of the structural and functional properties of two ligand-gated ion channels, the inhibitory γ-aminobutyric acid-A (GABA-A) neurotransmitter receptor and the excitatory subclass of glutamate neurotransmitter receptor, the *N*-methyl-D-aspartate (NMDA) receptor.

2. The choice of antigen

The choice of antigen for the production of antibodies against channel proteins is dictated by the purpose for which the antibody is required. For example, for a particular channel family, if there is no need to distinguish between subtypes, then the major antigenic determinants of the protein can be employed. These can be predicted from the respective primary structures by the method of Hopp and Woods which determines these epitopes by calculating the hydrophilicity of the protein sequence. This is readily carried out by available computer programs; the one in use in our laboratory is PC-GENE (IntelliGenetics Inc.). These peptide sequences can thus be utilized as antigen within either a synthetic peptide or a bacterially derived fusion protein (see below). However, if the aim is to produce an antibody which is channel-protein subtype-selective, or indeed, if the antibody is required to study defined domains within the protein primary stuctures, then the choice of the peptide sequence is dictated by the experiment. But again, the antigen could be either the synthetic peptide or a bacterially derived fusion protein. There are advantages and disadvantages to the use of both these types of antigen and these are discussed below.

2.1 Bacterially derived fusion proteins

For the production of a fusion protein, the DNA encoding the chosen region of receptor subunit is subcloned in the correct reading frame and orientation into a bacterial expression vector. Expression of the fusion protein or just the protein fragment itself, depending upon the characteristics of the vector, is induced and subsequently purified by affinity chromatography or preparative sodium dodecyl sulphate polyacrylamide gel electrophoresis (SDS-PAGE). There are now many bacterial expression vectors available commercially; an example of the use of one is given in *Protocol 1*.

Protocol 1. The production of bacterially derived fusion proteins for use as antigens in raising anti-receptor/channel antibodies

Equipment and reagents

- The GST gene fusion system is supplied to give three reading frames (pGEX1, 2, and 3) by Pharmacia Biotech. Ltd. It produces the required recombinant proteins as fusions with *Schistosoma japonicum* glutathione *S*-transferase (GST), thus facilitating their purification from bacterial lysates by glutathione affinity chromatography as described by Smith and Johnson (2).
- Glutathione agarose (Sigma)
- Phosphate-buffered saline (PBS): 8 g/l NaCl; 0.2 g/l KCl; 1.15 g/l Na_2HPO_4; 0.2 g/l KH_2PO_4
- PBS containing 1% (v/v) Triton X-100
- 50 mM Tris–HCl, pH 8.0 containing 5 mM glutathione
- Thrombin in PBS containing 5 mM $CaCl_2$
- Probe sonicator
- SDS-PAGE equipment

Method

1. Clone the fragment of the gene of interest into the multiple cloning site of the appropriate pGEX vector. Grow in the conditions recommended by the manufacturer for the production of the recombinant protein. Centrifuge the bacterial lysate for 20 min at 5000 *g* at 4°C. Discard the supernatant and resuspend the pellet with PBS containing 1% (v/v) Triton X-100. Sonicate the extract with a probe sonicator for 2 × 30 sec with cooling and freeze at −20°C. (Note that the efficiency for extraction may vary between different fusion proteins.) Thaw and purify the protein by glutathione affinity chromatography as described below.
2. For a 250 ml bacterial culture, swell 80 mg (1 ml final volume) glutathione agarose by heating at 50°C for 30 min in PBS. Wash the swollen gel three times by centrifugation and resuspension with PBS. Incubate the thawed bacterial lysate with the glutathione agarose for 20 min at 4°C.
3. Collect the glutathione agarose by centrifugation at 800 *g* for 20 min at 4°C. Wash the agarose three times, each with 50 ml PBS, by resuspension and centrifugation.
4. Incubate the agarose for 5 min at room temperature with 50 mM Tris–HCl, pH 8.0 containing 5 mM glutathione, prepared immediately before use. Collect the eluate by centrifugation as above and repeat the elution a second time.
5. Analyse the eluate, which should contain the glutathione *S*-transferase-fusion protein, by SDS-PAGE. We obtained yields of about 0.5–1 mg protein/250 ml culture.
6. The fusion protein can be used directly as the antigen or alternatively, the GST can be removed by thrombin digestion: incubate the fusion protein with thrombin in PBS containing 5 mM $CaCl_2$ at room temperature. The time of incubation and the concentration of thrombin are determined by experiment initially on a small scale. Samples are analysed by SDS-PAGE and used accordingly.
7. Additional points to note are that the recombinant protein can be verified by both size determination and partial N-terminal amino-acid sequencing. In our hands, the thrombin digestion was never 100% efficient, so GST was included routinely as part of the antigen. Anti-GST antibodies can be removed by preadsorption against a GST affinity column.

The main advantages of the fusion protein as an antigen are those of size and economy. It is relatively inexpensive to produce large quantities of purified polypeptide with a molecular weight up to at least 50 000 daltons. This has the advantage that the larger the protein fragment, the greater the probability that it will adopt the secondary structure of the channel, thereby eliciting the production of antibodies which recognize the native conformation of the protein. Also, once the correct bacterial clone has been identified, there is an infinite supply of antigen. However, the size of the fusion protein, although an advantage for the production of antibodies recognizing the native conformation of receptors, is a disadvantage with regard to the definitive specificity of the antibody.

2.2 Synthetic peptides

Synthetic peptides as antigens have the advantage of a greater predetermined specificity compared to fusion proteins. They are usually smaller in size and therefore require coupling to a carrier protein for antibody production. The lower molecular weight means that there is a reduced probability of reactivity of the antibodies with the native receptor and the use of a carrier protein requires that the anti-peptide antibodies be affinity-purified prior to use. The multiple antigen presentation system (MAPS) developed by James Tam has dispersed with the use of carrier protein (reviewed in ref. 3). In this system, multiple copies (usually eight) of the chosen peptide are coupled to a non-immunogenic lysine backbone and this is employed as the antigen-producing anti-peptide antibodies only. The disadvantages of the MAPS system are that MAPS peptides are expensive and the chemistry of synthesis is such that the C-terminal of the peptide is coupled to the lysine backbone. Thus if the C-terminal sequence of the protein is used, the peptide will not exactly mimic the native receptor. This can be circumvented by the use of spacer alanines between the C-terminal amino acid of the protein under study and the lysine backbone. The structure of a peptide is given in *Figure 1*.

Despite the disadvantages of the use of synthetic peptides as antigens, we have used them to considerable advantage for the study of the GABA-A receptors and, more recently, NMDA receptors. In summary, successful antibodies were raised with peptides of 11–18 residues in length. This was a compromise between the number of amino acids thought to be required for

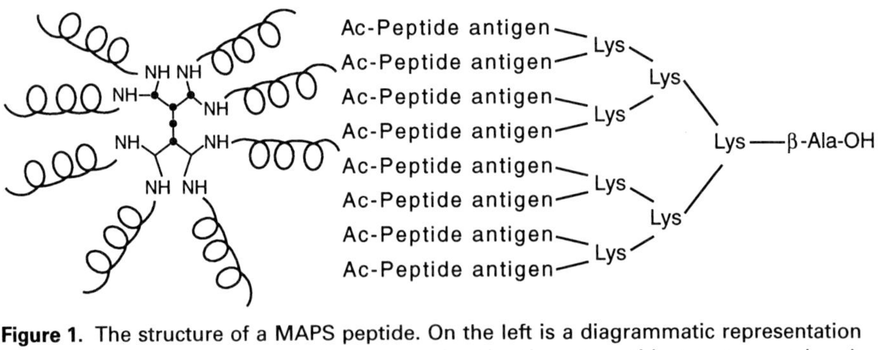

Figure 1. The structure of a MAPS peptide. On the left is a diagrammatic representation of the MAPS peptide with the helix representing the chosen peptide sequence and each ● is a lysine molecule. (Reprinted from ref. 3, with permission.)

a satisfactory immune response, a longer peptide having a greater probability of adopting the same conformation as that found within the native polypeptide, and cost. These peptides have been employed mostly for the production of antibodies for the identification of *in vivo* receptor subunit complements (see below). The sequences chosen were therefore from non-conserved regions of the deduced primary structures of the GABA-A receptor subunits; these corresponded to the extreme N- and C-terminal regions (*Figure 2*). A high success rate was achieved for the reaction of antibodies raised against these epitopes with the native receptor. These regions have also been successful for other proteins (4). Thus, they are good candidate regions for antibody production. *Figure 2* summarizes the peptide sequences that have been used for the raising of anti-GABA-A receptor α1 subunit antibodies. Of the eight peptides which were selected for various reasons, seven produced high titre anti-peptide antibodies, but of these only four were useful in that they recognized the GABA-A receptor α1 subunit.

Peptides were always coupled to a carrier protein which was either keyhole limpet haemocyanin (KLH) or, more recently, thyroglobulin (preferred now because of ease of solubility) The coupling method chosen was one that permitted covalent attachment at one end only in such a way that it would most closely resemble the parent subunit. Thus, for peptides corresponding to the N-terminal sequence of a GABA-A receptor subunit, coupling was always at the C-terminus and vice versa for C-terminal sequences. The preferred method of coupling was either glutaraldehyde, to utilize the terminal NH_2 (for C-terminal sequences), or 1-ethyl-3-(3-dimethylaminopropyl)-carbodi imide (EDAC)-activated COOH (for N-terminal sequences). However, if the internal sequence of the peptide contained lysine or glutamate/aspartate respectively, giving rise to the possibility of internal cross-linking between peptide and carrier protein, either cysteine or tyrosine was added to the end of the synthetic peptide to ensure unique terminal coupling. Peptide sequences taken from the internal peptide sequence were amidated at the

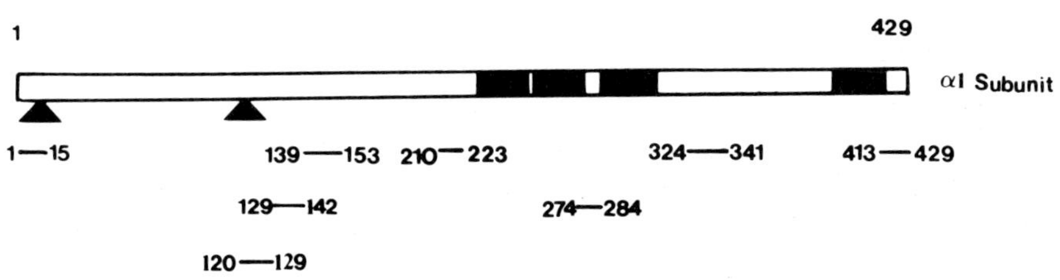

Figure 2. Schematic representation of peptide sequences used for the production of anti-GABA-A receptor α1 subunit antibodies. The figure represents the primary structure of the mature bovine GABA-A receptor α1 subunit where 1 is the N-terminal amino acid, the black arrowheads show predicted sites of *N*-glycosylation, and filled areas show the predicted transmembrane spanning regions, M1–M4. The peptide sequences used for antibody production are shown below the α1 polypeptide.

C-terminus, again to mimic the native subunit structure. The solubility of the synthetic peptide was considered, but generally this was just a check for the presence of charged amino acids. Additionally, the peptide sequences were checked for homology with proteins of known primary structure by screening a sequence database.

We have also successfully employed the use of a MAPS peptide for the production of a C-terminal anti-NMDAR2A subunit antibody (5). The method was successful, but in our hands and for this peptide sequence we still found it necessary to affinity purify the antibodies. The detailed methodologies for the different ways of coupling between peptide and carrier protein and the affinity purification of these respective anti-peptide antibodies can be found in ref. 1. For general protocols of animal immunization, etc., the reader is referred to ref. 6.

3. Antibody characterization

3.1 Antibody production

Antibody production is readily monitored and quantified by an enzyme-linked immunosorbent assay (ELISA), using the respective immunogen as the antigen, i.e. either the synthetic peptide or the bacterial fusion protein. However, although it is useful to know that antibodies are produced, the important point is to determine whether they react with the protein from which the antigen is derived. This is trivial if the purified receptor or channel is available, since this can be used as antigen in an ELISA assay and the antibody characterized further as described below. Alternatively, the cDNA encoding the receptor can be expressed in, for example, a mammalian cell line and the transfected and non-transfected cells screened for antibody reactivity by immunoblotting. The size of any immunoreactive band can be compared to the size of the protein deduced from the cDNA sequence with the caveat that there may be differences in post-translational modification, particularly *N*-glycosylation, between the natural and the cloned subunit. In the absence of either of these, then it is necessary to screen by immunoblotting using membranes prepared from tissues which are known to express the channel and with a negative control where there is no known tissue expression. Again, the size of any immunoreactive band is compared with that deduced from the respective cDNA sequence. For all these screening methods, antibody specificity controls should be carried out. In our laboratory this comprises overnight preincubation at 4°C of the antibody with the respective peptide used as antigen. A revised method for immunoblotting with increased sensitivity is given in *Protocol 2*. Examples of the different screening methods are given in *Figure 3*.

Protocol 2. Immunoblots using anti-peptide antibodies and the ECL detection system

Equipment and reagents

- Enhanced chemiluminescence (ECL) system (Amersham, UK)
- Transfer buffer: 25 mM Tris, 192 mM glycine, 20% (v/v) methanol, pH 8
- Tris-buffered saline (TBS): 50 mM Tris–HCl pH 7.4, 0.9% (w/v) NaCl
- Stripping buffer: 62.5 mM Tris–HCl pH 7.4, 2% (w/v) SDS, 100 mM β-mercaptoethanol
- TBS-T: TBS containing 0.2% (v/v) Tween 20
- Nitrocellulose membrane filter
- Dried skimmed milk powder
- Horseradish peroxidase conjugated anti-rabbit Ig (Amersham)
- Hyperfilm ECL (Amersham)
- SDS-PAGE equipment

Method

1. Run the antigen in the appropriate percentage polyacrylamide slab gel under reducing conditions according to standard methods of SDS-PAGE.
2. Transfer the antigens to a nitrocellulose membrane filter, which has been pre-soaked in transfer buffer, by electrophoresis at 50 V (constant voltage) for 2.5 h at room temperature (maintained with circulating water).
3. Rinse the membrane briefly with TBS and incubate the filter with stripping buffer for 30 min at 50°C. Wash the filter 2 × 10 min at 37°C with TBS-T.
4. Block the non-specific protein sites by incubation with TBS-T containing 5% (w/v) dried skimmed milk powder for 1 h at 37°C.
5. Incubate the nitrocellulose membrane filter overnight at 4°C with affinity-purified anti-peptide antibody diluted in TBS containing 2.5% (w/v) milk powder. Typical antibody concentrations are in the range 2.5–7.5 μg/ml.
6. Wash four times, 10 min each, at 37°C with TBS-T containing 2.5% (w/v) milk powder.
7. Incubate with a 1:1500 dilution of peroxidase-linked anti-rabbit Ig in TBS containing 2.5% (w/v) milk powder for 1 h at 37°C.
8. Wash as in step 6 and rinse twice with TBS-T.
9. Develop the immunoblot using the ECL detection system and Hyperfilm ECL according to manufacturer's instructions.
10. Immunoblots developed by this method can be reprobed with a different specificity anti-peptide antibody, as follows:
    (a) Immediately following the first development, rinse the nitrocellulose membrane with TBS then incubate with stripping buffer as in step 3 *but* at pH 11.5.

Protocol 2. *Continued*

(b) Wash twice with TBS-T then proceed as in steps 4–9. It is useful when reprobing immunoblots to have a control to ensure that the removal of the initial immunoreactivity is complete. This control is an antigen lane from the SDS-PAGE gel (step 1) which is incubated in the absence of the second anti-peptide antibody.

Figure 3. Examples of different screening methods for the reactivity of anti-peptide antibodies with the natural receptor protein. (a) and (b): demonstration of the specificity of anti-GABA-A receptor $\alpha1$, $\alpha2$, and $\alpha3$ subunit antibodies by immunoblotting using natural receptors. GABA-A receptors were purified from Na^+ deoxycholate extracts of bovine cerebral cortex by benzodiazepine affinity chromatography. Samples (50 µl) were subjected to characterization by immunoblotting as described in *Protocol 2* using 10% SDS-PAGE under reducing conditions and affinity-purified antibodies at 2.5–10 µg/ml final concentration. (a) Immunoblotting with anti-$\alpha1$ 324–341 (lane 1), anti-cys $\alpha2$ 414–424 (lane 2), and $\alpha3$ 454–467 (lane 3) antibodies alone. Note that each antibody recognizes a band with a distinct molecular weight, $\alpha1$ M_r 53 000, $\alpha2$ M_r 54 000, and $\alpha3$ M_r 58 000. (b) Lane 1, immunoblotting with all three specificity antibodies combined showing a summation of lanes 1–3 in (a); lane 2, same as lane 1 and preincubated overnight at 4°C with $\alpha1$ 324–341 (500 nM); lane 3, same as lane 1 but preincubated overnight with cys $\alpha2$ 414–424 (500 nM), and lane 4, same as lane 1 but preincubated overnight with cys $\alpha3$ 454–467 (500 nM). The position of pre-stained protein standards (kDa $\times$ 10^{-3}) are shown on the left. Further experimental detail can be found in ref. 9. (Reprinted from ref. 9 with permission.) (c) and (d): Demonstration of the reactivity of anti-NMDAR1 929–938 with the brain and the cloned NMDAR1 polypeptide by immunoblotting. Again immunoblotting was carried out as described in *Protocol 2* using 10% SDS-PAGE under reducing conditions, affinity-purified antibodies at 10 µg/ml final concentration and the antigens were either brain membranes or human embryonic kidney 293 cells transfected with the plasmid, pCISNMDAR1 (18). (c) Lane 1, non-transfected parent 293 cell homogenate (30 µg protein); lane 2, transfected 293 cell homogenate (30 µg protein); lane 3, as for lane 2 but antibody was preincubated overnight with NMDAR1 929–938 peptide prior to immunoblotting; lane 4, rat brain membranes (75 µg protein); lane 5, as for lane 4 but antibody was preincubated overnight with NMDAR1 929–938. (d) Lane 1, transfected 293 cell homogenate (30 µg protein); lane 2, transfected 293 cell homogenates after treatment with *N*-glycanase (30 µg protein); lane 3, rat brain membranes (75 µg protein), and lane 4, rat brain membranes (75 µg protein) after treatment with *N*-glycanase. Positions and molecular weights of pre-stained protein standards are shown on the left. The molecular weight of the *N*-deglycosylated peptide recognized by anti-NMDAR1 929–938 antibodies is M_r 97 000 compared to a value of M_r 105 000 predicted from the cDNA sequence. More details can be found in ref. 21. (Reprinted from ref. 21 with permission.) (e) and (f): Demonstration of anti-GABA-A receptor $\alpha1$ subunit antibody specificity by immunocytochemical staining of human embryonic kidney 293 cells expressing different GABA-A receptor α subunits. Human embryonic kidney 293 cells were grown on poly-D-lysine-coated coverslips and transfected with plasmids containing the different GABA-A receptor α subunits. The cells were fixed with methanol and probed, for the example shown here, with affinity-purified anti-$\alpha1$ 324–341 antibodies. Specific labelling was detected with the horseradish peroxidase/biotinylated Ig/diaminobenzidine system. (e) Shows 293 cells expressing the $\alpha1$ subunit and (f) shows non-transfected cells. Similar results showing background staining only were obtained when 293 cells expressing either the $\alpha2$, $\alpha3$, or $\alpha6$ GABA-A receptor subunits alone, thus demonstrating, in a different experimental paradigm to (a) and (c) above, the antibody specificity.

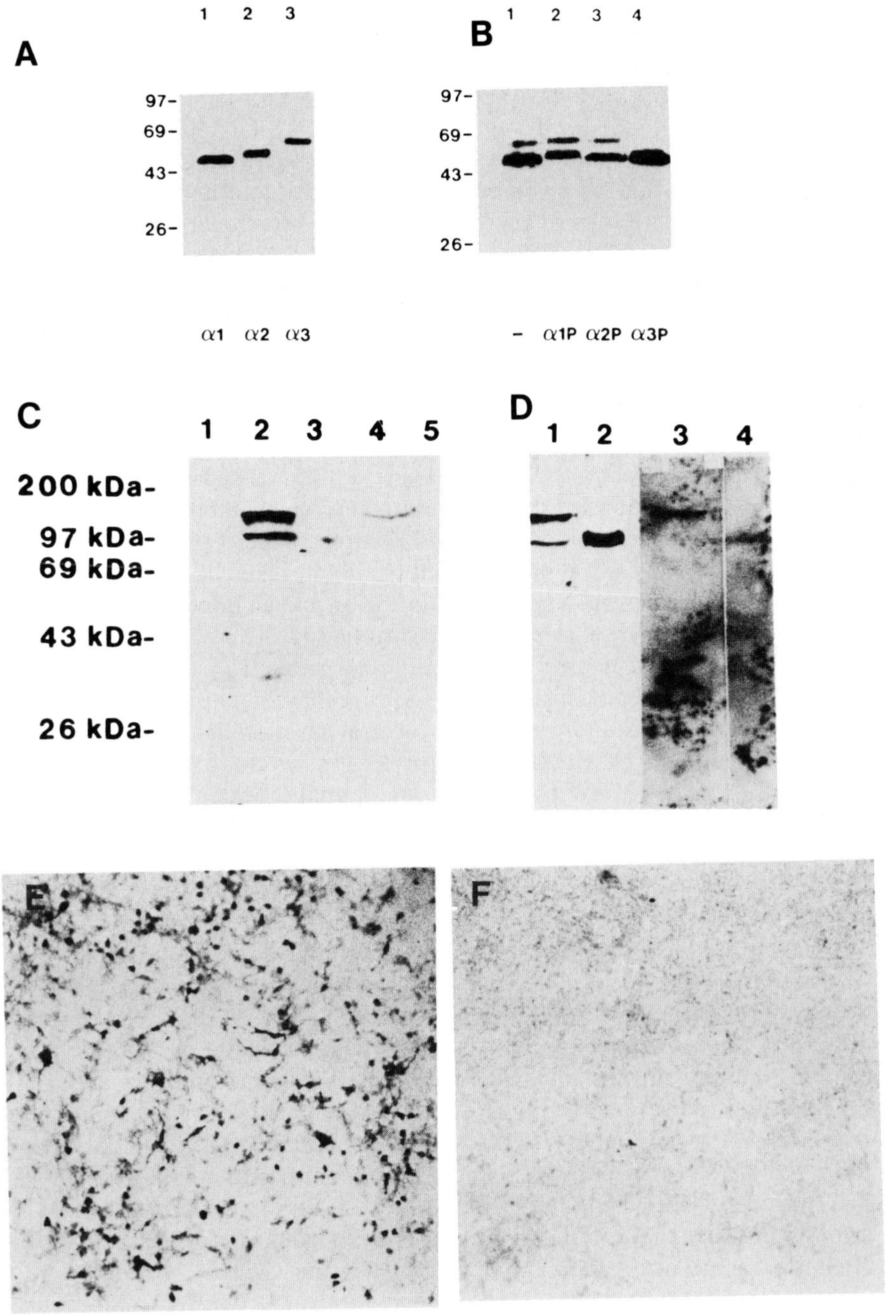

A
1 2 3
97-
69-
43-
26-
α1 α2 α3
B
1 2 3 4
97-
69-
43-
26-
- α1P α2P α3P
C
1 2 3 4 5
200 kDa-
97 kDa-
69 kDa-
43 kDa-
26 kDa-
D
1 2 3 4
E
F

4. Uses of antibodies

4.1 Identification of the subunit complements of native channels

Following the identification of the various cDNAs encoding the subunits of a particular gene family, an immediate goal is to determine which of the subunits co-assemble *in vivo* to form functional heteromeric proteins. The availability of a battery of subtype-specific antibodies, developed as described above, provides the tools with which to address this question.

One approach to the determination of native receptor subunit complements is that of immunoprecipitation. The discussion here relates to the GABA-A receptor, but the principles can be extrapolated to other channel proteins. GABA-A receptor genes comprise a large family which encode highly homologous polypeptides. Fifteen mammalian genes are known. These encode the GABA-A receptor subunits $\alpha1$–$\alpha6$, $\beta1$–$\beta3$, $\gamma1$–$\gamma3$, δ, and $\rho1$–$\rho2$. Different pentameric combinations of these are thought to coassemble *in vivo* to yield receptors with different biophysical and pharmacological properties. To investigate the co-association of the different subunits, a detergent-solubilized extract was prepared from a specific brain region and an immunoprecipitation assay carried out with a subunit-specific antibody (1, 7; *Protocol 3*). The abundance of the expression of this particular subunit in the chosen brain region at the defined developmental stage is readily quantified by the determination of the maximum percentage of specific radioligand binding sites immunoprecipitated by the antibody compared to control samples which contain the same concentration of non-immune immunoglobulin. For GABA-A receptors, the radioligands most frequently used are [^{3}H]muscimol and [^{3}H]flunitrazepam. The coexistence of this subunit with other GABA-A receptor subunit isoforms is then determined by the extraction of the immune complex from the immunoprecipitin precipitate followed by analysis by immunoblotting using a GABA-A receptor antibody of different specificity (*Protocol 3*).

One problem that we have encountered using this approach is that the receptor subunits are immunoprecipitated *and* the antibody is solubilized from the immune precipitate. Therefore, when the immunoblotting is carried out, the anti-rabbit Ig antibodies used for detection will recognize the Fc region of the primary anti-GABA-A receptor antibody. Since the antibody chain has a molecular weight, similar to GABA-A receptor subunits in SDS-PAGE under reducing conditions (about 50 000 Da), the experimental results are obscured by the antibody–antibody reactivity. This problem is readily circumvented by direct labelling of the second anti-GABA-A receptor antibody with biotin. For this we have used a biotinylation kit from Amersham International. Detection in the immunoblot now requires the use of streptavidin–peroxidase only. An alternative approach, to avoid the problems of antibody–antibody detection, is to use antibodies raised in different species. The availability of these antibodies is of particular value in double or even triple antibody immunofluorescent studies (see below).

Protocol 3. Quantitative immunoprecipitation of GABA-A and benzodiazepine binding sites from detergent extracts of mammalian brain

Equipment and reagents

- KP buffer: 10 mM potassium phosphate pH 7.4, 150 mM KCl
- Immunoprecipitin: 10% suspension of formalin-fixed *Staphylococcus aureus*
- Immunoprecipitin pretreatment buffer: PBS pH 7.4, 2% (w/v) SDS, 100 mM β-mercapto-ethanol
- KP-T buffer: KP buffer containing 0.5% (v/v) Triton X-100
- Electrophoresis sample buffer: 10 mM Na H_2PO_4 pH 7, 10% (v/v) glycerol, 0.02% (w/v) bromophenol blue, 20 mM DTT, 2.5% (w/v) SDS

Method

1. Incubate the detergent-solubilized brain extract with affinity-purified receptor subunit-specific antibody overnight at 4°C in a final volume of 500 μl. Typically, the volume of soluble extract is 200 μl, the volume of affinity-purified antibody is 200 μl containing a maximum of 125 μg protein, and the final assay volume is made up with KP buffer.
2. Pretreat Immunoprecipitin with pretreatment buffer at a ratio of 2:1 Immunoprecipitin:buffer for 30 min at 80°C. Centrifuge for 5 min at 10 000 *g* at room temperature. Wash the pellet three times by resuspension in PBS (three volumes) followed by centrifugation for 5 min at 10 000 *g* at room temperature. Resuspend to the original volume in PBS.
3. Add 100 μl pretreated Immunoprecipitin to the incubation mixture (step 1) and incubate for 1 h at room temperature.[a]
4. Add 400 μl KP-T buffer. Vortex and centrifuge for 15 min at 10 000 *g* at 4°C.
5. From the resultant supernatant, take 6 × 150 μl aliquots for the determination of total (3 × 150 μl) and non-specific (3 × 150 μl) radioligand binding activities.
6. Resuspend the pellet in 100 μl KP-T buffer by drawing through an air-displacement pipette several times. Add a further 900 μl of KP-T buffer and centrifuge for 10 min at 10 000 *g* at 4°C. Discard the supernatant and repeat the wash procedure twice more.
7. Resuspend the pellet as in step 6 to a final volume of 1 ml. Take 6 × 150 μl aliquots for the measurement of total (3 × 150 μl) and non-specific (3 × 150 μl) radioligand binding activities. Radioligand binding was measured by the polyethyleneimine filtration method (7).

Protocol 3. *Continued*

8. An essential control to take account of non-specific precipitation of proteins is to include an immunoprecipitation sample with the equivalent concentration of protein-A-purified non-immune IgG. Then proceed exactly as for the test samples (steps 3–7).
9. To determine the other receptor subunits co-precipitated with the original antibody, resuspend the pellet obtained at the end of step 6 in electrophoresis sample buffer (40 μl). Boil for 5 min and collect the supernatant by centrifugation for 5 min, 10 000 *g* at room temperature. The sample can then be analysed by immunoblotting as in *Protocol 2* using a different specificity antibody. Alternatively, a biotinylated primary antibody can be used and the immunoreactive species detected by incubation with streptavidin-linked horseradish peroxidase. Biotinylation of antibodies is readily carried out using a biotinylation kit (Amersham).

[a] Immunoprecipitation studies can be carried out under denaturing conditions because antigen:antibody reactions tolerate concentrations of SDS up to at least 0.2% (w/v). Use was made of this in our studies in the determination of the GABA-A-receptor subunit that was photoaffinity-labelled with [^{3}H]flunitrazepam. Immunoprecipitation studies with anti-α1 and anti-γ2 subunit antibodies showed that under both native *and* denaturing conditions the radioactivity was associated with only the α1 subunit (18).

Immunoprecipitation studies yield information regarding the coexistence of different subunit subtypes within a population of receptors immunoprecipitated by one specific antibody. However, this approach does not address whether this population of immunoprecipitated receptors is in fact homogeneous with respect to subunit complement, i.e. that all immunoprecipitated receptors do indeed have both these polypeptides. To answer this, it is necessary to immunoaffinity purify receptor subpopulations. The strategy is to immunoaffinity purify receptors using, for example, an anti-GABA-A receptor α1 subunit antibody-affinity column. The coexistence of subunits is thus determined by assaying the purified receptors by immunoblotting using antibodies of differing specificity. So far, this information is identical to that obtained by the immunoprecipitation method and therefore suffers the same limitations. However, if the purified receptor subpopulation is now subjected to a second round of immunoaffinity purification using a different antibody column, e.g. an anti-GABA-A receptor α2 subunit antibody column, more information is forthcoming. The double immunoaffinity-purified receptor population can be assayed by immunoblotting and the material applied and the material not bound to the second antibody column can also be assayed. This serial use of different specificity antibody affinity columns yields information regarding the homogeneity of the population of receptors immunoprecipitated by a single-specificity antibody. The experimental design is

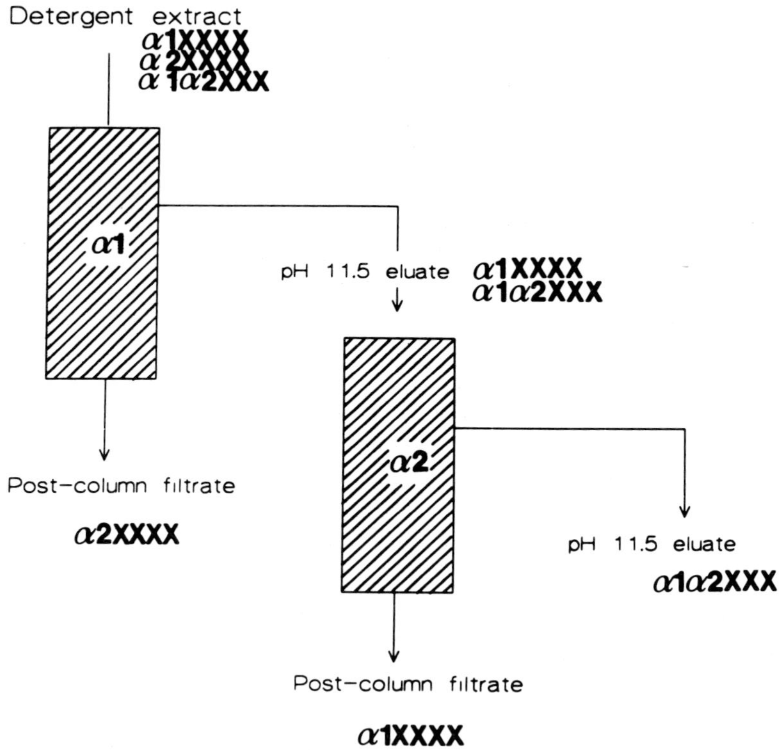

Figure 4. Schematic representation of the identification of GABA-A receptor subpopulations by immunoaffinity chromatography using different specificity antibody columns in series. α1xxxx, α2xxxx, and α1α2xxx represent different GABA-A receptor populations where 'x' is any subunit.

summarized in *Figure 4*. We have used this strategy successfully for the identification of GABA-A receptor subpopulations (9, 10).

The immunoaffinity chromatography purification procedure that we have used to isolate GABA-A receptor subpopulations is given in *Protocol 4*.

A disadvantage of this is that the method is based on the immunoreactivity of the receptor/channel and not the activity of the protein, defined here as the respective radioligand binding activity. Thus a potential criticism is that if there were a pool of precursor, unassembled receptor subunits present in the starting tissue, this would yield erroneous results with regard to the relative abundance of a particular receptor subunit subtype. The recovery of radioligand-binding activity in the purified samples suggested that this was probably not the case for the GABA-A receptor subunits. However, note that the recovery of activity from the given example is low for all types of elution (*Table 1*). The existence of nascent polypeptide chains is therefore a possibility.

Protocol 4. Purification of native GABA-A-receptor subpopulations

Equipment and reagents

- Dialysis buffer: 0.2 M $NaHCO_3$ pH 8.5, 0.5 M NaCl
- Cyanogen-bromide-activated Sepharose 4B gel (Pharmacia-Biotech)
- Disposable 10 ml column (BioRad)
- Wash buffer: 10 mM potassium phosphate pH 7.4, 0.5 M KCl, 1 mM EGTA, 2 mM magnesium acetate, 10% (w/v) sucrose, 0.2% (v/v) Triton X-100, 0.02% NaN_3
- Pre-elution buffer: 40 mM diethylamine pH 11, 0.5 M KCl, 1 mM EGTA, 2 mM magnesium acetate, 10% (w/v) sucrose, 0.2% (v/v) Triton X-100, 0.02% (w/v) NaN_3
- Collection buffer: 1 M glycine, 0.5 M KCl, 1 mM EGTA, 2 mM magnesium acetate, 10% (w/v) sucrose, 0.2% (v/v) Triton X-100, 0.02% (w/v) NaN_3

Method

1. Dialyse affinity-purified anti-peptide antibodies (5 mg in 1 ml) against 2 × 200 ml dialysis buffer. Couple the dialysed antibody[a] to cyanogen bromide-activated Sepharose 4B (1 ml swollen gel) by a standard method (11). (Our efficiency of antibody coupling was 95%, giving an immobilized antibody concentration of 15–30 μM.)
2. Pack the antibody affinity resin into a BioRad 10 ml disposable column and equilibrate with affinity wash buffer. Pre-elute the column with 10 ml pre-elution buffer and re-equilibrate with affinity wash buffer.
3. Dilute the solubilized brain extract 1:1 with affinity wash buffer. Recirculate the diluted extract (typically 15 ml; 20–40 mg protein) overnight at 4°C over the antibody affinity column at a rate of 10 ml/h.
4. Wash the antibody affinity column with 50 ml affinity wash buffer at a rate of 25 ml/h at 4°C.
5. Elute the column with 5 ml of the pre-elution buffer as in step 2 at a rate of 5 ml/h. Collect fractions of 1 ml into 200 μl collection buffer to give a final fraction volume of 1.2 ml and pH 8. Assay samples for radioligand binding activity and for antibody immunoreactivity as in *Protocol 2*. In our experience, the high-pH elution gave the highest yield of receptor as assayed by radioligand binding. Alternative elution conditions are summarized in *Table 1*.

[a] When Fab antibody affinity columns were used, the Fab antibody fragment was prepared from affinity-purified antibodies (6). It was immobilized at a concentration of 1.5 mg protein/ml resin and the purification procedure was carried out exactly as described here.

Another problem that we have encountered in the purification studies is that the immobilized immunoglobulin leaches from the affinity column. As for the immunoprecipitation studies, this can obscure the detection of antigen–antibody reactions for the less abundant populations of GABA-A receptors.

Table 1. Immunoaffinity purification of α1 subunit-containing GABA-A receptors from mammalian brain[a]

Sample	[^{3}H]Flunitrazepam binding sites (pmol)	Yield of [^{3}H]flunitrazepam binding sites (%)
Detergent extract	114.0 ± 58	100.0
α1 324–341 affinity column filtrate	49.0 ± 10	49.0 ± 16
α1 324–341 affinity column wash	4.0 ± 4	4.0 ± 4
0.14 mM α1 324–341 elution	0.8	1.1
0.5 M KSCN elution	0.1	0.1
3 M KSCN elution	0.1	0.1
pH 11.5 elution	4.9 ± 1.6	4.4 ± 1.2
pH 2.5 elution	0.8	2.5

[a] The table summarizes results obtained for the purification of GABA-A receptors containing an α1 subunit by anti-GABA-A receptor α1 324–341 subunit antibody affinity chromatography. Details of the experimental procedures can be found in ref. 9. Briefly, GABA-A receptors were solubilized from bovine cerebral cortex using Na^+ deoxycholate. The detergent extract was diluted 1:1 with 10 mM potassium phosphate, pH 7.4 containing 0.5 M KCl, 1 mM EGTA, 2 mM magnesium acetate, 10% (w/v) sucrose, 0.02% (w/v) NaN_3 and 0.2% Triton X-100 and applied to the anti-α1 324–341 antibody-affinity column. The column was washed to remove non-specifically bound proteins and then eluted using the conditions given above. Receptor activity was monitored by [^{3}H]flunitrazepam binding sites and the percentage yield calculated with respect to the initial starting activity.

More recently, we have prepared the lower molecular weight Fab antibody fragment and coupled this to sepharose. Alternatively, we have exploited the use of antibodies raised in different species. Other groups have employed a different approach. For example, in their studies of the immunoaffinity purification of non-NMDA glutamate receptors, Wenthold *et al.* (12) coupled the subunit-specific antibodies to protein A–sepharose by chemical cross-linking with the bifunctional reagent dimethyl pimelimidate hydrochloride. This has the added advantage that the immobilized antibody molecule is in the correct orientation for maximum antigen binding. In our studies, the immobilization of the antibody molecules to the affinity resin is not directed, as it is when using protein A, thus all Fab domains may not be accessible for antigen binding.

4.2 Use of antibodies to map ligand binding sites and functional domains of proteins

Again, following the availability of the primary structures of channel proteins, antibodies have been important tools in the localization of their functional domains. Here, the strategy has been once more to raise antibodies to defined regions of the respective channel subunits and to then use these antibodies to antagonize function. We have not used our antibodies for such studies, but the reader is referred to elegant work on the voltage-dependent Na^+ channel (summarized in ref. 13). Using amino acid sequence-directed antibodies, Catterall and colleagues identified residues important for the voltage-dependent binding of α-scorpion toxin to the extracellular domains I and IV of the Na^+ channel by antibody inhibition of radioligand binding (14). Additionally, sequence-directed antibodies identified parts of the intracellular domains III and IV responsible for the inactivation of the Na^+ channel. For these experiments, muscle cells in culture were perfused with the antibody prior to either electrophysiological recordings in the whole-cell clamp configuration (14) or for single-channel analysis (15). Importantly, it was shown that the inhibition of Na^+-channel inactivation by the antibody Ab_{SP19} did not occur at depolarizing potentials, thus demonstrating that the epitope is rendered inaccessible by inactivation and giving an insight into conformational changes undergone by Na^+ channels.

Antibodies are fairly large molecules and in the types of experiments described above it can be difficult to distinguish between steric effects and actual binding site inhibition. Therefore, as in the examples cited above and wherever possible, such studies should optimally use affinity-purified antibodies and preferably the Fab fragments thereof. Also, of course, control experiments using the corresponding non-immune Ig molecule should be carried out.

The mapping of ligand binding sites can also be determined using sequence-directed antibodies in conjunction with photoaffinity labelling studies. We used this approach for the localization of the binding site for the agonist benzodiazepine on the GABA-A receptor. It was known that the benzodiazepine, [^{3}H]flunitrazepam, specifically photoaffinity-labelled the α subunit of the GABA-A receptor (8, 16). The α1 subunit is the most abundant of this subunit class in adult brain and we had made antibodies to three regions of the α1 subunit: α1, 1–15; α1, 324–341; and α1, 424–439 (see *Figure 2*). All of these three antibodies recognized the denatured α1 subunit in immunoblots. Since the complete primary structure of the α1 polypeptide is known from the cDNA sequence, the fragments generated by cyanogen bromide cleavage can be predicted. Using partial and complete cleavage conditions, these peptides were identified using the anti-peptide antibodies. The [^{3}H]flunitrazepam photoaffinity-labelled cyanogen bromide peptides were identified by SDS-PAGE followed by fluorography. Further confirmation of the identity of peptides was obtained through knowledge of the predicted *N*-glycosylation sites of the α1 subunit and by [^{125}I] concanavalin-A overlay experiments. A typical mapping immunoblot is shown in *Figure 5*. From the results, we showed that the benzodiazepine-binding site must lie between residues 59 and 147 of the α1 subunit (17). It was later shown by site-directed

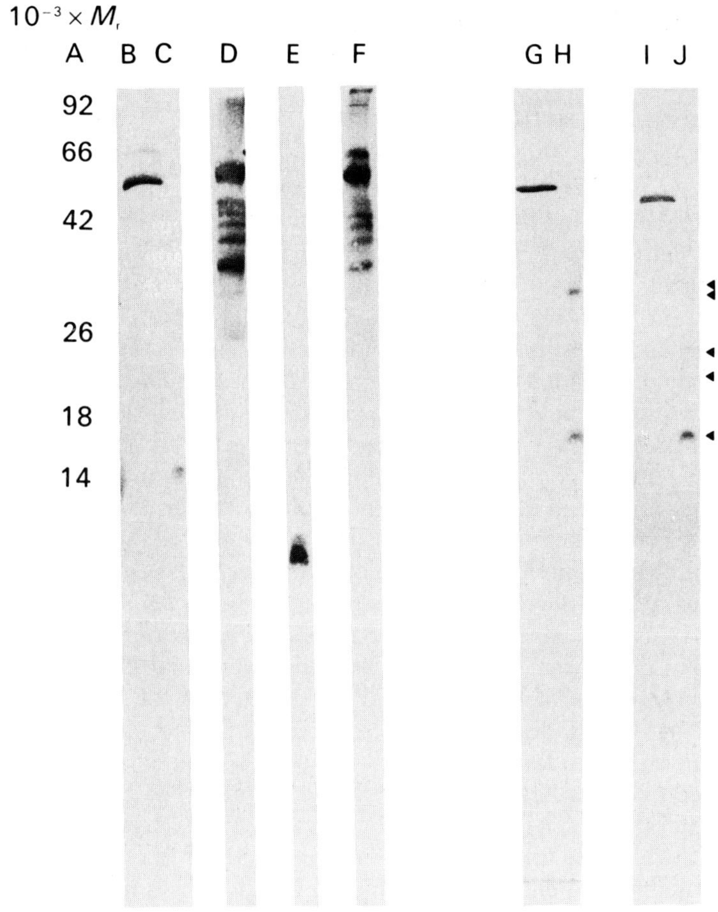

Figure 5. Partial delineation of the [^{3}H]flunitrazepam photoaffinity labelling site of the GABA-A receptor using a combination of immunoblotting with anti-peptide antibodies and fluorography. The purified native GABA-A receptor was photoaffinity labelled with [^{3}H]flunitrazepam and subjected to complete and limited cyanogen bromide (CNBr) chemical cleavage. The resultant peptides were analysed by immunoblotting as described in *Protocol 2* but using the diaminobenzidine detection system (1) after 15.5–20% SDS-PAGE under reducing conditions. Following development, the immunoblots were exposed to [^{3}H]-sensitive film (Amersham) at −80°C. Lanes are: A, pre-stained protein standards; B–F, blotting with anti-α1 1–15 antibodies with B, native purified GABA-A receptor; C, complete CNBr cleavage products; D, limited CNBr cleavage products; E and F, fluorographs of lane C and lane D respectively; G and H, immunoblotting with anti-α1 324–341 antibodies where G is native receptor; H, products of complete CNBr cleavage; I and J, immunoblotting with anti-α1 413–429 antibodies where I is whole receptor; and J, complete CNBr cleavage. The filled triangles denote the positions of five immunoreactive peptides found for both the anti-α1 324–341 and anti-α1 413–429 antibody immunoblots. The position of the photoaffinity labelling site was deduced to lie between residues α1 59–148 by the correlation of the experimental results with the predicted CNBr cleavage products from the α1 subunit primary structure. (Repinted from ref. 17 with permission.)

mutagenesis that histidine 101 (rat α1 sequence numbering) was critical for agonist benzodiazepine binding (18).

4.3 Immunocytochemical studies

I have focused so far on the use of antibodies for biochemical studies of channel proteins, which is our laboratory's expertise. Of course, receptor-subtype-specific antibodies are also extremely powerful tools for the immunocytochemical localization of channel proteins at the cellular and subcellular level. Indeed, there have been many reports of the use of both anti-peptide and anti-fusion protein antibodies for such studies. For example, we have used anti-peptide antibodies for the localization of the GABA-A receptor α1–α6 subunits in adult rat brain (19, 20). In these studies we always used the affinity-purified antibodies and the peptide sequences used were predicted to be extracellular. As in the immunoblotting experiments, controls were protein-A purified non-immune IgG and the preincubation of the antibody and the respective peptide prior to incubation with tissue. The most time-consuming part of these experiments was establishing and thereafter adhering *exactly* to the optimum fixation conditions for a particular antibody. Different batches of affinity-purified antibody from the same animal may vary, as indeed may fixation conditions between laboratories. In our studies, we used free-foating paraformaldehyde-fixed sections, affinity-purified antibodies at final concentrations of 0.025–4 μg/ml, and detection of immune reaction products by horseradish peroxidase/diaminobenzidine staining (19). We found it necessary in the peptide/antibody control experiments, for peptides conjugated to the carrier protein by cysteine, to preincubate with 0.2% (w/v) cysteine in 50 mM Tris, pH 11, to reduce heavy background, non-specific staining. Alternative fixation conditions are described in ref. 6.

Most channel proteins are heteromeric, therefore it is necessary to carry out multiple antibody incubations in colocalization studies for the determination of *in vivo* receptor structures. For these, at the light microscope level, it is useful to have the various specificity antibodies raised in different species concomitant with the use of different fluorescent substrates. At the electron microscope level, these different antibodies can be selectively labelled with different sizes of gold particles (e.g. ref. 20). The caveat with these types of experiments is that although the expression of the gene products themselves are studied rather than the RNA, as for *in situ* hybridization, immunoreactivity may not always be representative of functional, fully assembled receptor proteins (cf. Section 4.1).

5. Summary

In summary, antibodies with predetermined specificity are extremely useful probes for the investigation of channel protein structure and function. However, the production of the required antibody is not trivial. The point emphasized here is that judicious thought in the design of the antigen and thorough characterization of the resultant antibody are necessary in order to obtain unambiguous results. For receptor/channel localization studies, antibodies have the advantage over *in situ* hybridization of RNA in that they measure the gene products themselves, thus circumventing problems with possible post-transcriptional regulation. However, many channel protein subunit genes are alternatively spliced and the non-NMDA glutamate receptor genes are RNA edited. This differential processing gives rise to channels with important functional differences, but with minimal changes in protein primary structure. In the case of RNA editing, there is a single amino acid substitution. For these examples the production of specific antibodies would be extremely difficult and analysis at the DNA level is probably the only option.

Acknowledgements

The work described in this chapter was carried out by members of my research group who, past and present, include Dr Paul L. Chazot, Dr Miroslav Cik, Dr Michael J. Duggan, Simon Pollard, and Dr Christopher L. Thompson. It was supported by the Medical Research Council (UK) and the Science and Engineering Research Council (UK). The immunocytochemical studies were carried out in collaboration with Dr Jon D. Turner, Schering AG, Berlin, Germany. I gratefully acknowledge them all.

References

1. Stephenson, F. A. and Duggan, M. J. (1991). *Molecular neurobiology: a practical approach* (ed. J. Chad and H. Wheal), p. 183. IRL Press, Oxford.
2. Baude, A., Sequier, J.-M., McKernan, R. M., Olivier, K. R., and Somogyi, P. (1992). *Neurosci.* **51**, 739.
3. Tam, J. (1992). *Vaccine Res. Dev.* **1**, 51.
4. Van Regenmortel, H. V., Altschuh, D., and Klug, A. (1986). *Synthetic peptides as antigens: Ciba Foundation symposium 119* (ed. R. Porter and J. Whelan), p. 76. John Wiley, Chichester.
5. Cik, M., Chazot, P. L., and Stephenson, F. A. (1993). *Biochem. J.* **296**, 877.
6. Harlow, E. and Lane, D. (1988). *Antibodies: a laboratory manual.* Cold Spring Harbor Laboratory Press, Cold Spring Harbor, USA.
7. Duggan, M. J. and Stephenson, F. A. (1990). *J. Biol. Chem.* **265**, 3831.
8. Stephenson, F. A., Duggan, M. J., and Pollard, S. (1990). *J. Biol. Chem.* **265**, 21160.
9. Duggan, M. J., Pollard S., and Stephenson, F. A. (1991). *J. Biol. Chem.* **266**, 24778.
10. Pollard, S., Duggan, M. J., and Stephenson, F. A. (1993). *J. Biol. Chem.* **268**, 3753.
11. Smith, D. B. and Johnson, K. S. (1988). *Gene* **67**, 31.

12. Wenthold, R. J. (1991) *J. Biol. Chem.* **267**, 501.
13. Catterall, W. A. (1992). *Physiol. Rev.* **72**, S15.
14. Vassilev, P. M., Schauer, T., and Catterall, W. A. (1988). *Science* **241**, 1658.
15. Vassilev P., Scheuer, T., and Catterall, W. A. (1989). *Proc. Natl Acad. Sci. USA* **86**, 8147.
16. Sigel E., Stephenson, F. A., Mamalaki, C., and Barnard, E. A. (1983). *J. Biol. Chem.* **258**, 6965.
17. Stephenson, F. A. and Duggan, M. J. (1989). *Biochem. J.* **264**, 199.
18. Wieland, H. A., Luddens, H., and Seeburg, P. H. (1992). *J. Biol. Chem.* **267**, 1426.
19. Thompson, C. L., Bodewitz, G., Stephenson, F. A., and Turner, J. D. (1992). *Neurosci. Lett.* **144**, 53.
20. Turner, J. D., Bodewitz, G., Thompson, C. L., and Stephenson, F. A. (1993). *Anxiolytic β-carbolines: from molecular biology to the clinic* (ed. D. N. Stephens), p. 29. Springer, Berlin.
21. Cuatrecasas, P., Wilchek, M., and Anfinsen, C. B. (1968). *Proc. Natl Acad. Sci. USA* **61**, 636.
22. Chazot, P. L., Cik, M., and Stephenson, F. A. (1992). *J. Neurochem.* **59**, 1176.

7

Assembly, post-translational processing, and subcellular localization of ion channels

NEIL S. MILLAR, STEPHEN J. MOSS, and WILLIAM N. GREEN

1. Introduction

Most ion channels are complex, multi-subunit transmembrane proteins (1–3). A consequence of this structural complexity is that the correct functional expression of ion channels requires a diverse series of post-translational events including covalent modifications and subunit oligomerization. After protein synthesis and translocation into the endoplasmic reticulum (ER) membrane, newly-synthesized transmembrane polypeptides may undergo a variety of post-translational modifications (such as proteolytic cleavage, phosphorylation, glycosylation, fatty acylation, and disulphide bond formation). Individual polypeptides must also fold correctly into their native conformation and, in many cases, multiple polypeptide subunits must be accurately assembled into multi-subunit ion-channel complexes. Finally, the modified and assembled ion channels must be delivered to their appropriate subcellular location. Clearly, these subunit folding and oligomerization events must be carried out with great accuracy if ion channels are to perform their appropriate physiological function. How the processing, assembly, and subcellular localization of ion channels is achieved with sufficient accuracy is clearly of great interest and significance and has been the subject of a great deal of recent research.

1.1 Scope of the chapter

We have not attempted to describe all of the experimental techniques which have been used to investigate the assembly, post-translational processing, and localization of ion channels, or to provide a comprehensive review of the literature. Our aim instead has been to describe selected experimental techniques which have become, or are becoming, widely used in such studies. In addition, we have cited selected references to illustrate how these techniques have been used to address specific questions.

We have concentrated on research which has exploited developments in the molecular cloning of ion channels and their expression in heterologous expression systems such as the *Xenopus* oocyte (4) and cultured mammalian cells (5). The cloning and expression of ion channels is described in Chapter 5. Most of the experimental protocols described in this chapter were developed for the characterization of neurotransmitter-gated ion channels, in particular the nicotinic acetylcholine receptor (nAChR) and γ-aminobutyric acid receptor (GABA-AR). However, these experimental techniques should be applicable to the study of other ion channels and transmembrane proteins.

2. Structural organization of ion channels

In this section we review aspects of the structural organization of ion channels which are of relevance to their assembly and post-translational processing.

2.1 Neurotransmitter-gated ion channels

The most extensively studied ion channel (principally because of its great abundance) is the nAChR from the electric organ of the marine ray *Torpedo*. The *Torpedo* nAChR is a pentameric complex of four subunit types (α, β, γ, and δ) with the stoichiometry ($\alpha_2\beta\gamma\delta$). Direct evidence for this structure and stoichiometry has been provided by electron microscopy studies (6). It is generally assumed that other members of the super-family of neurotransmitter-gated ion channels (which includes neuronal and muscle nAChRs, GABA-ARs and ionotropic glutamate receptors) are also pentameric, although this is largely by analogy with the *Torpedo* nAChR. Most members of this diverse super-family, like the *Torpedo* nAChR, appear to form hetero-oligomeric ion channels. In some cases functional homo-oligomeric ion channels can be formed by expression in artificial expression systems (see, for example, ref. 7). There are, however, few convincing examples of homo-oligomeric neurotransmitter-gated ion channels expressed in native cells.

Despite the close homology among the four *Torpedo* nAChR subunits, there is no detectable variation in subunit composition, stoichiometry, or subunit position in the electric organ receptor. The fidelity of subunit assembly is remarkably consistent and suggests that mechanisms exist to ensure the accuracy of subunit assembly. In contrast to the *Torpedo* nAChR, many of the neurotransmitter-gated ion channels present in the central and peripheral nervous system (including neuronal nAChRs) are assembled from several subunit types into a variety of different ion-channel subtypes. For example, at least seven α subunits and four β subunits have been identified for neuronal nAChRs in rat and in other species (8). Since single neurones express multiple neuronal nAChR subunits, a mechanism must exist to select appropriate subunits into the required ion-channel subtypes.

2.2 Voltage-gated calcium and sodium channels

Most voltage-gated Ca^{2+} and Na^{+} channels are oligomeric complexes of several subunits (9, 10). The Na^{+} channel of the electric eel may, however, be an exception to this. The eel Na^{+} channel and the major pore-forming subunit of other Na^{+} or Ca^{2+} channels (the α or α1 subunits) contain four homologous repeated domains, each of which appears to be a 'pseudo-subunit', and these domains are believed to surround the ion-channel pore with four-fold symmetry. In many cases the pore-forming α and α1 subunits are associated with other subunits (for example, the β1 and β2 subunits of rat brain Na^{+} channels, and α2, β, γ, and δ subunits of L-type Ca^{2+} channels).

2.3 Voltage-gated potassium channels

The exact oligomeric composition of voltage-gated K^{+} channels has not been established biochemically. However, the similarity of individual K^{+} channel subunits to the four domains of the Na^{+} channel α subunit and the Ca^{2+} channel α1 subunit suggests that four subunits are required to form a functional K^{+} channel. This interpretation has been supported by electrophysiological data from the expression of mutated K^{+} channel subunits and of artificial cDNA constructs in which the coding sequences of multiple K^{+} channel subunits were linked into a single open reading frame. Evidence for the formation of hetero-oligomeric K^{+} channels has come from both electrophysiological studies and co-immunoprecipitation with subunit-specific antibodies (11, 12).

2.4 Other ion channels

Other families of ion channels (such as intracellular calcium release channels, cyclic nucleotide-gated channels, and gap junctions) will not be dealt with specifically in this chapter. There do, however, appear to be parallels between the assembly and processing of the neurotransmitter-gated and voltage-gated channels and these multisubunit transmembrane ion channels. For example, there is evidence for the assembly of hetero-oligomeric channels for each of these three ion channel families (13–15).

3. Post-translational processing and folding

3.1 Introduction

Several post- (and co-) translational modifications are essential for the generation of functionally active transmembrane polypeptides. These include protein folding, proline isomerization, the formation of hydrogen and disulphide bonds, phosphorylation, glycosylation, fatty acylation, and proteolytic cleavage. As in other sections of this chapter, rather than attempt to provide a comprehensive coverage of these subjects, we describe selected experimental techniques together with examples of the application of these techniques to specific questions and to particular ion-channel proteins.

3.2 Phosphorylation

Protein phosphorylation is widely regarded as being a pivotal mechanism in modulating the functional properties of many cellular proteins. The *in vivo* phosphorylation of several ion channels has been examined and shown to affect a variety of properties including desensitization, ion conductance, subunit assembly, and cell surface clustering (for a review see refs 9, 16).

The low abundance and heterogeneity of ion-channel subunits in their target tissues frequently makes the biochemical analysis of native ion-channel phosphorylation difficult to study. However, the availability of cDNAs encoding ion channel subunits, and of subunit-specific antisera, has greatly enhanced our understanding of ion-channel phosphorylation. By expressing cloned ion channels in mammalian cell lines, phosphorylation can be studied *in vivo* by metabolic labelling with [^{32}P]orthophosphate and immunoprecipitation with subunit-specific antibodies (*Protocol 1* and *Figure 1*). Techniques such as these have been used to study the phosphorylation of several ion channels, for example the nAChR (17), GABA-AR (18), glutamate receptors (19), and Na^+ channels (20).

Protocol 1. Phosphorylation of proteins expressed in mammalian cell lines

Equipment and reagents

- Cultured cells and cell-culture medium[a]
- PBS/Ca^{2+}: phosphate-buffered saline (PBS) containing 0.01% (w/v) $CaCl_2$
- Phosphate-free medium and [^{32}P]orthophosphate
- Lysis buffer[b]
- Pre-immune serum and relevant antisera
- Protein G or protein-A–sepharose
- Ultracentrifuge and low-speed centrifuge, rotary mixer (in cold room), heated water bath, and equipment for SDS-PAGE and fluorography

Method

1. Grow cultured cells expressing the ion channel of interest to near confluence in a 10 cm tissue culture dish[a] in conditions appropriate to the cell type (e.g. Dulbecco's modified Eagle's medium [DMEM] with 10% calf serum in an atmosphere of 5% CO_2 at 37°C).
2. Wash cell monolayer once with 5 ml warmed PBS/Ca^{2+}.
3. Optional: add warmed phosphate-free medium and incubate for 15 min.
4. Add warmed phosphate-free medium (2 ml) and [^{32}P]orthophosphate to a final concentration of 0.5–1 mCi/ml (18–37 MBq/ml) and incubate

cells in normal growth conditions for 4–12 h. If appropriate, add protein kinase activators (e.g. TPA, say 100 nM, for protein kinase C activation or forskolin, say 20–100 μM, for protein kinase A activation).

5. Wash cells in ice-cold PBS/Ca^{2+} (10 ml).
6. Scrape cells into 10 ml of ice-cold PBS/Ca^{2+} or, alternatively, ice-cold lysis buffer[b] without detergent and pellet cells (1000 *g*, 2 min).
7. Resuspend the pellet in 1 ml ice-cold lysis buffer[b] in a screw-cap Eppendorf tube, spin at 100 000 *g*, and transfer to fresh tubes for immunoprecipitation.
8. Pre-clear the lysate by incubation with pre-immune serum (1–100 μl) for 1 h at 4°C.
9. Add 50 μl of a 1:1 slurry of protein-G or protein-A–sepharose (see discussion in Section 4.3) in lysis buffer and incubate for an additional 30 min on a rotary mixer.
10. Centrifuge the lysate at 10 000 *g*, 1 min, and transfer the supernatant into a clean screw-cap tube.
11. Repeat steps 8 and 9 with antisera against the protein of interest.
12. Centrifuge the lysate, discard the supernatant, and wash the protein-G or protein-A–sepharose. Pellet three times with an excess of lysis buffer supplemented with 0.5 M NaCl and then finally with lysis buffer.
13. Resuspend the pellet in 50 μl of 2× sample buffer (see *Protocol 6*), heat to 40°C to elute the precipitated proteins, and load on an SDS-polyacrylamide gel.
14. After electrophoresis, fix the gel for 30 min in 25% methanol/10% acetic acid, dry, and expose to X-ray film at −70°C with intensifying screens.

[a] Transiently transfected cells are typically labelled 48 h after transfection.
[b] To maintain the phosphorylation state of the proteins of interest, it is necessary to use a lysis buffer containing phosphatase inhibitors. The lysis buffer described in *Protocol 5*, supplemented with phosphatase inhibitors: sodium fluoride (50 mM) and sodium pyrophosphate (5 mM), has been used successfully in studies of the nAChR expressed in transfected cells. The buffer includes protease inhibitors where appropriate and detergents for efficient solubilization. An alternative lysis buffer comprises 6 mM Na_2HPO_4, 4 mM NaH_2PO_4, 5 mM EDTA, 5 mM EGTA, 50 mM NaCl, protease inhibitors (0.1 mM PMSE, 10 mg/ml leupeptin, 10 mg/ml antipain, 10 mg/ml pepstatin, 5 mg/ml chymostatin, 0.1 mg/ml aprotinin), phosphatase inhibitors (50 mM NaF, 1 mM sodium orthovanadate, 5 mM sodium pyrophosphate), solubilizing agents (2% (v/v) Triton X-100, 0.5% (w/v) sodium deoxycholate, 0.2% (w/v) SDS). This lysis buffer will efficiently solubilize many membrane proteins and maintain conformational epitopes and subunit interactions. If the antiserum to be used for immunoprecipitation only recognizes denatured epitopes, resuspend the cell pellet from step 6 in a small volume (200 μl) of lysis buffer supplemented with 1% (w/v) SDS, boil for 5 min, and then add five volumes of lysis buffer (containing 2% (v/v) Triton X-100 and 0.5% (w/v) deoxycholate). This treatment will produce a viscous lysate which will require sonication before further treatment, but has been found to give very clean immunoprecipitations.

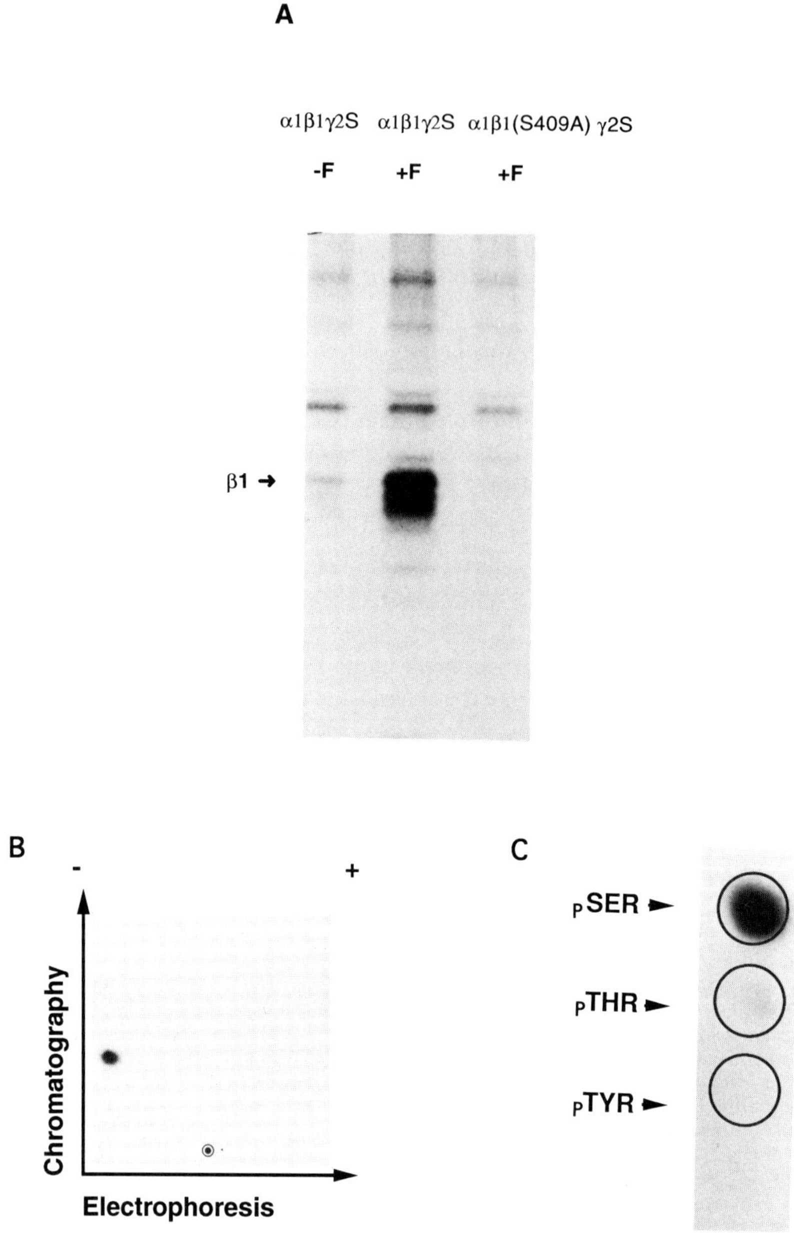

Once subunit phosphorylation of an ion channel has been established, the specific sites of phosphorylation can be determined. Heterologous expression systems allow the precise sites of phosphorylation to be determined by site-directed mutagenesis (see Chapter 5). Typically, putative phosphorylation sites at serine or threonine residues are altered by site-directed mutagenesis

Figure 1. Phosphorylation of GABA-A receptors expressed in HEK 293 cells. Mouse GABA-AR subunits α1, β1, and γ2S were transiently expressed in human embryonic kidney 293 cells and labelled with [^{32}P]orthophosphate (see *Protocol 1*). Cells were then treated with (+F) or without (−F) forskolin (20 μM) for 20 min. GABA-ARs were isolated by immunoprecipitation and separated by SDS-PAGE. The resulting autoradiogram is shown (a). Forskolin treatment resulted in phosphorylation of only the β1 subunit (lane 2). Mutagenesis of a single serine residue (serine 409) in the peptide sequence (^{405}R-R-R-A-*S*-L-Q-K^{412}) to an alanine residue abolished forskolin-stimulated phosphorylation of the β1 subunit (lane 3). The phosphorylation of the β1 subunit was further investigated by tryptic phosphopeptide mapping (b). The area of the gel containing the phosphorylated β1 subunit was excised, digested with trypsin, and the resulting phosphopeptides subjected to two-dimensional phosphopeptide mapping (*Protocol 2*). A single basic phosphopeptide was detected, as shown in (b). Phosphoamino acid analysis (*Protocol 3*) of the β1 subunit clearly indicates the presence of only phosphoserine.

---

to alanine residues, whilst putative tyrosine phosphorylation sites are usually mutated to phenylalanine residues. The functional effects of phosphorylation can then be determined precisely by comparing the properties of wild-type phosphorylated ion channels with mutant non-phosphorylated channels. An inspection of the putative intracellular domains of many ion-channel subunits will reveal numerous potential sites for phosphorylation by a wide variety of both serine and tyrosine protein kinases. Although it may be possible to predict the sites of phosphorylation in this way, there are important limitations to this approach. The existence of an apparent consensus sequence for a protein kinase does not guarantee that a protein can be phosphorylated, nor is it a foolproof indicator of which protein kinase is responsible for phosphorylation (if, indeed, phosphorylation does occur at the site). Therefore, before attempting mutational analysis of ion-channel phosphorylation, it is often desirable to gain information concerning the number of phosphorylation sites and the identity of the phosphoacceptor residues.

Two-dimensional peptide mapping of phosphorylated polypeptide subunits (*Protocol 2*) can provide information about the number of phosphorylation sites and the charge environment surrounding the site of phosphorylation. The phosphoprotein of interest, from an excised SDS-polyacrylamide gel slice (*Protocol 1*), is proteolytically digested and the resultant phosphopeptides separated on the basis of charge by electrophoresis at pH 3.5 then subjected to ascending chromatography (21). At pH 3.5, the migration of the peptide is largely determined by the charge environment around the phosphoacceptor group (22). If proteases are used which have specific cleavage sites (such as trypsin or thermolysin), this analysis provides information on the likely site(s) of protein phosphorylation.

Two-dimensional phosphopeptide mapping will usually demonstrate the number of phosphorylation sites in the protein of interest (unless more than one site is phosphorylated on a single proteolytic fragment). The identity of individual phosphorylated amino acids can be determined by acid hydrolysis and thin-layer chromatography (23). This technique is described in *Protocol 3*.

Protocol 2. Two-dimensional phosphopeptide mapping

Equipment and reagents

- 10% Acetic acid/20% methanol and 50% methanol (all v/v)
- Protease solution: 0.3 mg/ml TPCK-treated trypsin[a] in 50 mM NH_4HCO_3
- Cellulose TLC plate (Kodak 13255); wetting solution: pyridine:acetic acid:H_2O 1:19:89, pH 3.5; chromatography buffer: pyridine: butan-1-ol:acetic acid:H_2O 15:10:3:12; tank; Whatman 3MM paper
- Tracking dyes: 0.5 mg/ml basic fuschin and phenol red in H_2O
- Scintillation counter
- Speed-Vac concentrator/drier (optional)
- 0.2 μm filters and filter manifold (optional)
- High-speed centrifuge
- Water bath or heating block
- High-voltage thin-layer electrophoresis apparatus (e.g. Pharmacia or BioRad)
- X-ray film, cassette, and screens

Method

1. Excise the phosphorylated protein band of interest from a dried SDS-polyacrylamide gel (the band of interest is identified by autoradiography, as described in *Protocol 1*) and determine the amount of incorporated radioactivity by Cherenkov counting.
2. Remove SDS from the gel slice by washing three times for 30 min in 10% acetic acid/20% methanol (50 ml).
3. Wash a further three times for 30 min in 50% methanol (50 ml).
4. Dry the gel slice for one hour at room temperature (e.g. Speed Vac).
5. Digest the phosphoprotein by resuspending the dried gel slice in 1 ml protease solution[a] and incubating at 37°C for 20 h.
6. Add an additional 0.5 ml of protease solution and incubate at 37°C for a further 2 h.
7. Filter the supernatant through a 0.2 μm filter and dry the sample.
8. Determine the recovery of radioactivity (usually in excess of 90%) by Cherenkov counting.
9. Dissolve the dried sample in 10–20 μl of water by heating at 65°C for 2 min.
10. Centrifuge at 10000 *g* (max) for 5 min.
11. Spot the sample (200 c.p.m. is sufficient for an overnight exposure) on to the centre of a 20 cm^2 cellulose TLC plate, 4 cm up from one edge (see *Figure 2*). The sample should be added in 0.5 μl aliquots and allowed to dry between sample applications. Also spot 1 μl samples of the tracking dyes.
12. Wet the TLC plate with wetting solution. This is best achieved by

placing wetting-solution-soaked filter paper, out of which a 2 cm circle has been cut, over the area of sample application on to the TLC plate. Add further wetting solution from the outside edges of the filter paper. All areas of the plate should be wet, but puddles should be avoided.

13. Place the TLC plate on a high-voltage thin-layer electrophoresis apparatus and electrophorese at 500 V in the wetting solution until the tracker dyes are within 4 cm of the plate edge. This will take approximately 1 h. Remove the plate and allow to dry completely in a fume hood.
14. Subject the dried TLC plate to ascending chromatography (see *Figure 2* for orientation). Prepare a conventional TLC tank by soaking two wicks of filter paper in chromatography buffer and add 200 ml of the buffer to the tank. It is advisable to leave the tank overnight to equilibrate. Allow the chromatogram to run until the buffer reaches 4 cm from the top of the plate.
15. Dry the TLC plate in a fume hood, cover completely with cling film, and expose to X-ray film at −70°C with an intensifying screen.

[a]TCPK is an inhibitor of chymotrypsin, a common contaminant of trypsin preparations. Alternatively, the phosphoprotein can be digested with thermolysin (0.3 mg/ml) supplemented with 2 mM $CaCl_2$.

Protocol 3. Phosphoamino-acid analysis

Equipment and reagents

- 10% Acetic acid/20% methanol and 50% methanol (all v/v)
- Protease solution: 0.3 mg/ml TPCK-treated trypsin[a] in 50 mM NH_4HCO_3
- Cellulose TLC plate (Kodak 13255); wetting solution: pyridine:acetic acid:H_2O 1:19:89, pH 3.5; tank; Whatman 3MM paper
- Tracking dye: 0.5 mg/ml phenol red in H_2O
- Scintillation counter
- Speed-Vac concentrator/drier (optional)
- 0.2 μm filters and filter manifold (optional)
- Heating block or oven set at 105°C
- High-voltage thin-layer electrophoresis apparatus (e.g. Pharmacia or BioRad)
- X-ray film, cassette, and screens
- 6 M HCl
- N_2 gas
- Sample solution: formic acid:acetic acid: H_2O 1:10:89
- Phosphoamino-acid standards: 10 mg/ml each of phosphoserine, phosphothreonine, and phosphotyrosine
- 1% (w/v) ninhydrin

Method

1. Process an SDS-PAGE gel slice containing the phosphoprotein of interest as described in *Protocol 2* as far as step 8.
2. Resuspend the dried gel slice in 1 ml of 6 M HCl and transfer to a 15 ml glass tube with a screw-cap fitting. Blow nitrogen gas over the sample, cap the tube, and incubate at 105°C for 1 h.

Protocol 3. *Continued*

3. Dry the sample (e.g. Speed Vac).
4. Resuspend in 10 μl sample solution and spot the sample on to a TLC plate as described above (*Protocol 2*, step 11). Spot the sample 4 cm from the bottom edge and side (up to four different samples can be applied to the same plate, leaving 4 cm between samples). Add 1 μl of non-radioactive phosphoamino-acid standards and 1 μl of tracking dye to each sample. Wet the TLC plate by overlaying with filter paper soaked in the sample solution, transfer the plate to a TLC electrophoresis apparatus, and electrophorese at 500 V until the phenol red migrates 5 cm from the origin (as described in *Protocol 2*).
5. Remove the TLC plate, soak in wetting solution (see *Protocol 2*, step 12) and electrophorese at 500 V until the phenol red has migrated a further 7 cm (see *Figure 2*).
6. Dry the TLC plate in a fume hood.
7. Submerge the plate in 1% ninhydrin dissolved in acetone and dry again. As the plate dries purple spots will appear where the phosphoamino-acid standards are.
8. Circle the phosphoamino-acid standard spots lightly with pencil, cover with cling film and expose the TLC plate to X-ray film at −70°C with intensifying screens to detect the position of the radioactive phosphoamino acids.

3.3 Glycosylation and fatty acylation

The role of glycosylation in ion-channel expression and function can be investigated *in vivo* by the use of inhibitors of glycosylation such as the antibiotic tunicamycin. Glycosylation at asparagine residues can be inhibited by the addition of tunicamycin directly to cell growth medium (typically 1–10 μg/ml for several hours). The effect of tunicamycin on protein glycosylation can be observed by changes in the apparent molecular weight of proteins on polyacrylamide gel electrophoresis (see *Figure 3*). Tunicamycin has been used to study the glycosylation of several ion channels (e.g. 24, 25).

The use of endoglycosidases can provide information on the carbohydrate composition of proteins. Simple asparagine-linked carbohydrate side chains (which are added to proteins in the ER) are sensitive to treatment with both endoglycosidase H and F, whereas complex-type carbohydrates (modified in the medial stacks of the Golgi apparatus) are insensitive to endoglycosidase H but remain sensitive to endoglycosidase F. This differential sensitivity of carbohydrate side chains to endoglycosidases has been used to identify the subcellular location of assembled and unassembled nAChR subunits (26). The effect of *in vitro* endoglycosidase F treatment on GABA-ARs purified from transiently transfected cells is shown in *Figure 3*.

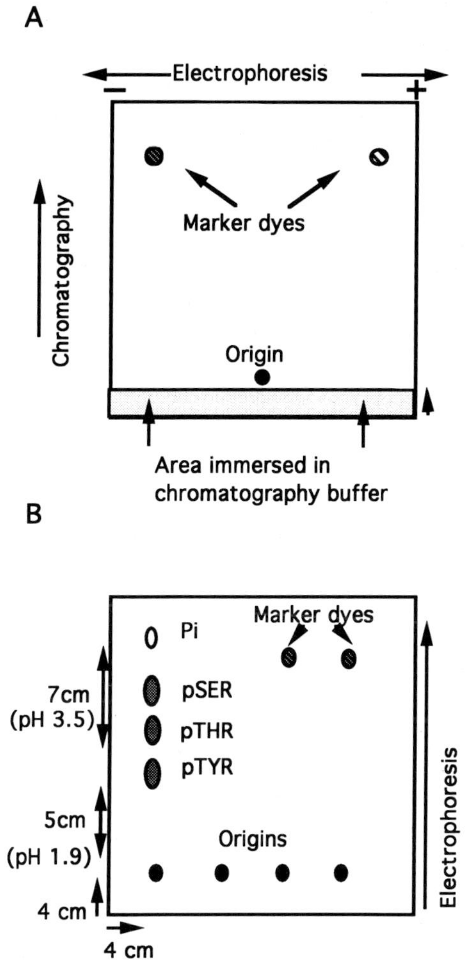

Figure 2. Phosphopeptide mapping and phosphoamino acid analysis. (a) Sample position and TLC plate orientation for two-dimensional peptide mapping (*Protocol 2*). The position of the origin and the migration of the marker dyes during electrophoresis is shown. After electrophoresis the plate was allowed to dry and subjected to ascending chromatography in the direction shown. (b) Sample position and TLC plate orientation for phosphoamino acid analysis (*Protocol 3*). The samples and standards were applied to the origins 4 cm from the bottom edge (up to four samples can be run on the same TLC plate). The samples were electrophoresed in formic acid:acetic acid solution (pH 1.9) for 5 cm, followed by electrophoresis in pyridine:acetic acid solution (pH 3.5) for 7 cm, dried, and then soaked in 1% (w/v) ninhydrin, as described in *Protocol 3*. The relative positions of phosphoserine (pSER), phosphothreonine (pTHR), and phosphotyrosine (pTYR) are shown.

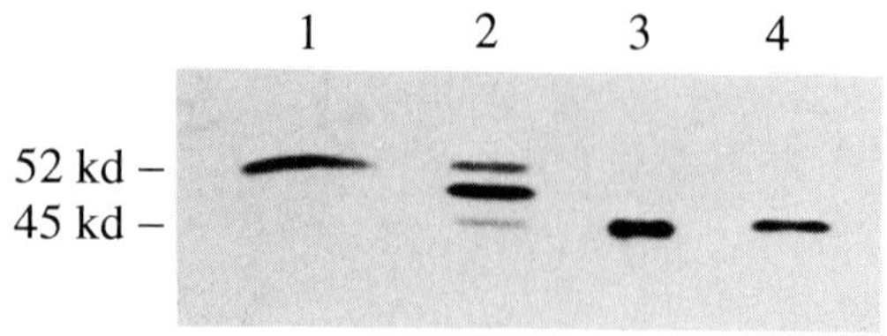

Figure 3. Effect of tunicamycin and endoglycosidase treatment on the glycosylation of GABA-AR subunits. Mouse GABA-AR subunits α1, β1, and γ2S were transiently expressed in human embryonic kidney 293 cells. Transfected cells were grown for 18 h either in the absence (lane 2) or presence (lane 3) of tunicamycin (10 μg/ml). The cells were then harvested, lysed by sonication, separated by SDS-PAGE, and transferred to a nitrocellulose membrane. The α1 subunit was visualized by Western blotting using a rabbit polyclonal antiserum against the α1 subunit. In addition, membranes from transfected cells which had not been treated with tunicamycin were treated with endoglycosidase F for 18 h prior to SDS-PAGE and Western blotting (lane 3). Lane 1 contains GABA-ARs purified from mouse brain.

Site-directed mutagenesis (see Chapter 5) provides an alternative approach to the study of glycosylation and enables the importance of glycosylation at specific sites to be investigated. The role of asparagine-linked glycosylation in the assembly, surface expression, and functional properties of the nAChR have been studied extensively by the elimination of specific glycosylation target sites by mutagenesis (e.g. 27, 28, 29).

Some ion-channel polypeptides are modified by the covalent addition of fatty acids, which may help to anchor or to orientate subunits into lipid membranes. Radioactively labelled fatty acids can be incorporated into polypeptides expressed in cultured cells and then the radioactively labelled proteins identified by solubilization, immunoprecipitation, SDS-PAGE, and autoradiography (*Protocols 4–6*). This approach has been used to investigate the incorporation of [^{3}H]palmitate into nAChRs in a muscle cell line (30) and into rat brain Na^+ channels in cultured neurones (31).

3.4 Disulphide bond formation

The presence of disulphide bonds within a protein can usually be demonstrated by the difference in migration of a protein on SDS-PAGE in the presence and absence of a reducing agent such as dithiothreitol (10 mM) in the gel-loading buffer (see footnote to *Protocol 6*). Site-directed mutagenesis (see Chapter 5 for details) has been used to investigate the role of specific disulphide bonds in various aspects of ion-channel expression and function, as has been described for studies of both phosphorylation and glycosylation (see Sections 3.2 and 3.3). For example, by combining site-directed mutagenesis with metabolic labelling (*Protocol 4*) and immunoprecipitation (*Protocol 6*), it has been possible to identify the role of particular disulphide bonds in the ligand binding, assembly, and cell surface expression of the nAChR (28, 29, 32).

In contrast to the generally rapid formation of intra-chain disulphide bonds, the formation of disulphide bonds between two polypeptide chains may be a later event in protein maturation. Immunoprecipitation studies of Na^+ channels in rat primary embryonic neuronal cultures indicates that only a third of the α subunits are disulphide bonded to β2 and that only disulphide linked α and β2 subunits appear on the cell surface (33).

3.5 Protein folding

Many of the experimental techniques which have found widespread use in the study of protein folding (e.g. circular dichroism, nuclear magnetic resonance spectroscopy) have been applied principally to soluble proteins *in vitro*. Nevertheless, techniques have been developed which enable some aspects of the folding of membrane proteins to be followed *in vivo*.

The early events in the folding of the nAChR α subunit have been studied by following the formation of both ligand binding sites and conformational epitopes for monoclonal antibodies. Immediately after synthesis the α subunit is unable to bind either acetylcholine or α-bungarotoxin, a peptide antagonist of acetylcholine binding. Pulse-chase metabolic labelling (*Protocol 4*) of the muscle nAChR expressed in cultured cells has shown that the α subunit does not acquire the ability to bind α-bungarotoxin until 15–30 min after synthesis (34). The kinetics of the process by which the *Torpedo* nAChR acquires a conformation which is able to bind either α-bungarotoxin or antibodies which recognize conformation-dependent epitopes have been followed by immunoprecipitation of metabolically labelled proteins (see *Protocols 4* and *6* and ref. 35).

The ER contains several chaperone proteins (of which the best characterized is the immunoglobulin-binding protein, BiP), which recognize and stabilize intermediates during polypeptide folding and oligomerization (36). The interaction of the *Torpedo* and mouse muscle nAChR subunits with BiP has been investigated in cultured cell lines by co-immunoprecipitation (37–39).

4. Subunit assembly

Many of the fundamental questions concerning the assembly of individual polypeptide subunits into multi-subunit complexes such as ion channels can be investigated by experimental techniques such as metabolic labelling, together with immunoprecipitation and sucrose-gradient centrifugation. These techniques are described here, together with selected examples of how they have been applied to particular aspects of ion-channel assembly. In particular, we have focused upon studies which have combined these techniques with molecular cloning and expression (described in Chapter 5).

4.1 Metabolic labelling of proteins expressed in mammalian cells

Many ion channels are expressed at relatively low levels in their native tissues or in artificial expression systems. It is, therefore, frequently necessary to use a radioactive label to aid detection of ion channel proteins. Ion channel subunits can be radioactively labelled by a variety of methods, for example by the use of specific radioligands, or by the direct incorporation of radioactive amino acids into newly synthesized polypeptides. Perhaps the most versatile technique is to 'pulse-chase' label proteins during protein synthesis *in vivo* (*Protocol 4*). Newly synthesized polypeptides can be labelled by the addition of one or more radioactive amino acids (e.g. [^{35}S]methionine or [^{3}H]leucine) to the cell culture medium. Typically, this is done for a short 'pulse' which is then followed by a 'chase' period of variable length in the absence of radioactive amino acids. In this way, the proteins synthesized during the pulse-labelling can be followed over a period of time as they are processed, assembled, and directed to their appropriate cellular location. An alternative approach is to give an extended labelling and no chase. This will maximize the amount of radioactively labelled protein and will label all 'pools' (intracellular unassembled subunits, intracellular assembled subunits, and cell-surface assembled subunits).

Protocol 4. Metabolic labelling

Reagents

- Cultured cells and PBS/Ca^{2+} (see *Protocol 1*)
- Amino-acid deficient medium and relevant radiolabelled and unlabelled amino acid(s)

Method

1. Grow cultured cells to confluence in a 10 cm tissue culture dish in conditions appropriate to the particular cell type (e.g. Dulbecco's modified Eagle's medium [DMEM] with 10% calf serum in an atmosphere of 5% CO_2 at 37 °C).
2. Wash the cell monolayer once with 5 ml warmed PBS/Ca^{2+}.
3. Add 5 ml warmed medium lacking the amino acid which is to be used for the metabolic labelling (e.g. methionine-free (met^-) DMEM). Incubate cells in normal growth conditions (e.g. 5% CO_2 and 37 °C) for 15 min to starve cells of the appropriate amino acid.
4. Remove medium and replace with warmed met^- medium (2 ml) and 333 μCi (12 MBq) [^{35}S]methionine and incubate in normal growth conditions for the desired time, usually 5–60 min. Ensure that the incubator shelf is level to avoid cells drying out.

5. Stop the labelling by washing three times with pre-warmed culture medium containing 5 mM methionine. Add additional 'cold' methionine to dilute any remaining [^{35}S]methionine.
6. Incubate cells in pre-warmed culture medium containing methionine at 5% CO_2 and the appropriate temperature for the desired chase time.

4.2 Cell lysis and solubilization of ion channel proteins

Adequate cell lysis and ion-channel solubilization are critical for many of the experimental techniques described in this chapter. An appropriate solubilizing agent may need to be selected empirically for a particular ion-channel type and may depend upon, for example, whether it is necessary to maintain the native protein conformation and subunit interactions. The solubilizing agent must be strong enough to ensure the quantitative release of the ion channels, yet mild enough to preserve interactions between subunits (for studies of subunit assembly) and to maintain the correct conformation of the antibody epitope (for immunoprecipitation studies with antibodies directed against conformational epitopes). Alternatively, if an antibody is being used which only recognizes a denatured (linear) epitope, harsher solubilization conditions would normally be appropriate. The use of antibodies directed against denatured proteins is discussed further in *Protocol 1*.

The importance of using sufficiently mild conditions to maintain subunit interactions has been illustrated by studies of subunit assembly performed with the *Torpedo* nAChR (35). This investigation demonstrated that nAChR assembly intermediates (although not mature pentameric nAChRs) dissociate in 1% Triton X-100, but that these subunit interactions are preserved if a milder detergent (Lubrol PX) together with phospholipid (phosphatidylcholine, PC) were used. This solubilizing agent (LPC) is composed of 1.83 mg/ml PC and 1% (w/v) Lubrol, resulting in a molar ratio of PC to Lubrol of approximately 1:7. A more extensive discussion of the uses of different detergents for protein solubilization can be found in a companion book in this series (40).

Protocol 5. Cell lysis and solubilization of ion channels

Equipment and reagents

- Cultured cells and PBS/Ca^{2+} (see *Protocol 1*)
- Rubber policeman (cell scraper)
- Eppendorf-type microcentrifuge tubes and microcentrifuge capable of reaching 10–12 000 r.p.m. (in cold room)
- Vortex mixer
- Crushed ice
- Lysis buffer: 150 mM NaCl, 5 mM EDTA, 50 mM Tris–HCl pH 7.4, 0.02% (w/v) NaN_3, protease inhibitors (e.g. 2 mM PMSF[a], 2 mM *N*-ethylmaleimide), solubilizing agent (e.g. 1% (v/v) Triton X-100)

Protocol 5. *Continued*

Method

1. After the pulse-chase labelling (*Protocol 4*), wash the cell monolayer twice with 5 ml ice-cold PBS/Ca^{2+}. Remove the final wash by aspiration.
2. Dislodge the cells by scraping the dish with a rubber policeman and harvest the cell suspension by tilting the plate. Load the cells into a 1.5 ml Eppendorf tube. Harvest any remaining cells by washing with ice-cold PBS/Ca^{2+} (400 μl).
3. Pellet cells by centrifugation at 1000 *g* for 3 min and resuspend by vortexing in 100–500 μl ice-cold lysis buffer.
4. Store resuspended cell lysates on ice for several hours, vortexing occasionally. We have found that the long incubation on ice minimizes contamination of the supernatant (after step 5) with actin which can otherwise co-precipitate in subsequent immunoprecipitation experiments.
5. Pellet any non-solubilized material by centrifugation at 10 000 *g* for 10 min at 4°C and transfer the cell lysate to a fresh Eppendorf tube.

[a] AEBSF may be a possible less toxic alternative to PMSF (Editor's note).

4.3 Immunoprecipitation

Immunoprecipitation is a commonly used technique for the isolation and purification of proteins and is used frequently in studies of ion-channel assembly and post-translational modification. Monoclonal and polyclonal antibodies which are specific for single subunits of a hetero-oligomeric ion-channel protein complex can be generated by standard techniques (see Chapter 6). *Figure 4* shows an autoradiograph of an SDS-polyacrylamide gel in which all four (α, β, γ, and δ) [^{35}S]methionine-labelled *Torpedo* nAChR subunits have been immunoprecipitated by subunit-specific antibodies.

One of the key tests for subunit assembly is co-immunoprecipitation. Under conditions that do not dissociate subunit complexes, subunits are assumed to be specifically associated when antibodies specific for one subunit co-precipitate other subunits. In order to assay for co-immunoprecipitation, subunits must be labelled or tagged, for example by metabolic labelling with [^{35}S]methionine (see Section 4.1). After the addition of a subunit-specific antibody to an appropriately solubilized complex (see Section 4.2), the antigen–antibody complex is precipitated by protein G or protein A conjugated to a large sugar molecule such as agarose or sepharose. Both protein G and protein A bind to the constant regions of the heavy chain of a variety of antibodies. Protein G is more versatile than protein A in its binding since it recognizes antibodies from a larger variety of species (including most IgG types in rat, which are not recognized by protein A). For more details of the uses of proteins G and A, see ref. 41.

Immunoprecipitation can be used to isolate and estimate the sizes of different subcellular ion channel 'pools' (e.g. cell surface, intracellular, or total). For example, with metabolically labelled cells (*Protocol 4*) expressing the *Torpedo* nAChR, it is possible to isolate independently the surface (assembled), intracellular (assembled), and intracellular (unassembled) nAChR pools. Surface-assembled nAChR can be isolated by binding α-bungarotoxin to intact cells, followed by cell lysis (*Protocol 5*) and immunoprecipitation with anti-bungarotoxin antisera. Intracellular-assembled nAChR can be isolated by immunoprecipitation of cell lysates (from which surface nAChRs have been immuno-depleted) with an antibody which recognizes assembled nAChR, and finally, unassembled intracellular subunits can be immunoprecipitated with subunit-specific antisera (42).

Protocol 6. Immunoprecipitation

Equipment and reagents

See also *Protocol 2* (for cell lysis) and *Protocol 5* (pre-clearing the lysate)

- Cell-lysis buffer (see *Protocols 2* and *5*)
- Protein G or protein A–sepharose
- Specific antibodies
- Sample buffer and tracking dye (2 × SBTD): 4% (w/v) SDS, 20% (v/v) glycerol, 0.125 M Tris–HCl (pH 6.8), reducing agent (optional[a]) –10 mM DTT, and tracking dye–0.01% (w/v) bromophenol blue
- Rotary mixer (in cold room)
- Vortex mixer
- Microcentrifuge (in cold room)
- Aspirator (use a 23 gauge needle attached to a vacuum aspirator)
- Equipment and reagents for SDS-PAGE
- Equipment and reagents for fluorography or immunoblotting (depending on detection method)

Method

1. (Optional) If using polyclonal antibodies, pre-clear the cell lysate (see *Protocol 5*) by addition of pre-immune serum (10–50 μl) and rotate at 4°C. After 1 h add 50 μl protein G or protein A–sepharose solution (see discussion above) resuspended 1:1 in lysis buffer (*Protocol 5*) and rotate again at 4°C for 1 h. Spin for 30 sec and remove supernatant. For monoclonal antibodies, add protein-G or protein A–sepharose solution (50 μl), rotate at 4°C for 15 min, spin for 30 sec, and remove supernatant.
2. Add ion-channel specific (or individual subunit-specific) antibody or antibodies to the cell lysate (see *Protocol 2*). The amount of antibody will depend on the amount of antigen in the lysate and the affinity of the antibody for the antigen. An estimate can be made by titrating the antibody against a constant amount of the protein (ion-channel) antigen. Rotate at 4°C for 3–6 h.
3. Add protein G or protein A–sepharose solution (50 μl) and rotate at 4°C for a further 2–3 h.

Protocol 6. *Continued*

4. Spin for 20 sec and aspirate supernatant.
5. Add 700 μl lysis buffer (see *Protocol 5*), vortex, spin for 20 sec, and aspirate supernatant. Repeat three more times.
6. After the final spin, remove as much of the supernatant as possible and add 35 μl 2 × sample buffer and tracking dye (SBTD). Incubate at room temperature for 10 min, vortexing occasionally. Samples may be boiled to decrease aggregation, but this can lead to problems with protein degradation (this is a problem, for example, with the nAChR γ subunit). Spin for 1 min to pellet debris, remove supernatant, taking care not to remove any of the sepharose, and either store at −20°C or load the supernatant directly on to an SDS-polyacrylamide gel (a 7.5% polyacrylamide gel has been found to give good separation of the four *Torpedo* nAChRs, see *Figure 4*).
7. If the immunoprecipitated subunits are radioactively labelled with weak β-emitting isotopes such as ^{35}S, ^{14}C, or ^{3}H (see *Protocol 1*), they can be detected by fluorography: fix the gel for 30 min in 25% methanol 10% acetic acid, immerse the gel in an appropriate fluorographic reagent, e.g. Amplify (Amersham) for 30 min, dry, and expose to X-ray film at −70°C. Alternatively, non-radioactive proteins can be detected by Western blotting.

[a] The addition of a reducing agent such as dithiothreitol (DTT) to the sample buffer (e.g. 200 mM in the 2 × SBTD, 10 mM final concentration) will reduce both inter-subunit and intra-subunit disulphide bonds. The DTT should be added to the sample buffer immediately prior to use (a 1 M stock of DTT should be prepared in 0.01 M sodium acetate, pH 5.2, and stored in aliquots at −20°C). The decision to add DTT will depend on whether or not one wants disulphide bonds eliminated. A possible disadvantage of adding DTT is that the disulphide bonds between the heavy and light chains of the immunoprecipitating antibodies will also be reduced. This can lead to overloading of proteins in the region where the immunoglobulin heavy chain migrates (approximately 50 kDa), leading possibly to sample band distortion. This is a particular problem if the proteins are to be Western blotted due to the anti-IgG second antibody to the 50 kDa immunoglobulin band.

4.4 Sucrose gradient centrifugation

Sucrose gradient centrifugation is a widely used technique for estimating the size of polypeptides and protein complexes and for following the assembly of multi-subunit proteins. For example, it is possible to separate and resolve unassembled nAChR subunits from fully assembled (pentameric) complexes and even partially assembled subunit complexes by sedimentation of the metabolically labelled and solubilized ion channel on an appropriate (5% to 20%) sucrose gradient (35). An example of a sucrose gradient profile showing fully assembled, cell surface, *Torpedo* nAChR is shown in *Figure 5*.

An estimate of the sedimentation coefficient of a solubilized protein (ion

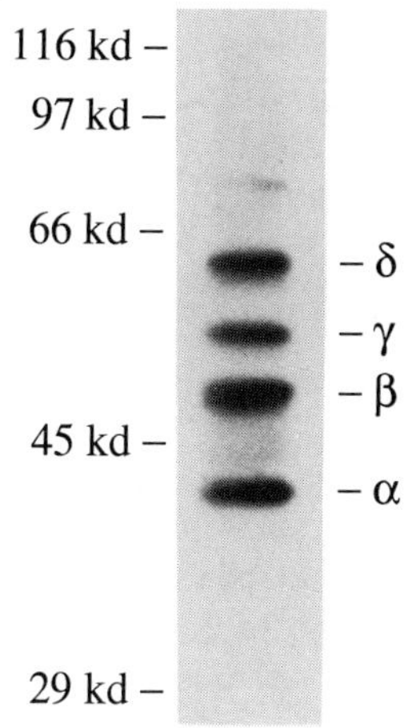

Figure 4. Immunoprecipitation of the four (α, β, γ, and δ) [^{35}S]methionine labelled *Torpedo* nAChR subunits. A single 10 cm culture of mouse fibroblast cells expressing *Torpedo* nAChR α, β, γ and δ subunits (the 'All-11' cell line, see ref. 5) was incubated in DMEM containing 0.33 μCi [^{35}S]methionine for 15 min (*Protocol 4*). The cells were solubilized in lysis buffer containing 1% (v/v) Triton X-100 (*Protocol 5*) and the subunits were precipitated with subunit-specific antibodies (*Protocol 6*): polyclonal serum against the α and δ subunits, and monclonal antibodies MAb 148 (anti-β) and MAb 168 (anti-γ). Precipitated subunits are shown after electrophoresis on 7.5% (w/v) SDS polyacrylamide gels. The subunits α, β, γ, and δ migrate at molecular weights of approximately 40, 50, 60, and 65 kDa, respectively, consistent with the protein standards: of 29 kDa (carbonic anhydrase), 45 kDa (egg albumin), 66 kDa (bovine plasma albumin), 97.4 kDa (phosphorylase B), and 116 kDa (β galactosidase).

channel) sample can be determined by sedimentation through a linear sucrose gradient. Sedimentation coefficient values can be obtained by running protein standards in parallel. The units of sedimentation coefficient are seconds but, since values in the region of 1×10^{-13} sec are common, the Svedberg unit (S) is more commonly used (1 Svedberg unit = 1×10^{-13} sec).

Protocol 7. Sucrose gradient centrifugation

Equipment and reagents

- Cell-lysis buffer (see *Protocol 5*)
- 5% (w/v) and 20% (w/v) sucrose in cell-lysis buffer
- 5 ml Linear gradient former (e.g. Hoefer SG5) and peristaltic pump (e.g Ranin Rabbit)
- Access to an ultracentrifuge with a swing-out rotor (e.g. Beckman SW 50.1 or similar)
- Automated fraction collector (e.g. BioRad 2110) (optional)

Method

1. Prepare 5% (w/v) and 20% (w/v) sucrose solutions in lysis buffer with an appropriate solubilizing agent (see *Protocol 5*).

Protocol 7. *Continued*

2. Connect the outlet of a 5 ml linear gradient maker to a peristaltic pump by silicone tubing. Place a magnetic bar in the chamber closest to the outlet of the gradient maker (the mixing chamber) and place the gradient maker on a magnetic stirrer.
3. With the taps on the outlet and between the two chambers of the gradient maker closed, add 2.7 ml of the 5% (w/v) sucrose solution to the chamber closest to the outlet (the mixing chamber), and 2.7 ml of the 20% (w/v) sucrose solution into the chamber furthest from the outlet.
4. While stirring the 5% (w/v) sucrose chamber, open the two taps and switch on the peristaltic pump (a flow rate of about 50–100 μl/sec is suitable). Pour a 5 ml linear sucrose gradient into an ultracentrifuge tube (e.g. 13 × 51 mm polyallomer Beckman No 326819 tube). The end of the silicone tubing is placed just above the bottom of the centrifuge tube in a fixed position and removed carefully after the gradient has been poured. Often a jack is used to raise and lower the centrifuge tube (on a rack) while the silicone tubing is maintained in its fixed position.
5. Layer the cell lysate (*Protocol 5*) on to the top of the gradient. The volume of the lysate should be no more than 250 μl for a 5 ml gradient. If larger lysate volumes have to be analysed, a larger gradient is needed.
6. Pre-chill the rotor and centrifuge to 4°C. Centrifuge for an appropriate time[a] at 4°C in a swinging bucket ultracentrifuge rotor.
7. Remove sample fractions of an appropriate volume (e.g. 100–350 μl) to give the desired number of fractions. This can be done either by careful removal by hand from the top of the gradient (with, for example, a Gilson P1000 pipette) or from the bottom of the tube using an automated fraction collector.

[a] Centrifugation at 40 000 r.p.m. to $\omega^2 t = 9.0 \times 10^{11}$ (14.25 h in a Beckman SW 50.1) has been found to be suitable for analysis of nAChR subunit monomer and assembled subunit complexes.

The distribution of proteins within the sucrose gradient can be determined either by immunoprecipitation of metabolically labelled proteins (*Protocols 4–6*) followed by separation by SDS-PAGE and autoradiography or by use of immunoprecipitation together with specific radioligands.

Immunoprecipitation of metabolically labelled proteins from individual sucrose-gradient fractions provides information about the subunit composition of ion-channel complexes within each sucrose-gradient fraction. For example, immunoprecipitation of fractions containing fully assembled *Torpedo* nAChR with an α-subunit-specific antibody will co-precipitate the β, γ, and

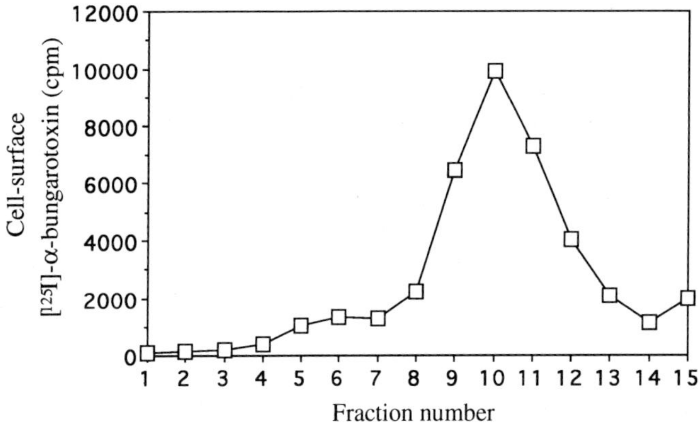

Figure 5. Sucrose gradient sedimentation profile of cell-surface expressed *Torpedo* nAChR. A single 10 cm culture of mouse fibroblast cells expressing *Torpedo* nAChR α, β, γ, and δ subunits (the 'All-11' cell line, see ref. 5) was cell-surface labelled with 4 nM of [^{125}I]α-bungarotoxin and solubilized in lysis buffer containing 1% Triton X-100 (*Protocol 5*). The cell lysate was layered on a 5 ml 5–20% (w/v) linear sucrose gradient made from 1% (v/v) Triton X-100 lysis buffer and either 5 or 20% (w/v) sucrose. Gradients were centrifuged in a Beckman SW 50.1 rotor at 40 000 r.p.m. to $\omega^2 t = 9.0 \times 10^{11}$. Approximately 15.5 fractions (320 μl) were collected from the top of the gradient and the ^{125}I counts measured on a gamma counter. Most of the [^{125}I]α-bungarotoxin counts migrate in a single peak at 9S, precisely where the native monomeric form of the *Torpedo* nAChR migrates (5).

δ subunits. If an essentially irreversible radioligand is available with which the ion channel can be pre-incubated and which will remain bound during sedimentation (such as [^{125}I]α-bungarotoxin in the case of the *Torpedo* and muscle nAChRs) the bound radioligand can be assayed by counting the radioactivity within each fraction (see *Figure 5*). This approach has the advantage that fully assembled cell-surface nAChRs may be isolated by binding [^{125}I]α-bungarotoxin to intact cells prior to receptor solubilization.

If a suitable irreversible radioligand is unavailable it may be possible to perform an immuno-immobilization assay (43) with radioligands with faster off-rates, for example [^{3}H]nicotine binding to the neuronal nAChR (44). In this technique proteins within each fraction are preincubated with an appropriate antibody together with protein G–sepharose (or a second antibody coupled to sepharose) and then incubated with an appropriate radioligand. Bound radioligand is separated from unbound by rapid centrifugation and washing of the radioligand–protein–antibody–sepharose complex and samples counted by scintillation counting (for further details see ref. 43).

4.5 Overview of approaches to the study of subunit assembly

The ability to express functional multi-subunit (hetero-oligomeric and homo-oligomeric) ion channels from cloned genes in heterologous expression systems (reviewed in Chapter 5), together with techniques such as those described in this chapter, has been critical to the investigation of many aspects of ion channel assembly. Selective examples of the many diverse approaches to the study of ion channel assembly are described below.

Specific regions and amino acids which are important in determining the specificity of subunit interactions are beginning to be identified by techniques in which artificially modified subunits are expressed. The expression of truncated muscle nAChR subunits together with intact subunits has demonstrated that recognition signals for subunit interactions are present within the extracellular N-terminal domains (45). The importance of the N-terminal domain in subunit assembly has also been demonstrated by the expression of chimeric β/ϵ nAChR subunits (46). Individual amino acids have been identified in the α and ϵ subunits of muscle AChR which are important in the efficiency of subunit assembly by site-directed mutagenesis and the expression of chimeric subunits (47, 48). A similar approach using truncated and chimeric K^+ channels suggests that the intracellular hydrophilic amino-terminal region of K^+ channel polypeptides is a critical structural element in determining the assembly of K^+ channel subunits into multimeric channels (49). The expression of artificially truncated cDNAs in an *in vitro* expression system has been used to identify regions which are important in the assembly of K^+-channel subunits into tetrameric complexes (50).

The expression of single subunits and partial subunit combinations of the cloned *Torpedo* nAChR (35, 51) or muscle nAChR (52–54) in mammalian cell lines has enabled the contribution of individual subunits to nAChR assembly to be investigated by co-immunoprecipitation (*Protocol 6*) and sucrose gradient centrifugation (*Protocol 7*). An alternative approach to identifying the contribution of individual subunits to functional ion channels is the use of antisense oligonucleotides to prevent the expression of specific subunits in cultured cells (55).

The stoichiometry of neuronal nAChRs formed by the co-expression of the α4 and β2 subunits has been estimated by careful quantification of the amount of [^{35}S]methionine incorporated by metabolic labelling of subunits expressed in *Xenopus* oocytes (56). The stoichiometry of K^+ channels has been studied by the expression of cDNAs in which the coding sequence of multiple (2–5) subunits were linked into a single open reading frame (57). The stoichiometry of neuronal nAChRs and K^+ channels has also been inferred from electrophysiological studies of mutated channel subunits (58, 59).

5. Subcellular localization of ion channels

Ion-channel subunits are synthesized in the ER and then targeted, via the Golgi body, to either the cell surface or to various intracellular locations. Several techniques are available with which to investigate both the subcellular localization of transmembrane protein and the efficiency of subcellular targeting. Techniques such as radioligand binding and immunoprecipitation can be used to estimate the efficiency with which proteins are expressed on the cell surface (by immunoprecipitation or binding assays with intact and solubilized cells). Similar information can be obtained by immunofluorescence microscopy of both permeabilized and non-permeabilized cells. In addition, immunofluorescence microscopy can also be used to study the intracellular localization and cell surface distribution of proteins.

5.1 Surface expression of subunit combinations

The level of cell surface expression of proteins can be assayed by metabolic labelling and immunoprecipitation (*Protocols 4* and *6*) and also, although less quantitatively, by immunofluorescent microscopy (see Section 5.2). An alternative approach is to use a specific radioligand (or radiolabelled antibody) to assay surface expression. This can be done easily for cells expressing the *Torpedo* and muscle nAChR by incubating cell monolayers with $[^{125}I]\alpha$-bungarotoxin (e.g. add 5 nM $[^{125}I]$toxin in PBS/Ca^{2+} to the monolayer and rock gently for 2 h at room temperature, wash three times in PBS/Ca^{2+}, harvest cells by addition of 1% (v/v) Triton X-100, and count in a gamma counter). This has been used to assay the ability of partial combinations of cloned nAChR subunits to reach the cell surface when expressed in cultured cells (35, 53, 60). Details of binding assays with alternative radioligands can be found in a companion volume in this series (61).

5.2 Immunofluorescence microscopy

Immunofluorescence microscopy (*Protocol 8*) is a powerful technique with which to investigate many aspects of the subcellular localization and targeting of ion channels. Many ion channels which are expressed on the cell surface have a non-uniform distribution. A striking example of this is the synaptic clustering of the nAChR at the neuromuscular junction. Immunofluorescence microscopy has been used to investigate the cell surface clustering of the nAChR and nAChR-associated proteins at the neuromuscular junction, in cultured muscle cells and in transfected cells (62–64). Immunofluorescence can also be used to identify the intracellular distribution of proteins. It has been possible to demonstrate that nAChR assembly occurs in the ER membrane by immunofluorescence microscopy with an antibody which preferentially binds to assembled, rather than unassembled, nAChR (26).

Cell surface proteins are frequently targeted to specific regions of the cell's plasma membrane, for example to the cell body or to the dendrites or to the axons of neuronal cells. The targeting of GABA-AR subunits has been investigated by their expression in cultured polarized epithelial cells. There is evidence that the sorting of proteins to either the apical or basolateral surface of the polarized epithelial cells occurs by a similar mechanism to the axonal and dendritic sorting in neuronal cells. Immunofluorescence microscopy can be used to distinguish between subunits which have been targeted to either the apical or basolateral surface of the polarized cells (for further details see ref. 65).

Protocol 8. Immunofluorescence microscopy

Equipment and reagents

- PBS/Ca/Mg: PBS containing 0.005% (w/v) $CaCl_2$ and 0.005% (w/v) $MgCl_2$. To avoid precipitation, prepare freshly each day by adding $CaCl_2$ and $MgCl_2$ (e.g. from 2% (w/v) stocks) to PBS.
- PBS/Ca/Mg/BSA: PBS/Ca/Mg containing 1% (w/v) BSA
- 2% (w/v) paraformaldehyde in PBS (freshly made): heat paraformaldehyde to 60°C and add a few drops (per 100 ml) of conc. NaOH to dissolve. Allow to cool and filter (e.g. Whatman 3MM paper) before use.
- 0.1% (w/v) saponin in PBS
- Primary antibodies and fluorescently labelled secondary antibodies
- Fluorosave (Calbiochem)
- Parafilm
- Fluorescence microscope, slides, and coverslips
- Humid chamber (e.g. plastic slide box with lid in which slides may be laid horizontally, on a support, above soaked tissue or filter paper, to prevent them from drying out).

Method

1. Grow cells under normal conditions on a glass microscope coverslip. This can be done conveniently by plating cells into a 35 mm tissue culture dish containing a sterile 22 × 22 mm coverslip.
2. Wash cells by dipping gently in PBS/Ca/Mg (all washing steps in the protocol can be done conveniently by gently dipping the coverslip (held by fine tweezers) into about 40 ml of wash solution in a medium-sized plastic weigh-boat (Sigma). After each wash remove excess liquid by touching the edge of the coverslip to a paper tissue but do not allow cells to dry out).
3. Fix cells by an appropriate method.[a,b] This can be done by placing the coverslip in a 35 mm tissue culture dish containing 2 ml fixative. For surface labelling use 2% (w/v) paraformaldehyde in PBS and fix for 10 min at room temperature. For intracellular labelling use 2% (w/v) paraformaldehyde, 0.1% (w/v) saponin in PBS and fix for 10 min at room temperature.
4. Wash cells by dipping coverslip three times in PBS/Ca/Mg and once in PBS/Ca/Mg/BSA.
5. Label cells with the primary antibody. Cover a glass slide with a layer of Parafilm, add 100 μl of an appropriate dilution of the primary antibody in PBS/Ca/Mg/BSA, and place Parafilm-covered slides in humid chamber. Close the lid and incubate at room temperature for 1–2 h.
6. Wash three times in PBS/Ca/Mg and once in PBS/Ca/Mg/BSA.

7. Label with a fluorescent secondary antibody (e.g. goat anti-mouse IgG (Calbiochem) if a mouse primary antibody was used in step 5) by adding 100 μl of an appropriate dilution of the fluorescent secondary antibody (e.g. 1:100) in PBS/Ca/Mg/BSA to a glass microscope slide covered in Parafilm and place the cover slip on the drop. Labelling should be done in the dark and in a humidified chamber for 1–2 h at room temperature.
8. Wash four times in PBS/Ca/Mg and once in water.
9. Blot excess water by touching the edge of the coverslip to a paper tissue and mount on a glass microscope slide in an anti-fade medium (e.g. FluorSave, Calbiochem), and store in the dark.
10. Examine by fluorescence microscopy.

[a] Numerous cell fixation methods have been described. It may be necessary to vary fixation conditions for particular antigens and primary antibodies. We have described a simple method which has been found to give successful results with several antibodies.
[b] Alternatively, for surface labelling, cells may be fixed after labelling with the primary antibodies (i.e. after step 5).

6. Concluding remarks

Many aspects of the post-translational processing, assembly, and localization of even the most well characterized ion channels are still rather poorly understood. However, the techniques described in this chapter and elsewhere in this book have provided answers to many important questions concerning these processes. The sites at which several post-translational modifications occur have been clearly identified in many ion channels. The ability of different subunits to co-assemble into functional channels and the distribution of ion channels within the cell have also been demonstrated in many cases. Despite these advances, much remains to be explained. In most cases the structural and functional consequences of post-translational modifications are not fully understood. The mechanism by which the assembly of ion channel subunits is accurately regulated and by which ion channels are localized to particular subcellular locations is also largely unknown.

Acknowledgements

We would like to thank Dr Toni Claudio for the All-11 cell line and polyclonal antibodies (directed against the *Torpedo* α and δ subunits) and Dr Jon Lindstrom for MAbs 148 and 168 (directed against the *Torpedo* β and γ subunits) which were used in *Figure 4*.

References

1. Unwin, N. (1989). *Neuron* **3**, 665.
2. Barnard, E. A. (1992). *Trends Biochem. Sci.* **17**, 368.
3. Catterall, W. A. (1993). *Trends Neurosci.* **16**, 500.
4. Sumikawa, K., Houghton, M., Emtage, J. S., Richards, B. M., and Barnard, E. A. (1981). *Nature* **292**, 862.
5. Claudio, T., Green, W. N., Hartman, D. S., *et al.* (1987). *Science* **238**, 1688.
6. Unwin, N. (1993). *J. Mol. Biol.* **229**, 1101.
7. Couturier, S., Bertrand, D., Matter, J. M., *et al.* (1990). *Neuron* **5**, 847.
8. Sargent, P. B. (1993). *Ann. Rev. Neurosci.* **16**, 403.
9. Catterall, W. A. (1992). *Physiol. Rev.* **72**, S15.
10. DeFelice, L. J. (1993). *J. Membrane Biol.* **133**, 191.
11. Jan, L. Y. and Jan, Y. N. (1992). *Ann. Rev. Physiol.* **54**, 537.
12. Pongs, O. (1992). *Physiol. Rev.* **72**, S69.
13. Sorrentino, V. and Volpe, P. (1993). *Trends Pharmacol. Sci.* **14**, 98.
14. Chen, T. Y., Peng, Y. W., Dhallan, R. S., Ahamed, B., Reed, R. R., and Yau, K. W. (1993). *Nature* **362**, 764.
15. Dermietzel, R. and Spray, D. C. (1993). *Trends Neurosci.* **16**, 186.
16. Swope, S. L., Moss, S. J., Blackstone, C. D., and Huganir, R. L. (1992). *FASEB J.* **6**, 2514.
17. Miles, K., Greengard, P., and Huganir, R. L. (1989). *Neuron* **12**, 1517.
18. Moss, S. J., Doherty, C. A., and Huganir, R. L. (1992). *J. Biol. Chem.* **267**, 14470.
19. Raymond, L. A., Blackstone, C. D., and Huganir, R. L. (1993). *Nature* **361**, 637.
20. Costa, M. R., Casnellie, J. E., and Catterall, W. A. (1982). *J. Biol. Chem.* **257**, 7918.
21. Nairn, A. C. and Greengard, P. (1987). *J. Biol. Chem.* **262**, 7273.
22. Cooper, J. A., Sefton, B. M., and Hunter, T. (1983). In *Methods in enzymology* (ed. J. D. Corbin and J. G. Hardman), Vol. 99, p. 387. Academic Press, London.
23. Hirano, A. A., Greengard, P., and Huganir, R. L. (1988). *J. Neurosci.* **50**, 1447.
24. Merlie, J. P., Sebbane, R., Tzartos, S., and Lindstrom, J. (1982). *J. Biol. Chem.* **257**, 2694.
25. Bar-Sagi, D. and Prives, J. (1983). *J. Cell Physiol.* **114**, 77.
26. Ross, A. F., Green, W. N., Hartman, D. S., and Claudio, T. (1991). *J. Cell Biol.* **113**, 623.
27. Gehle, V. M. and Sumikawa, K. (1991). *Brain Res. Mol. Brain Res.* **11**, 17.
28. Blount, P. and Merlie, J. P. (1990). *J. Cell Biol.* **111**, 2613.
29. Mishina, M., Tobimatsu, T., Tanaka, K., *et al.* (1985). *Nature* **313**, 364.
30. Olsen, E. N., Glaser, L., and Merlie, J. P. (1984). *J. Biol. Chem.* **259**, 5364.
31. Schmidt, J. D. and Catterall, W. A. (1987). *J. Biol. Chem.* **262**, 13713.
32. Sumikawa, K. and Gehle, V. M. (1992). *J. Biol. Chem.* **267**, 6286.
33. Schmidt, J., Rossie, S., and Catterall, W. A. (1985). *Proc. Natl Acad. Sci. USA* **82**, 4847.
34. Merlie, J. P. and Lindstrom, J. (1983). *Cell* **34**, 747.
35. Green, W. N. and Claudio, T. (1993). *Cell* **74**, 57.
36. Munro, S. and Pelham, H. R. B. (1986). *Cell* **46**, 291.
37. Forsayeth, J. R., Gu, Y., and Hall, Z. W. (1992). *J. Cell Biol.* **117**, 841.

38. Paulson, H. L., Ross, A. F., Green, W. N., and Claudio, T. (1991). *J. Cell Biol.* **113**, 1371.
39. Blount, P. and Merlie, J. P. (1991). *J. Cell Biol.* **113**, 1125.
40. Haga, T., Haga, K., and Hulme, E. C. (1990). In *Receptor biochemistry: a practical approach* (ed. E. C. Hulme), p. 1. IRL Press, Oxford.
41. Harlow, E. and Lane, D. (1988). *Antibodies: a laboratory manual.* Cold Spring Harbor Laboratory, New York.
42. Green, W. N., Ross, A. F., and Claudio, T. (1991). *Neuron* **7**, 659.
43. Whiting, P. and Lindstrom, J. (1986). *J. Neurosci.* **6**, 3061.
44. Whiting, P., Schoepfer, R., Lindstrom, J., and Priestley, T. (1991). *Mol. Pharmacol.* **40**, 463.
45. Verrall, S. and Hall, Z. W. (1992). *Cell* **68**, 23.
46. Yu, X. M. and Hall, Z. W. (1991). *Nature* **352**, 64.
47. Chavez, R. A., Maloof, J., Beeson, D., Newsom-Davis, J., and Hall, Z. W. (1992). *J. Biol. Chem.* **267**, 23028.
48. Gu, Y., Camacho, P., Gardner, P., and Hall, Z. W. (1991). *Neuron* **6**, 879.
49. Li, M., Jan, Y. N., and Jan, L. Y. (1992). *Science* **257**, 1225.
50. Shen, N. V., Chen, X., Boyer, M. M., and Pfaffinger, P. J. (1993). *Neuron* **11**, 67.
51. Claudio, T., Paulson, H. L., Green, W. N., Ross, A. F., Hartman, D. S., and Hayden, D. (1989). *J. Cell Biol.* **108**, 2277.
52. Blount, P., Smith, M. M., and Merlie, J. P. (1990). *J. Cell Biol.* **111**, 2601.
53. Gu, Y., Forsayeth, J. R., Verrall, S., Yu, X. M., and Hall, Z. W. (1991). *J. Cell Biol.* **14**, 799.
54. Blount, P. and Merlie, J. P. (1988). *J. Biol. Chem.* **263**, 1072.
55. Listerud, M., Brussaard, A. B., Devay, P., Colman, D. R., and Role, L. W. (1991). *Science* **254**, 1518.
56. Anand, R., Conroy, W. G., Schoepfer, R., Whiting, P., and Lindstrom, J. (1991). *J. Biol. Chem.* **266**, 11192.
57. Liman, E. R., Tytgat, J., and Hess, P. (1992). *Neuron* **9**, 861.
58. Cooper, E., Couturier, S., and Ballivet, M. (1991). *Nature* **350**, 235.
59. MacKinnon, R. (1991). *Nature* **350**, 232.
60. Sine, S. M. and Claudio, T. (1991). *J. Biol. Chem.* **266**, 19369.
61. Hulme, E. C. (ed.) (1992). *Receptor–ligand interactions: a practical approach.* IRL Press, Oxford.
62. Hartman, D. S., Millar, N. S., and Claudio, T. (1991). *J. Cell Biol.* **115**, 165.
63. Dubinsky, J. M., Loftus, D. J., Fischbach, G. D., and Elson, E. L. (1989). *J. Cell Biol.* **109**, 1733.
64. Kuromi, H., Brass, B., and Kidokoro, Y. (1985). *Dev. Biol.* **109**, 165.
65. Perez-Velazquez, J. L. and Angelides, K. J. (1993). *Nature* **361**, 457.

8

Ion channels and signal transduction systems in plant and animal cells

I. McFADZEAN and C. BROWNLEE

1. Introduction

The advent of the patch-clamp technique in the late 1970s brought about a revolution in the study of ion channels. Whilst in the main this arose as a result of being able to resolve single channel currents, an equally important development was that it became possible to manipulate the intracellular environment of the cell, allowing the control of ion currents by cytoplasmic messengers to be studied in far greater detail than hitherto conceivable. The vast majority of such studies have been performed in animal cells, in which the transduction pathways coupling neurotransmitter receptor activation to changes in ion channel activity have been elucidated in great detail in a wide range of cell types. More recently the patch-clamp technique has been used to identify ion channels in plant cells and to study their control by intracellular constituents, in particular Ca^{2+}.

This chapter has two overlapping aims. The first is to review the methodology involved in studying the control of ion currents by intracellular mediators and the second is to serve as an introduction to patch clamping plant cells (in the context of changes in intracellular calcium as a regulator of ion channel function). It is assumed throughout that the reader has some basic knowledge of electrophysiological techniques, in particular patch-clamping (see Chapter 1).

2. Recording from plant cells

2.1 Special considerations for recording from plant cells

Plant cells present a special set of problems for studies of ion channels and intracellular signalling, related principally to cell structure and specific physiological properties. In many cases, however, methods have been developed to overcome these problems and virtually all electrophysiological and related techniques applied to animal cells are now possible with many plant cell types.

Special considerations when recording from plant cells include the following.

- The presence of the cell wall makes direct recording with a patch electrode impossible and impalement with intracellular electrodes problematic, particularly with thick-walled cells.
- The presence of a large intracellular vacuolar compartment means that in higher plant cells especially, the cytosol often comprises only a small fraction of the total cell volume. This can create uncertainties in the location of intracellular electrodes for recording or microinjection purposes. The relatively large vacuolar membrane area can also lead to uncertainties in the origin of a particular membrane in protoplasts prepared for patch clamping.
- Cell turgor must be eliminated to prevent bursting of protoplasts when the cell wall is removed. This requires incubation of protoplasts in an unphysiological 'osmoticum'. Turgor pressure of intact cells can also displace ion selective sensors or fluorescent dyes from the electrode tip unless appropriate back pressure is applied to the electrode.
- The lack of a rigid cytoskeleton in walled plant cells tends to make gigaohm seal formation during patch-clamp experiments more difficult than with many animal cells.

2.2 Voltage clamping plant cells

Both single (switch-clamp) and double-electrode voltage-clamp methods have been used in a variety of plant cells. Voltage-clamp with a double-barrelled electrode has been used effectively on the stomatal guard cell (e.g. ref. 1). This cell is one of the most intensively studied with respect to signal transduction mechanisms in higher plants. In addition to playing a major role in plant–water relations, guard cells show very clear responses (stomatal opening or closure) corresponding to changes in cell shape resulting, in turn, from changes in ion and water fluxes across the plasma membrane in response to well defined signals, including plant hormones (abscisic acid, auxin), light, and CO_2.

The single-electrode discontinuous voltage-clamp has been used with eggs of the marine alga, *Fucus serratus* (2). With this method, a single electrode performs both the voltage recording and current injection tasks with a rapid switching frequency, determined by the time constants of the electrode (τ_e) and cell membrane (τ_c). This allows voltage associated with charging of the electrode capacitance during the current injection cycle to decay before membrane potential is sampled. The relatively large size (~70 mm diameter) of *Fucus* eggs ensured a sufficiently high τ_c despite a relatively low input resistance (100 MΩ). Electrode switching frequency can be optimized by bevelling the tip of the electrode (3) and by dipping it in mineral oil to minimize stray capacitance. Limitations on the range of usable voltage-clamp potentials arose from the relatively small τ_c/τ_e ratio (typically around 150), resulting in clamp error or oscillations induced by extreme command voltages.

2.3 Patch clamping plant cells

The primary and most crucial step in patch clamping walled plant cells is the removal of the cell wall to reveal a clean plasma membrane on which gigaohm seals can be obtained with a patch pipette. To date, most patch-clamp studies of plant cells have utilized protoplasts from which the cell wall has been removed by enzyme digestion.

2.3.1 Enzymatic methods

A variety of enzyme methods have been developed for the production of viable plant cell protoplasts. These vary considerably between different cell and tissue types. A major problem with protoplast production is the isolation of particular identified cell types from tissue which may contain several different kinds of cell. Isolation of a particular cell type may involve initial dissection of the tissue, followed by preferential disruption of the unwanted cells, and selective enzymatic digestion of the walls of the cells of interest. A good example of the successful isolation of a particular protoplast type is the preparation of stomatal guard cell protoplasts from leaf tissue (*Protocol 1*). Loss of turgor and a non-physiological osmotic environment are unavoidable factors in the application of patch-clamp techniques to most plant cells.

Protocol 1. Enzymatic method for isolating stomatal guard cell protoplasts from mechanically prepared epidermis of *Vicia faba* leaves (4, 5)

Equipment and reagents

- Homogenizer, rotating shaker, and water bath or incubator
- Razor blades, nylon mesh, paper towels
- Homogenization medium: Ficoll (type 400) 10% (w/v), 5 mM $CaCl_2$, 0.1% (w/v) PVP 40
- Digestion medium A: 0.25 M D-mannitol, 0.7% (w/v) cellulysin (from Behring Diagnostics), 1.0 mM $CaCl_2$, 0.1% (w/v) PVP 40, 1 mg/ml pepstatin A, pH 5.5
- Digestion medium B: 0.35 M mannitol, 1.0 mM $CaCl_2$, 1% (w/v) cellulase 'Onozuka' RS (from Yakult, Honsha Co.), 0.01% (w/v) pectolyase (Y-23, from Seishin Pharmaceutical Co.), 1 mg/ml pepstatin A
- Rinsing solution: 0.25 M mannitol, 1.0 mM $CaCl_2$, pH 5.0
- 0.35 M mannitol

A. *Mechanical disruption of leaf tissue*

1. Harvest leaflets from the second node of 3–4 week old plants. Store in Petri dishes with moistened paper towels.
2. Remove midveins with a razor blade and cut each leaf into eight pieces.
3. Transfer pieces from 10 leaves into 100 ml homogenization medium and homogenize at 20–22 000 r.p.m. for 50–70 sec at 20 °C.

Protocol 1. *Continued*

B. *Enzymatic digestion of mesophyll cells*

1. Collect and combine successive homogenizations and filter through a 220 mm nylon mesh.
2. Rinse 2–3 min with tap water.
3. Transfer to 100 ml digestion medium A and shake at approximately 135 r.p.m at 27 °C for 45 min or until no mesophyll tissue remains (up to 6 h).

C. *Guard cell protoplast collection*

1. Collect epidermis on a 220 mm nylon mesh and rinse with 500 ml rinsing solution.
2. Transfer to 50 ml digestion medium B and incubate at 26–27 °C at 44 r.p.m. Guard cell protoplast release should occur after approximately 2 h. Observe epidermal fragments microscopically for release of protoplasts.
3. Pour through 30 mm nylon mesh, centrifuge filtrate at 60 *g* for 7 min, and wash three times in 0.35 M mannitol.

The success rate of gigaohm seal formation in plant protoplasts prepared by standard enzyme methods varies but is generally low compared with most animal cells. Significant improvements in seal rates have, however, been produced by modifications of the enzyme treatment, particularly by reducing this to a minimum. Brief enzyme treatment to weaken the cell wall, followed by osmotic adjustment to make the protoplasts swell and pop out of the weakened wall appears to improve sealing rates in several cell types (6, 7) (*Protocol 2*). Gigaohm seal (>10 MΩ) formation can be obtained usually within 2 min of the application of suction to the pipette.

Protocol 2. Quick enzymatic method for protoplast isolation from pea epidermal cells (6, 7)[a]

Equipment and reagents

- Equipment as in *Protocol 1*
- Digestion solution: 1.7% (w/v) cellulase (see *Protocol 1*), 1.75% (w/v) cellulysin (see *Protocol 1*), 0.02% (w/v) pectolyase Y-23 (see *Protocol 1*), 0.2% (w/v) BSA, 2.35% (w/v) Gamborg's BS (Gibco), 10 mM $CaCl_2$, 10 mM MES-KOH, mannitol to adjust osmolarity to 610 mOsm, pH 5.5
- Wash solution: as digestion solution but without enzymes
- Patch-clamp solution: 5 mM $CaCl_2$, 2 mM $MgCl_2$, 10 mM potassium citrate, mannitol to give an osmolarity of 210 mOsm, pH 5.5

Method

1. Peel epidermis from stem or underside of young unfolding leaves.

2. Shake (50 r.p.m.) for 5–10 min at 30°C in hydrolytic digestion solution.
3. Transfer tissue to 5 ml wash solution and wash for 5 min at 30°C.
4. Transfer tissue to patch-clamp solution.

[a] Also applicable to *Phaseolus vulgaris* leaf mesophyll, *Avena sativa* coleoptile cortex, and *Arabidopsis thaliana* cotyledon mesophyll.

2.3.2 Laser ablation

A UV laser can be used to ablate regions of the cell wall through which either selected regions of the plasma membrane can be exposed or protoplasts can be completely extruded (8, 9), allowing gigaohm seal formation with a patch pipette. Advantages of this technique include the following.

- Enzyme treatment is not required, reducing damage or modification of cell surface properties.
- Protoplasts can be patch clamped immediately following release.
- Individual cell types can be selected.
- The method can be applied to cell types from which protoplasts cannot be prepared enzymatically.

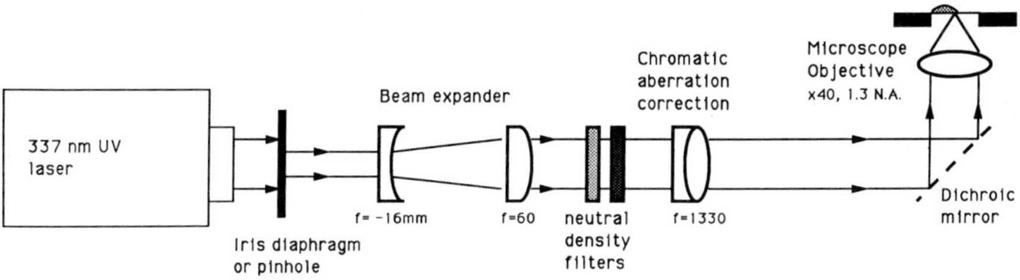

Figure 1. Optical arrangement for delivery of a UV laser and focusing to a small spot by the microscope objective for microsurgery.

Protocol 3. Laser ablation for protoplast production or localized patch clamping

Equipment and reagents

- Microscope with fluorescence port and UV transmitting optics (e.g. Nikon Diaphot, Zeissaxiovert)
- Pulsed nitrogen UV laser (e.g. VSL 337 from Laser Science Inc.)
- Appropriate laser delivery and focusing optics (see *Figure 1*), e.g. from Spindler and Hoyer Ltd
- Plasmolysis medium A: filtered sea water containing 0.6 M mannitol, 3–6% (w/v) polyethylene glycol 3500
- Growth medium: 5 mM MES buffer, pH 6.0, 1% (w/v) bactoagar (Difco)
- Plasmolysis medium B: 5 mM MES buffer, 0.35 M sorbitol or glycine

Protocol 3. *Continued*

Method

1. Plasmolyse cells in an appropriate plasmolysis medium:
   (a) Transfer *Fucus serratus* zygotes (8) to plasmolysis medium A on a coverslip.
   (b) For *Medicago sativa* root hairs (9), grow seedlings for 48 h at 21 °C in the dark on growth medium then transfer to plasmolysis solution B on a coverslip.
2. Align cell wall in region of interest and use focused laser spot to ablate a hole in the wall. One or more pulses is usually sufficient.
3. Adjust the osmolarity of the external solution so that a protoplast is partially or wholly extruded.
4. Approach with patch-clamp electrode as soon as possible.

3. Measurement of cytosolic Ca^{2+}

3.1 Methods for measuring cytosolic Ca^{2+}

(a) Ca^{2+}-selective microelectrodes (10) can be used with cells which are sturdy enough to withstand impalement with two electrodes or a double-barrelled electrode. This is probably the least expensive method for cytosolic $[Ca^{2+}]$ measurement. The main drawbacks of this method include the relatively slow response time of the electrode and calibration problems (re-calibration of the electrode after impalement of the cell is often difficult). These problems can be particularly acute with plant cells where turgor can result in partial sensor displacement and subsequent loss of sensitivity. *In vivo* calibration is usually not possible. If the location of the electrode is known, then this method gives a signal which reflects only the $[Ca^{2+}]$ in that compartment.

(b) Luminescent photoproteins, the most commonly used of which is aequorin (11). The luminescence signal increases with the cube of the Ca^{2+} concentration and can be measured with photometry or imaged using a sensitive photon detector. Aequorin reliably reflects free $[Ca^{2+}]$ and does not chelate Ca^{2+} or interfere with the kinetics of Ca^{2+} responses. It is, therefore, potentially a good method for monitoring small, transient, and localized excursions from resting levels. A particular advantage is that cells can be transformed with the aequorin gene allowing long-term, non-invasive monitoring of Ca^{2+} (12). Aequorin can also be targeted to particular organelles (e.g. mitochondria), allowing study of intra-organellar Ca^{2+} dynamics (13).

(c) Fluorescent indicators. Currently, these represent the most widely used

Table 1. Some commonly used fluorescent indicators for intracellular calcium measurements and their applications[a]

Dye	Type	Excitation λ (nm)	Emission λ (nm)	Application
Fura-2	Dual excitation	340, 380	530	Ratio photometry, ratio imaging
Indo-1	Dual emission	360	405, 480	Ratio photometry, ratio imaging, ratio confocal imaging (UV)
Fluo-3	Single	480	530	Single wavelength photometry with UV flash photolysis, confocal imaging
Ca green	Single	480	530	As Fluo-3 but with large signal at low [Ca^{2+}]
Ca crimson	Single	593	620	Long wavelength and confocal imaging
Ca orange		550	575	As calcium crimson
Fura red		545	580	Long wavelength measurements, confocal imaging
Rhod-2		553	576	As Fura red
Mag Fura-2		340, 370	500	Ratio photometry and imaging of [Ca^{2+}] in the high mM range

[a] More details of these and other fluorescent Ca^{2+} indicators can be obtained from Molecular Probes Inc., Eugene, OR, USA.

method. There are many variations of this technique, each reflecting the particular cell type and kind of information sought. A wide range of dyes is available with varying affinities and optical properties, including dual- and single-wavelength excitation or emission and UV and longer wavelength dyes (*Table 1*).

3.2 Use of fluorescent dyes

3.2.1 Advantages and disadvantages of fluorescent Ca^{2+} indicators

Many cell types can be loaded with Ca^{2+}-sensitive dye by incubation with the appropriate acetoxymethyl (AM) ester (see Section 3.2.2). In comparison with luminescent indicators, the fluorescent signal from most available dyes is strong, rendering them ideal for single-cell studies and high-resolution imaging. The most commonly used dyes, fura-2 and indo-1, show a spectral shift on binding the Ca^{2+} (14) and can be used in ratiometric mode (*Table 1*), so that the dye reports [Ca^{2+}] irrespective of factors such as the dye concentration or cell thickness. Not all cells will load with AM ester and in some cells, intracellular ester hydrolysis may be incomplete or hydrolysis may occur externally. AM esters are generally very insoluble, but solubility and loading can often be improved by incubating cells with dye in the presence of the non-ionic surfactant pluronic (Molecular Probes Inc.). Intracellular compartmentalization or loss of the dye from the cytosol can occur to varying degrees in different cell types, resulting in decreased cytoplasmic signals and contamination by intra-organelle signals. One way of avoiding this is to microinject the free-acid form of the dye directly into the cell. This is usually effective where compartmentalization is due to intracellular sequestration of uncleaved ester or when ester is hydrolysed preferentially in subcellular compartments. However, many cell types, particularly plant cells, are able to transport the free-acid form of the dye across intracellular membranes. This can be prevented by treating the cells with the anion transport inhibitor probenecid (15). A better method is to microinject a dextran-conjugated form of the dye. Most Ca^{2+}-sensitive dyes are now available with M_r of 10 and 70 kDa (Molecular Probes Inc.).

The other major potential problem with the use of fluorescent indicators is the fact that they are Ca^{2+} buffers and bind Ca^{2+} effectively. Thus, if intracellular dye concentrations are high, significant buffering of cytosolic Ca^{2+} may occur. For example, use of the original dye, Quin-2, occasionally required intracellular concentrations of up to 1 mM for reasonable signals, and since its K_d (for Ca^{2+}) is >0.1 μM, the probe would actually bind >0.5 mM cation at normal resting cellular levels. Any buffering can slow and smooth out rapid transients, and dyes can also potentially act as *mobile* Ca^{2+} chelators, smoothing out *spatial* variations in Ca^{2+} distribution. To minimize buffering, intracellular dye concentrations should clearly be kept as low as possible (of the order of tens of μM or less). For a dual-wavelength indicator, intracellular dye concentration can be monitored by measuring the fluorescence at the isosbestic (Ca^{2+} independent) excitation (Fura-2) or emission (Indo-1) wavelength. In spectrofluorimetric measurements in cell suspensions, intracellular dye concentration can be estimated by permeabilizing cells to release dye and monitoring the isosbestic fluorescence.

3.2.2 Methods for loading cells with fluorescent Ca^{2+} indicators

i. Ester loading

Cells are incubated in culture medium in the presence of the AM form (up to 1–5 mM) of the indicator. The AM ester will, in theory, enter the cell by diffusion, where it can be cleaved by the cell's own esterases to release the free-acid form which is charged and therefore trapped intracellularly. The following general method applies:

(a) Prepare stock solution of 1 mM Fura-2 in dry DMSO and add to culture medium to give a final concentration of 1–5 μM.

(b) Incubate cells in culture medium with dye, either in suspension or attached to coverslips, and observe fluorescence of sub-samples at 10 min intervals until fluorescence is visible or can be monitored with the microscope recording system (Section 3.2.4).

(c) When loading is complete (usually after 30–90 min), wash cells and allow to settle on coverslips for fluorescence monitoring.

ii. Microinjection

This method (10) can be used where cells cannot be ester loaded, where dextran-conjugated dyes are used, or where co-loading with other indicators, buffers, or other compounds is required. Microinjection can be done either *via* ionophoresis (of free-acid forms of the dye), whereby small negative current pulses (0.1–100 nA) are passed through the microelectrode, or by pressure pulses from a pressure injection system.

iii. Patch-pipette loading

Where patch clamping of cells is routine, it is often convenient to include a Ca^{2+} indicator (free-acid or dextran-conjugated form) in the patch-pipette containing the whole-cell recording solution. This allows the combined monitoring of Ca^{2+} and whole-cell currents during patch-clamp experiments. (See, for example, ref. 16.)

3.2.3 Calibration of the fluorescence signal

Theoretical and practical aspects of intracellular $[Ca^{2+}]$ measurement are discussed in detail in ref. 10. There is no single method for calibration of the dye fluorescence in terms of $[Ca^{2+}]$. Some examples of the most commonly used methods are given below:

(a) *In vitro* calibration. This is the simplest but least informative method of calibration. Fluorescence of droplets of different Ca^{2+} buffers containing 10–50 mM dye can be monitored. The method can only give an approximation of the cytosolic $[Ca^{2+}]$ since no account is taken of the optical or other properties of the cell which may strongly influence the dye signal.

(b) *In vivo* calibration. The dual wavelength indicators Fura-2 and Indo-1 show a sigmoidal ratio response to increasing $[Ca^{2+}]$ with saturation of the ratio in the Ca^{2+}-bound (R_{max}) and Ca^{2+}-free (R_{min}) forms (14). $[Ca^{2+}]$ can be calculated from any cell ratio value (R) if R_{max} and R_{min} are known. Thus

$$[Ca^{2+}]=bK_d[R-R_{min}/R_{max}-R],$$

where $b = Ca^{2+}{}_{free}/Ca^{2+}{}_{bound}$ dye fluorescence at the longer excitation wavelength (Fura-2). R_{min} and R_{max} can, in principle, be obtained by treatment of the cells with Ca^{2+} ionophore (up to 10 nM A23187 or ionomycin) in the presence of high external Ca^{2+} (for R_{max}), followed by

Ca^{2+}-free medium (for R_{min}). Practical difficulties may be encountered due to loss of dye signal during ionophore treatment, inability to reach a steady R_{min} or R_{max}, or very slow or zero response to the ionophore. In the absence of a usable response to ionophore treatment, an alternative *in vivo* calibration method involves microinjection of Ca^{2+} buffers into the cell at concentrations sufficient to override the cell's buffering capacity. This requires pressure injection of buffer while recording [Ca^{2+}] until a new steady-state fluorescence ratio is obtained.

In whole-cell patch-clamp experiments, when cells have been loaded with dye through the patch-pipette, the cells can be perfused internally through the patch-pipette at the end of the experiment with intracellular recording solution containing either 10 mM EGTA to give R_{min} or 10 mM $CaCl_2$ for R_{max}.

3.2.4 Equipment for fluorescence measurement of cytosolic [Ca^{2+}] in single cells

The equipment required for the measurement of cytosolic [Ca^{2+}] in single cells varies depending on the type of response studied, the type of cell, and the information required. The simplest and least expensive method for single- or dual-wavelength measurements employs photomultipliers, while more sophisticated imaging methods make use of high-sensitivity cameras or confocal microscopy. The essential equipment (*Figure 2*) and most common methods of fluorescence monitoring are listed below:

- Fluorescence microscope (with UV conducting optics if UV dyes are to be used). Several suppliers now provide fluorescence microscopes specially adapted for Ca^{2+} monitoring.
- Fluorescent light source, usually a mercury vapour or xenon lamp. This can be part of the microscope's fluorescence illuminator or a remote light source delivered to the microscope by fibre optics. This removes electrical noise and mechanical vibration from the microscope, improving electrical recordings and the quality of images.
- If a dual-wavelength excitation dye is to be used (e.g. Fura-2), a filter changer is required in the excitation light path. This is controlled by a computer and synchronized with data acquisition from the photomultiplier or imaging device. Dual-wavelength emission dyes (e.g. Indo-1) require either two photomultipliers or a filter changer in front of the imaging device.
- Photomultipliers can be used to monitor fluorescence in single cells, in groups of cells, or in cell suspensions. The photomultiplier signal gives an average fluorescence value without any spatial information. Photomultipliers, in combination with appropriate hardware or software, can provide very highly resolved temporal information on average, whole-cell Ca^{2+} dynamics.

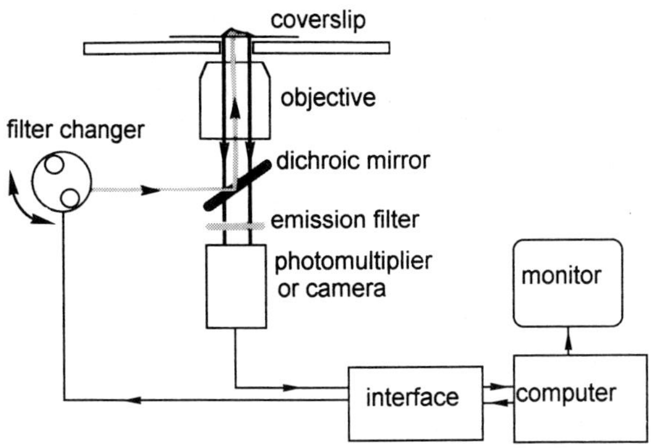

Figure 2. Schematic diagram of the essential components of a fluorescence microscope system for use with dual or single wavelength indicators.

- Imaging cameras. Intensified SIT video cameras provide information at video rate, but this needs to be digitized before processing. Images can be averaged for noise reduction with consequent reduction in temporal resolution, but noise levels depend ultimately on dye signal strength. Cooled CCD cameras can provide very low noise digital images and have a large dynamic range (usually 12 bits). They are especially useful for the application of deconvolution algorithms for removal of out-of-focus information.
- Confocal laser scanning microscopy (CLSM) is being used increasingly for dynamic imaging of cytosolic Ca^{2+}. Though confocal microscopes are available with UV lasers for use with UV excitable dyes (particularly Indo-1), most CLSMs are equipped with an argon laser which emits light at 487 nm. *Table 1* provides examples of Ca^{2+}-sensitive dyes which can be used at longer wavelengths with the confocal microscope. In general, these are not ratiometric, and it is important to be aware of potential artefacts due to laser and fluorescence light attenuation, particularly with large pigmented cells. This problem can be overcome largely by sequential ratio imaging whereby a sequence of fluorescent images is divided by an initial image to give ratio images representing the relative change in dye fluorescence.

The basic sequence for obtaining ratiometric measurements of cytosolic $[Ca^{2+}]$ in dye-loaded cells is as follows:

(a) Monitor cell autofluorescence at each excitation wavelength, and store these values for subsequent subtraction (the method used will depend on the computer software employed). For ester-loaded cells it is usually easier to obtain autofluorescence values after the experiment by a treatment which quenches the intracellular dye fluorescence (see, for example, ref. 18).

(b) Load cells (see Section 3.2.2). Select cells for photometry or imaging and obtain a sequence of images or photometric measurements at each excitation wavelength.

(c) Ratioing can usually be carried out on-line for photometric measurements, either by using analogue circuitry or through software. For imaging, ratioing can be performed on-line if the computer is fast enough, otherwise images must be stored for later analysis.

4. Animal cells

Whilst there is no doubt that the use of electrophysiological methods to study plant cell physiology is a rapidly expanding field, it is very much in its infancy when compared to the wealth of knowledge that exists on the electrophysiological responses to neurotransmitters, hormones, and drugs in excitable cells derived from animal tissues. In general, the neurotransmitter or hormone acts on a receptor, activating it, and setting in motion a series of events which leads to the opening or closing of specific ion channels. Electrophysiological techniques can be used to study the properties of both the receptor and the ion channel in great detail, but here we are concerned primarily with the way in which these techniques can be used to study the link between the two, i.e. signal transduction mechanisms. In particular, we have attempted to suggest how such investigations may be planned from 'first principles' when tackling a specific biological problem.

4.1 The use of patch clamping to determine the involvement of a diffusible second messenger

Often the first step in elucidating the transduction pathway between receptor and channel is to determine whether it employs a second messenger which diffuses through the cell cytoplasm. In many instances this is achieved by a process of elimination, whereby the investigator carries out a series of experiments designed to interfere sequentially with the various potential transduction pathways (Section 4.4) in the hope that, at worst, a number of pathways can be eliminated, and at best, the operative pathway or a component of it can be positively identified. This approach is thc only one available when the effect of a neurotransmitter or hormone on whole-cell currents is being studied and where the single channel currents underlying the total current have not been identified. If the single channels have been identified and, more importantly, can be detected in cell-attached patches, then the involvement of a diffusible messenger can be tested by applying agonist in the bath solution. Since the gigaohm seal formed between the patch-pipette and the cell membrane acts as a physical barrier, preventing agonist reaching receptors on the extracellular face of the patch, any change in channel activity detected is most likely to occur as a result of a cytoplasmic messenger (formed following activation of receptors outside the patch) diffusing 'under' the patch pipette to modulate the channel activity. Lack of effect of bath-applied agonist points to a more direct, 'membrane-delimited' coupling between receptor and channel, and this can be confirmed by comparing channel activity in cell-attached recordings made with and without agonist added to the patch-pipette filling solution.

One of the most intriguing questions in neurophysiology at the moment is, what couples activation of muscarinic receptors to inhibition of the 'm-current', a time and voltage-dependent potassium current? A number of attempts have been made to elucidate the coupling mechanism, and a series of experiments performed in neuroblastoma × glioma hybrid (NG108–15) cells by Brown and co-workers illustrates the approach well (19, 20). Initial experiments (19) provided evidence for which systems were *not* involved, after over 30 inhibitors of known transduction pathways were tested, none of which reduced the inhibition of the m-current produced by acetylcholine. More recent experiments (20) showed that agonist applied outside the patch pipette could inhibit channel activity in cell-attached patches, suggesting the involvement of an as yet unidentified diffusible messenger.

4.2 Ligand-gated ion channels

If experiments such as those outlined above suggest that there is no soluble second messenger coupling the receptor to a channel, it may be possible to confirm this using excised, outside-out patches. For example, 5-HT_3 receptor-mediated currents have been recorded in outside-out patches excised from both cell lines and dissociated rat superior cervical ganglion neurones (21, 22). The 5-HT_3 receptor belongs to the family of ligand-gated or receptor-operated channels, where the receptor and ion channel are part of the same macromolecular complex. It is perhaps therefore not surprising that single channel currents can be recorded in excised patches since, leaving aside for the moment regulation of the channel properties by, for example, protein phosphorylation, if the receptor is present in the patch then so too is the channel. This is not to say, however, that if single channel currents can be evoked by agonists in outside-out patches, the agonist is acting via a ligand-gated channel. Take, for example, the situation in atrial-muscle cells. Here application of muscarinic receptor agonists activates an inwardly rectifying potassium conductance. Bath-applied agonist is ineffective in experiments using cell-attached patches (23), suggesting a tight coupling between receptor and channel. This, however, is not a ligand-gated channel since interposed between the receptor and channel is a guanine-nucleotide binding protein (G protein).

4.3 G proteins: membrane-delimited pathways

G proteins play a pivotal role in coupling receptors to ion channels, whether by coupling receptor to channel directly (membrane-delimited pathways) or by being the first step in the transduction pathway leading to the formation of another second messenger. A description of the electrophysiological approaches used to study G-protein function, including how the involvement of a G protein in an electrophysiological response is determined and methods of identifying which G protein is involved, has been published recently in another volume in the *Practical approach* series (24). What follows is effectively a summary of that chapter; for a more detailed description of the methodology involved, the reader is referred to the original.

4.3.1 Determining the involvement of a G protein

The first step is determining whether a G protein is involved in a response. This can be achieved in several ways, and these are listed below.

- Is the response GTP-dependent? Compare responses obtained in whole-cell patch-clamp experiments using micropipette solutions containing either 100 μM GTP or no GTP.
- Is the response to agonist mimicked or rendered irreversible by the inclusion of a non-hydrolysable analogue of GTP (GTPγS or Gpp(NH)p) or AlF_3 in the micropipette? The GTP analogues are added directly to the micropipette-filling solution at concentrations of 100–500 μM. AlF_3 is formed *in situ* by the addition of 10 mM KF, CSF, or NaF to the micropipette-filling solution. The fluoride ions react with the aluminium in the glass of the micropipette to produce the reactive AlF_4^-.
- Is the response blocked by GDPβS? This GDP analogue competes with GTP for the guanine nucleotide binding site of the G protein, and thus prevents its activation. GDPβS is added directly to the micropipette-filling solution at concentrations up to 1 mM, depending on the cells under study.

It should be noted that, while a positive result using any of the agents listed above is good evidence for the involvement of a G protein, the converse does not hold—i.e. the lack of effect of, for example, GDPβS on a response to agonist does not necessarily mean the lack of involvement of a G protein.

4.3.2 Identifying the G protein

Having confirmed that a G protein is involved in mediating the response, the next step is to try and identify which G protein is being employed. There are several approaches to this. The first, and most straightforward, is to determine whether the response is sensitive to pertussis toxin (Ptx). This toxin ADP-ribosylates the α subunit, and in so doing inactivates G proteins of the G_i/G_o, and G_t subtypes. Recordings can be made from cells pre-treated with Ptx (200–500 ng/ml) for several hours. Alternatively, tissues can be prepared from animals treated with the toxin, e.g. 1–1.2 μg Ptx intracerebroventrically, 1–5 days prior to making recordings (25).

More rigorous identification of the G protein has relied up until now on three main approaches:

- Reconstitution experiments, where the ability of purified G proteins to reconstitute the responses to agonists in, for example, cells pretreated with Ptx, is tested.
- Antibody studies, where specific antibodies raised against the α subunits of different G proteins are introduced into cells and tested for their ability to block responses to agonists.
- More recently, the injection of antisense oligonucleotides into cells has been used to block selectively expression of the different G-protein subunits.

Reconstitution experiments can be performed using either the whole-cell or inside-out configurations of the patch-clamp technique. In whole-cell experiments, the endogenous G proteins must be Ptx-sensitive since they must first be inactivated by pre-treatment with the toxin. The purified G-protein α subunits are then added back to the cell via the patch-pipette filling solution at concentrations of 15–100 nM. When the single channel currents under study are stable in excised patches, pre-activated α subunits can be applied to the cytoplasmic face of inside-out patches, and the relative potency of the α subunits from different G proteins at activating the channel can be compared.

In studies using antibodies, the antibodies are loaded into the cell either by their inclusion in the micropipette-filling solution (when making whole-cell recordings) or by microinjection. The former method has been used successfully to study the coupling between D_2 receptors and K^+ and Ca^{2+} currents in pituitary cells (26), the purified antibodies being incorporated in the micropipette-filling solution at a concentration of 180 μg/ml. Microinjection of antibody has been used in NG108–15 cells (27) and, more recently, in superior cervical ganglion neurones (28), in both cases to study coupling between neurotransmitter receptors and voltage-dependent Ca^{2+} currents.

More recently, an alternative method of loading cells with antibody has been described (29). This involves a 'scrape-loading' procedure based on that first used by McNeil and co-workers (30), and involves transiently severing the neurites from cultured neurones by washing them gently from the coverslips to which they are attached using a small volume of culture medium and a fire-polished Pasteur pipette. The severing of the neurites allows the antibody, present in the culture medium, access to the cell interior, with a calculated efficiency of around 25% for a 150 kDa molecule (30). A similar technique has been used to load the oncogene protein T25-*ras* into embryonic chick neurones. Whilst the scrape-loading technique as used by Menon-Johansson and Dolphin (29) has proved successful in cultured rat dorsal root ganglion neurones, a far lower success rate has been achieved in superior cervical ganglion cells, with loading (or lack of it) being verified immunocytochemically (M. P. Caulfield, personal communication). In these cells, microinjection remains the method of choice.

The main advantage that antibodies have over reconstitution methods is that the antibodies block the functional transduction pathway, whereas reconstitution studies give information on what coupling *can* occur as opposed to what *does* occur. An alternative method of blocking the transduction pathway was employed by Kleuss *et al.* (31) in a study of the coupling between muscarinic and somatostatin receptors and voltage-dependent calcium currents in rat pituitary GH_3 cells. These workers used a microinjection technique to introduce antisense DNA oligonucleotides into cells. These were complimentary to unique sequences of the mRNA encoding the α-subunits of G_o subtypes. Control cells were microinjected with nonsense DNA, and the ability of the anti-sense DNA to inhibit the production of the respective G proteins over a 72 h period was followed immunocytochemically. A combination of the immunohistochemical and electrophysiological studies showed that maximum inhibition of G-protein production occurred around 40 h after microinjection of the anti-sense DNA. More recently, the same method has been used in cultured dorsal root ganglion neurones, again to study the coupling between receptors—in this case GABA-B receptors—and voltage-dependent calcium channels (32).

4.4 Pathways involving soluble second messengers

As well as coupling directly to ion channels, G proteins couple receptors to a number of effector enzymes which produce biologically active second messengers. These second messengers include cAMP, IP_3, DAG, and arachidonic acid metabolites, all of which have been shown to be involved in the regulation of ion-channel activity in excitable cells. G proteins do not appear to be involved directly in the activation of guanylate cyclase to produce cGMP, but this cyclic nucleotide has also been shown to regulate ion-channel activity, most notably in the retina.

4.4.1 Adenylate cyclase

The production of cAMP by adenylate cyclase can be both stimulated and inhibited following receptor activation. The stimulatory action is mediated by the pertussis toxin-*insensitive* G protein, G_s (of which at least four variant forms have been isolated) and the inhibitory action by the pertussis toxin-*sensitive* G protein, G_i, (of which at least three forms have been isolated). If a response to an agonist is thought to be mediated by an increase in cAMP, then the response should be mimicked by a receptor-independent increase in the intracellular concentration of cAMP. This can be achieved by the extracellular application of a membrane-permeable analogue of the nucleotide or the intracellular application of cAMP itself, or by the receptor-independent activation of adenylate cyclase using forskolin.

The membrane-permeable analogues of cAMP most commonly used are dibutyryl and 8-bromo- derivatives. Both are freely soluble in aqueous solution and are effective at concentrations of around 0.1–0.3 mM, although some workers have used as much as 1 mM. cAMP can be applied intracellularly by including it in the patch-pipette filling solution. As with the internal diffusion of all molecules, the effective pipette concentration of cAMP will depend to some extent on the geometry of the patch electrodes being used (see, for example, ref. 33). In cardiac myocytes a pipette concentration of 200 μM has proved effective, although the extracellular addition of the phosphodiesterase inhibitor *iso*butylmethyl xanthine (IBMX) can reduce the maximally effective cAMP concentration to around 30 μM. cAMP can also be introduced into cells using the technique of flash photolysis. Here a photolysable 'caged' analogue of cAMP is allowed to diffuse into the cell via the patch pipette. The caged compound is biologically inert, but active cAMP can be released by a brief flash of UV light (34).

Forskolin is a diterpene which stimulates adenylate cyclase directly and so can be used to increase the intracellular concentration of cAMP in a receptor-independent fashion. Forskolin can be made up as a stock solution in DMSO at a concentration of 10 mM and is effective when applied extracellularly at concentrations of 0.1–50 μM. The response to any agent which acts to increase the intracellular concentration of cAMP should be enhanced by inhibition of phosphodiesterase. This can be achieved using IBMX, which is effective when applied extracellularly at concentrations of 20–100 μM.

As cAMP produces its effects within cells by activating cAMP-dependent protein kinase (protein kinase A), which goes on to phosphorylate numerous cell proteins, including for example the L-type calcium channel, the whole-cell variant of the patch-clamp technique can be used to study the transduction pathway at the level of the kinase. For example, the catalytic subunit of cAMP-dependent protein kinase has been introduced into cardiac myocytes (33) as has the inhibitor of the enzyme Rp-adenosine 3′,5′-monothionophosphate (Rp-cAMPS) (35). In addition, analogues of ATP which are either non-hydrolysable (e.g. AMP-PNP), or produce irreversible thiophosphorylation (ATP-γ-S), can be tested for their ability to abolish the response to agonist or render it irreversible, respectively. The effective concentrations of these analogues are determined to some extent by trial and error, but in each case a concentration of 1 mM is a good starting point.

As mentioned above, a number of receptors can act to inhibit the activity of adenylate cyclase. If an electrophysiological response is mediated by such an action it should be sensitive to pertussis toxin. In addition, flooding the cell with cAMP, either by applying it or one of its analogues directly, or by using a combination of forskolin and IBMX, should prevent the response to agonist. This approach has been used in the study of the inhibitory effects of α-adrenoceptor agonists on calcium currents in NG108–15 cells (36).

4.4.2 Guanylate cyclase

Recently there has been a resurgence of interest in the guanylate cyclase/cGMP signalling pathway, due mainly to the realization that nitric oxide, which is a potent activator of soluble guanylate cyclase, is formed in a number of cell types in both the periphery and the central nervous system. Experimental protocols used to study the role of cGMP in mediating electrophysiological responses are similar to those described above for cAMP. The membrane-permeable analogue of cGMP used most commonly is 8-bromo cGMP (at concentrations in the range 10–500 μM). Alternatively, cGMP can be added to the patch-pipette during whole-cell experiments (at similar concentrations). Although a number of drugs (e.g. the methylxanthines) inhibit cGMP phosphodiesterase, very few are selective in their action, since they also inhibit cAMP phosphodiesterase and thus may raise cAMP and cGMP levels simultaneously. The drug MB 22948 (Zaprinast) is a preferential inhibitor of cGMP phosphodiesterase. It can be obtained directly from Rhone-Poulenc Rorer (Dagenham, Essex, UK) and is active at concentrations around 1–10 μM. The drug is dispersed initially in triethanolamine (10% (v/v) in water) at a concentration of 10 mM, before being diluted to the required concentration in extracellular solution.

4.4.3 Phospholipase C

Activation of phospholipase C leads to the hydrolysis of the membrane phospholipid phosphatidyl inositol 4,5-bisphosphate (PIP_2) to produce two second messengers, inositol (1,4,5)-trisphosphate (IP_3), and diacylglycerol (DAG). IP_3 is a soluble, cytoplasmic messenger which acts on a specific receptor to mobilize calcium from intracellular stores. A number of IP_3 receptors have been identified, and these (along with the family of ryanodine receptors first identified in skeletal muscle) act as intracellular calcium ion channels. A number of studies have been carried out on the electrophysiological properties of these channels (e.g. see Chapter 2) and in the case of the IP_3 receptor it has been shown that the channel is regulated both by IP_3 and by the cytoplasmic calcium concentration. In addition, IP_3 is metabolized to a wide range of inositol phosphates, such as inositol 1,3,4,5-tetrakisphosphate (IP_4) which has been postulated to stimulate calcium influx across the plasma membrane. DAG activates protein kinase C, and both arms of the phospholipase-C signalling pathway have been widely reported to be involved in the coupling between receptors and ion channels.

i. The IP_3 pathway

Since responses mediated by IP_3 occur as a result of a rise in the concentration of free intracellular Ca^{2+}, the ability of intracellular Ca^{2+} buffers to inhibit a response to agonist is often the first clue that IP_3 might be involved. Either EGTA or BAPTA, at intracellular concentrations of 1–10 mM, should be effective in blocking such responses (also see remark in Section 3.2.1). If the response is Ca^{2+} dependent, then a more selective interference with the IP_3 pathway can be achieved by the intracellular application of heparin ($M_r \simeq 5000$ Da) which blocks the IP_3 receptor at concentrations in the range 1–5 mg/ml. It should be noted, however, that heparin may also block the production of IP_3 by inhibiting G-protein activation (37) so care must be taken when interpreting results obtained with this agent.

In theory, it should be possible to apply IP_3 intracellularly using patch-pipettes. In practice, this has proved difficult, probably because the diffusion of IP_3 into the cell causes a *gradual* rise in intracellular calcium as opposed to the *transient*, synchronized release produced by the physiological stimulus. One approach to overcome this problem has been to use caged IP_3, where the active inositol phosphate is released from an inactive, photolysable precursor by a brief flash of UV light (cf. remarks in Section 4.4.1 on 'caged' cAMP). Recently this approach has been used to show that IP_3 can release stored Ca^{2+} to inhibit voltage-dependent calcium currents in smooth muscle cells (38). The response to IP_3 was blocked by 10 mM EGTA and 2 mg/ml heparin.

ii. The DAG pathway

DAG activates protein kinase C (PKC), which in turn phosphorylates a number of intracellular target proteins. The main pharmacological tools used in the study of this pathway are direct activators or inhibitors of PKC. Phorbol esters are powerful inflammatory and tumour-promoting compounds which activate protein kinase C, and should thus be able to mimic responses mediated by the DAG pathway. The β-phorbol esters, 12-*O*-tetradecanoyl-phorbol-13-acetate (TPA), phorbol 12-myristate 13-acetate (PMA), and phorbol 12,13-dibutyrate (PDBu) are membrane-permeable and activate the kinase when applied extracellularly at concentrations around 100 nM. The three phorbols mentioned above can be made up as 10 mM stock solutions in DMSO prior to being diluted in extracellular buffer. Appropriate control experiments should be performed using DMSO or an inactive phorbol such as 4-α-phorbol 12,13-didecanoate (4α-PDD)—which does not activate protein kinase C—dissolved in DMSO (see below). If solvent effects prove to be a problem, more water-soluble phorbols are available such as phorbol 12,13-diacetate (PDAc), but generally higher concentrations of these are required, presumably due to restricted diffusion across the cell membrane. An alternative to using phorbol esters to activate protein kinase C is to use the structural analogue of DAG, 1,2-oleoylacetyl glycerol (OAG). This is again active when applied extracellularly, the concentrations required being around 1–10 μM.

Phorbol esters, as well as being used acutely to activate protein kinase C, can also be used over longer periods to downregulate the enzyme. In such experiments cells are incubated with the phorbol esters for typically 18–24 h prior to testing agonist responses. This approach has been used recently in NG108–15 cells to study the coupling between bradykinin receptors and the m-current (39). The most commonly used inhibitors of protein kinase C are staurosporin and H-7. Staurosporin is active at low-nM concentrations and can be applied either in the pipette or added extracellularly after being dissolved in DMSO. At higher concentrations, staurosporin will also inhibit other kinases (see ref. 40 and references therein). H-7 is water soluble and active when applied extracellularly at concentrations of around 10–100 μM. In addition to these two inhibitors, pseudosubstrate peptides such as protein kinase C (19–36) have also been used to inhibit the enzyme when applied intracellularly (41).

4.4.4 Arachidonic acid metabolites (eicosanoids)

Arachidonic acid is produced from the breakdown of membrane phospholipids by the enzyme phospholipase A_2. This enzyme can be activated by G-protein coupled receptors (42) and the arachidonic acid so produced is subsequently metabolized to a range of biologically active oxidation products, known collectively as the eicosanoids. Two main enzymatic pathways have been shown to be involved in the production of eicosanoids from arachidonic acid: the cyclo oxygenase pathway giving rise to prostaglandins (PG) and thromboxanes (TX), and the lipoxygenase pathway giving rise to leukotrienes (LT) and hydroxyeicosatetraenoates (HETE). Eicosanoids released extracellulary have long been known to be involved in the inflammatory response. More recently they have also been shown to act as intracellular messengers, and the products of the lipoxygenase pathway in particular have been shown to be capable of regulating ion-channel activity.

In most cases, the first clue to the involvement of eicosanoids in a response comes from the ability of exogenously applied arachidonic acid to mimic the response to agonist. The arachidonic acid can be prepared at a stock concentration of 10 mM in DMSO or 100% ethanol and then diluted in either extracellular or intracellular medium to a final concentration of 20–50 μM. Assuming that the arachidonic acid produces a response, specific inhibitors of the lipoxygenase and cylo-oxygenase enzymes can be used to probe the involvement of these different pathways. The lipoxygenase pathway can be inhibited by nordihydroguaiaretic acid (NDGA) with an EC_{50} of around 5 μM. The inhibitor is again made up in DMSO or ethanol as a 10 mM stock solution prior to being diluted in extracellular medium. The cyclo-oxygenase pathway can be inhibited by the non-steroidal anti-inflammatory (NSAI) class of drugs, of which indomethacin is the most commonly used, being effective at concentrations between 2 and 5 μM. Stock solutions (10 mM) are prepared in 0.5% Na_2CO_3 or in ethanol.

If inhibitors of neither pathway inhibit the response to arachidonic acid (or agonist), then a final possibility which should be considered is that the fatty acid is acting directly on the ion channel. In cases such as this, other fatty acids which are *not* substrates for either the cyclo-oxygenase or the lipoxygenase enzymes should also be effective. One example is myristic acid, which is a fully saturated fatty acid and is therefore not a substrate for the enzymes. The direct regulation of ion channels by fatty acids has been the subject of a recent review (43).

5. Conclusions

In this chapter we have attempted to review the methodology used to study second-messenger regulation of ion channels in both plant and animal cells. It has not been our intention to include every experimental protocol employed in such studies, but rather to provide a basic set of protocols for free [Ca^{2+}] measurements (and for preparing part cells and membranes) and to outline the types of experiments that can be attempted using electrophysiological techniques to elucidate signal-transduction mechanisms. It is likely that in the time between preparation of the manuscript and the appearance of the book, there will have been a number of advances in the biochemistry of signal transduction. For example, at the time of writing there is great excitement about the possibility that cyclic ADP ribose might play a second-messenger role during calcium homeostasis. It is hoped that the principles underlying the experiments described above can be put to good use in probing further the complexities of signal-transduction pathways.

References

1. Blatt, M. R. (1991). In *Methods in plant biochemistry* (ed. K. Hostettmann), p. 281. Academic Press, London.
2. Taylor, A. R. and Brownlee, C. (1993). *Planta* **189**, 109.
3. Kaila, K. and Voipa, J. (1985). *J. Physiol.* **369**, 8P.
4. Kruse, T., Tallman, G., and Zeiger, E. (1989). *Plant Physiol.* **90**, 1382.
5. Raschke, K. and Hedrich, R. (1989). In *Methods in enzymology* (ed. S. Fleischer and B. Fleischer), Vol. 174, p. 312. Academic Press.
6. Elzenga, J. T. M., Keller, C. P., and van Volkenburg, E. (1991). *Plant Physiol.* **97**, 1573.
7. Vogelzang, S. A. and Prins, H. B. A. (1992). *Protoplasma* **171**, 104.
8. Taylor, A. R. and Brownlee, C. (1992). *Plant Physiol.* **99**, 1686.
9. Kurkdjian, A., Leitz, G., Manigault, P., Harim, A., and Greulich, K. O. (1993). *J. Cell Sci.* **105**, 263.
10. Ogden, D. (ed.) (1994). *Microelectrode techniques: the Plymouth workshop handbook* (2nd edn). Company of Biologists Ltd, Cambridge.
11. Cobbold, P. H. and Rink, T. J. (1987). *Biochem. J.* **248**, 313.
12. Knight, M. R., Campbell, A. K., Smith, S. M., and Trewavas, A. J. (1991). *Nature* **352**, 405.
13. Rizzuto, R., Simpson, A. W., Brini, M., and Pozzan, T. (1992). *Nature* **358**, 325.
14. Grynkiewicz, G., Poenie, M., and Tsien, R. Y. (1985). *J. Biol. Chem.* **260**, 3440.
15. Di Virgilio, F., Steinberg, T. H., Swanson, J. A., and Silverstein, S. C. (1988). *J. Immunol.* **140**, 915.
16. Almers W. and Neher, E. (1985). *FEBS Lett.* **192**, 13.
17. Gillot, I. and Whitaker, M. (1993). *J. Exp. Biol.* **184**, 213.
18. Malgaroli, A., Milani, D., Meldolisi, J., and Pozzan, T. (1987). *J. Cell Biol.* **105**, 2145.

19. Robbins, J., Marsh, S. J., and Brown, D. A. (1993). *J. Physiol.* **469**, 153.
20. Selyanko, A. A., Stansfeld, C. E., and Brown, D. A. (1992). *Proc. R. Soc. Lond. B* **250**, 119.
21. Lambert, J. J., Peters, J. A., Hales, T. G., and Dempster, J. (1989). *Br. J. Pharmacol.* **97**, 27.
22. Yang, J., Mathie, A., and Hille, B. (1992). *J. Physiol.* **448**, 237.
23. Soejima, M. and Noma, A. (1984). *Pflügers Arch. Eur. J. Physiol.* **400**, 424.
24. McFadzean, I. and Brown, D. A. (1992). In *Signal transduction: a practical approach* (ed. G. Milligan), pp. 167–79. IRL Press, Oxford.
25. Aghajanian, G. K. and Wong, Y. Y. (1986). *Brain Res.* **371**, 390.
26. Lledo, P. M., Homburger, V., Bockaert, J., and Vincent, J.-D. (1992). *Neuron* **8**, 455.
27. McFadzean, I., Mullaney, I., Brown, D. A., and Milligan, G. (1989). *Neuron* **3**, 177.
28. Caulfied, M. P., Vallis, Y., Mullaney, I., Milligan, G., and Brown, D. A. (1993). *J. Physiol.* **473**, 41P.
29. Menon-Johansson, A. S. and Dolphin, A. C. (1992). *J. Physiol.* **452**, 177P.
30. McNeil, P. L., Merphy, R. F., Lanni, F., and Taylor, D. L. (1984). *J. Cell Biol.* **98**, 1556.
31. Kleuss, C., Hescheler, J., Ewel, C., Rosenthal, W., Schultz, G., and Wittig, B. (1991). *Nature* **353**, 43.
32. Campbell, V., Berrow, N., and Dolphin, A. C. (1993). *J. Physiol.* **470**, 1.
33. Kameyama, M., Hofmann, F., and Trautwein, W. (1985). *Pflügers Arch. Eur. J. Physiol.* **405**, 285.
34. Kaplan, J. H. and Somlyo, A. P. (1989). *Trends Neurosci.* **12**, 54.
35. Hescheler, J., Tang, M., Jastorff, B., and Trautwein, W. (1987). *Pflügers Arch. Eur. J. Physiol.* **410**, 23.
36. McFadzean, I. and Docherty, R. J. (1989). *Eur. J. Neurosci. Eur. J. Physiol.* **1**, 141.
37. Ito, H., Takikawa, R., Iguchi, M., Hamada, E., Sugimoto, T., and Kurachi, Y. (1990). *Pflügers Arch. Eur. J. Physiol.* **417**, 126.
38. Komori, S. and Bolton, T. B. (1991). *Pflügers Arch. Eur. J. Physiol.* **418**, 437.
39. Schäfer, S., Béhé, P., and Meves, H. (1991). *Pflügers Arch. Eur. J. Physiol.* **418**, 581.
40. Bosma, M. M. and Hille, B. (1989). *Proc. Natl Acad. Sci. USA* **86**, 2943.
41. Rane, S. R., Walsh, M. P., McDonald, J. R., and Dunlap, K. (1989). *Neuron* **3**, 239.
42. Axelrod, J., Burch, R. M., and Jelesema, C. L. (1988). *Trends Neurosci.* **11**, 117.
43. Ordway, R. W., Singer, J. J., and Walsh, J. V. (1991). *Trends Neurosci.* **14**, 96.

9

Channel structure

JOHN B. C. FINDLAY and DEREK MARSH

1. Introduction

It is now well appreciated that integral membrane proteins in general and ion channels in particular present formidable problems for the determination of their 3D structure. The first is that most proteins of this class are naturally present at low levels and can be isolated only with considerable difficulty. Thus far, expression systems have not really solved this problem in that, unlike water-soluble proteins, the amounts which can be produced and purified are in most cases still very low. Although there are prospects that heterologous expression may eventually provide the answer, progress at present is slow. The second major problem is that proteins with substantial intramembranous domains do not, with a few notable exceptions, readily form 3D crystals suitable for X-ray analysis using present-day methods. Much more hope lies with 2D systems where ordered arrays are more easily produced and electron microscopy, although more difficult, can be used to provide evidence at low to medium resolution. This topic is dealt with extensively in Chapter 10.

All hope is not lost, however, for there are a variety of biochemical and biophysical techniques which can be employed to generate data which will allow some structural appreciation of the molecule. Although the resultant models are terribly crude, they can provide a useful basis for further experiments designed to refine or correct the model. This will still generate only a low resolution appreciation of the structure, however, and the next steps, such as the identification of mechanisms of action and ligand docking, are highly speculative in the absence of firm structural data. In the long run, it is likely that the procedures and rules governing modelling studies will gradually develop to the point where the models will become more accurate. At present, however, the general dictum must be 'the more data, the more realistic the representation'.

This chapter will deal, mostly in general terms, with the kinds of approaches and data which can help in the production of sensible representations of integral membrane proteins. For that reason, it lacks the usual detailed protocols but it serves the purpose of identifying particular experimental strategies which may provide the required information, although we do cover spectroscopic techniques, including sample preparation, in some detail. Information on protein purification can be found in many other textbooks, including other books in the *Practical approach* series (e.g. *Biological membrane: a practical approach* (ed. J. B. C. Findlay and W. H. Evans) and *Protein purification applications: a practical approach* (ed. E. L. V. Morris and S. Angal).

2. General modelling principles

A number of basic pieces of information are required before the construction of any sensible model can be contemplated. Assuming that the complete protein sequence is available, it is still too speculative to attempt modelling unless analysis of this sequence indicates firm homology with systems for which 3D structural information has been obtained. One now usually has to accumulate steadily a body of evidence using a variety of approaches.

2.1 Topography

Currently, the number and approximate position in the sequence of transmembrane segments is first elucidated by the application of hydropathy plots of various kinds (1–4). These are useful to give a first approximation of the topography and several scales should be applied in order to get some indication of potential variability. Where the protein in question can be identified as a member of a previously known family by the use of feature analysis (e.g. ref. 5), then the delineation of the polypeptide chain in the bilayer can be facilitated. Analysis of a series of aligned sequences is often a more reliable indicator, particularly when used with the determination of hydrophobic and conserved faces (see later). Newer methods based on these approaches look to be a significant improvement (6, 7). In all these situations, however, it should be appreciated that the scope for error is considerable. Hydrophobic stretches are not necessarily located in the bilayer, while charged regions such as occur in Ca^{2+} ATPase can be membrane-inserted and hydrophilic regions can be found in the midst of transmembrane segments (e.g. some G-protein coupled receptors). In addition, we may face a particular ambiguous situation with oligomeric associations. In these cases, some intramembrane sections may be buried in a completely protein environment and therefore may not be subjected to the hydrophobic influences provided by the fatty acid milieu of a lipid bilayer. An example of this may already be seen in the H5 segment in a wide range of ion channels (8, 9).

Recourse should be made to experimental strategies to establish topography. There are several of these and they have different strengths and weaknesses. The first of these is protein modification by either biological or non-biological processes. The identification of glycosylated residues, for example, is almost always symptomatic of a non-cytosolic orientation. Conversely, phosphorylated or methylated residues, particularly if these arise by transient phenomenia, suggest intracellular locations. The identification of such sites or regions is best achieved by submitting the relevant part of the protein to amino acid sequencing (10) and/or analysis by mass spectrometry (11). Inspection of sequences alone for the presence of such sites is a very insecure basis for actual biological modification. Potential glycosylation and phosphorylation sites (e.g. that for myosin kinase, see ref. 12) can be introduced by genetic methods and the engineered construct can be examined with regard to *in vivo* or *in vitro* modification. One other interesting, but by no means reliable, indicator is the observation that the cytoplasmic surface of an emergent transmembrane segment is often occupied by positively charged residues, particularly lysine and arginine (13). This observation has been made with respect to bacterial membrane proteins and although not universally true, there are indications that it may also apply to a limited number of eukaryotic transmembrane segments, e.g. helix 1 of vertebrate rhodopsins. An even less universal situation is the presence of aromatic residues, particularly tyrosine, at the hydrophobic/hydrophilic boundary (14).

Chemical labelling with a variety of hydrophilic probes has been used to identify particularly exposed residues (15). Although very effective, such an approach is very demanding given the problems associated with the amount of material available and with the difficulty in manipulating and purifying hydrophobic peptides. Modern protein engineering techniques can, however, substantially circumvent such problems. Thus, the substitution of single cysteine residues at defined positions in the sequence and subsequent treatment of the protein whilst in the bilayer with radiolabelled, fluorescent or spin-labelled sulphydryl reagents (e.g. ref. 16) can pinpoint the location of particular residues with a high degree of accuracy and are perhaps amongst the most definitive of the topographic probes. Such approaches can reveal nearest-neighbour interactions by monitoring the formation of disulphide bonds (17). Distance relationships can also be calculated by fluorescence quenching measurements.

A quite different approach makes use of the splicing of the transmembrane regions to reporter enzymes such as β-lactamase or PhoA (18, 19). Depending upon whether the resulting construct expresses the reporter enzyme on the intra- or extracellular surface, one can identify the location of the immediately adjacent sequence. This process can be repeated for each of the putative loop regions. The final approach also relies on the prediction of the extramembranous domains to allow the generation of anti-peptide antibodies to these regions (20). The hope here is that such antibodies will recognize the native protein and in that way localize the particular section to an intra- or extracellular position. This strategy has been successful with the β-adrenoceptor but in many cases is frustrated by the inability of anti-peptide antibodies to recognize the native protein.

In these ways one may be able to arrive at a firm estimate of a very fundamental piece of information—how many transmembrane segments are there in the protein and approximately where are the bilayer boundaries located? It is quite surprising that this basic item of information is known for only a small number of proteins. The recent structural data published for the acetylcholine receptor illustrate dramatically how misleading the sole use of hydropathy indices can be (21).

2.2 Conformation

The acetylcholine receptor also illustrates the importance of determining the conformation, at least in crude terms, of the intramembranous sector of a protein. The identification of a 20–22 residue hydrophobic region has generally been taken to indicate that the segment adopts an α-helical conformation, largely on the basis of the structure of the light-harvesting complex where such a stretch of residues is required to traverse the hydrophobic portion of the bilayer. Less than half that number are required for a β-sheet type structure and then only every other residue need be hydrophobic. The suggestions made with respect to the acetylcholine receptor tend to shatter that presumption for, with this protein, there are three 20–22 residue hydrophobic sections which may in fact adopt a largely β-sheet conformation. Presumably, there will be at least six transmembrane passes in the β-sheet conformation.

Secondary structure predictions either solely or in the form of a composite analysis (22) can give some indication of the conformation of a transmembrane segment but these predictions are highly unreliable and are often incapable of deciding between α-helical and β-sheet forms. A more recent approach makes use of aligned sequences and Fourier-transform analysis of hydrophobic and conserved positions (6). The position of the peak in the Fourier-transform power spectrum is indicative of the periodicity in the sequence alignment and can be compatible with either a helix or β-sheet structure. The method can be used further to indicate whether there are any discontinuities in the regular structure.

When the protein can be obtained in sufficient quantities, however, the best approach is an experimental one using either circular dichroism (CD) or Fourier-transform infrared (FTIR) spectroscopy. These techniques offer the best prospects of obtaining a rough estimate of the proportion of α-helix and β-sheet in a protein or in the transmembrane domain. This has been carried out for bacteriorhodopsin (23) and the proton channel of the vacuolar H^+ ATPase (24). Potentially, nuclear magnetic resonance (NMR) spectroscopy could be very useful but the technique at present requires work on peptides or with proteins dissolved in organic solvent. Both situations raise concerns as to whether perturbation of the native structure may result, and for these reasons the approach is not that of first choice. Some small proteins, however, may be addressed in mild detergent, and in such cases, NMR may produce valuable information. Finally, ESR with spin-labelled residues can indicate periodicity in the availability of the probe to quenching and this has been interpreted to indicate secondary structure (25). The approach is ingenious but relies on a combination of protein engineering, expression of relatively large quantities of material, and efficient reconstitution of active protein.

2.3 'Sidedness'/orientation

The orientation, i.e. 'sidedness', of each of the transmembrane segments is the next important parameter to ascertain. Since these regions will be largely regular, this kind of information for only a small proportion of the total residues can be used largely to orient most of the segment. Again, one can adopt a chemical labelling approach by determining the availability of individual side chains to photosensitive or sulphydryl hydrophobic probes which partition into the bilayer (26, 27). Alternatively, the environment of spin labels attached to particular residues can be determined (25). Both approaches are labour intensive but should give consistent results and have the potential to indicate regions which may depart from structural regularity.

The chemical labelling strategy has one other potentially important use, particularly for ion channels which, because of their likely oligomeric organization, may have segments which are not exposed to the lipid milieu. In such cases, it should not be possible to modify any residue by probes which partition into the lipid phase. Thus if one produces a series of sulphydryl-containing mutants representing a contiguous series of six to eight residues and none of these mutants becomes modified then it is not unreasonable to assume that the segment is not in contact with the lipid phase.

The Fourier-transform prediction method mentioned above will indicate the probable internal, hydrophobic, and conserved faces of the transmembrane segments. For the G-protein linked receptors, one face of each helix is more conserved than the other and this conserved face is likely to be sequestered within the structure (28). The buried face appears to contain all the intramembranous charged residues and most of the polar side chains while the variable face is more hydrophobic, all useful indicators of a non-lipid orientation. Interestingly, if, for example, the helical structure extends beyond the confines of the hydrophobic phase of the bilayer into the phospholipid head group regions and beyond, the hydrophobic and conserved vectors now lie in approximately the same direction, consistent with the sequestration of hydrophobic residues away from the now largely hydrophilic external environment. This change of direction of the hydrophobic vector is also likely to be a useful indicator of where the polypeptide chain emerges from the bilayer (29).

Both the labelling and predictive procedures can also give a rough estimate of the proportion of the circumference of a helix that is likely to be exposed and/or buried. When one comes to packing these helices into a 3D structure, this information can be useful in deciding both the connectivity and position of a segment of the protein, particularly if low resolution data are available from electron microscopy. But, before moving to the final phase of model building, it is vital to appreciate the quaternary structure of the protein. The classical approach is chemical cross-linking, particularly in detergent-soluble form when intermolecular collision frequency is low (30). Such knowledge could have a profound interpretation on lipid exposure and hence on the packing arrangement.

2.4 Folding

The final, most definitive but also most difficult element of structure to obtain is some idea of the 3D folding pattern or packing arrangement of the transmembrane segments. This can at present really only be obtained from X-ray or electron diffraction studies. Such information is known in a few cases to quite high resolution (31–33) but even at low resolution, the data can be very valuable. It will, for example, provide firm evidence for the quaternary organization of a protein and/or for its overall shape. This may in turn allow information such as the kind indicated in the above sections to be amalgamated into a reasonable appreciation of the structure, including the connectivity of the transmembrane segments. Without some indication of the folding pattern, it is this author's fear that all models may be too speculative to be credible.

This chapter has so far attempted to give a guide to the kinds of information required for any realistic modelling of integral membrane proteins and to provide a survey of how the necessary data might be obtained. What follows is a more detailed description of the biophysical approaches which can be utilized to provide some of this information.

3. Spectroscopic characterization of channel structure and assembly

The spectroscopic methods that can be employed for channel studies divide into roughly two types: those such as Fourier-transform infrared (FTIR), circular dichroism (CD), and nuclear magnetic resonance (NMR) spectroscopies which give direct information on the protein structure, and those such as spin-label electron spin resonance (ESR) and fluorescence spectroscopies from which the structural information is obtained by rather more indirect methods. Practically, FTIR and ESR spectroscopies are well adapted to membrane studies, since they are undisturbed by large, scattering particulate systems. High-resolution NMR spectroscopy, on the other hand, is applicable only to relatively small peptide stretches either in solution or in a micellar environment. The optical methods, particularly CD, are potentially susceptible to scattering artefacts when used with large particulate systems. For the above reasons, FTIR and ESR spectroscopies are considered in detail below and the other spectroscopies are given a briefer, more general treatment. Additionally, although NMR spectroscopy is potentially an extremely powerful method, it is technically too complex for us to give a comprehensive coverage here of all its instrumental aspects.

3.1 Fourier-transform infrared (FTIR) spectroscopy

FTIR spectroscopy is especially suitable for determining the secondary structure of channel peptides, both in solution and in opaque or highly scattering (i.e. membranous) samples, by using the amide I vibrational bands of the peptide backbone (cf. *Table 1*). In addition, vibrational bands may also be resolved from certain of the peptide side chains. Identification of the individual bands characteristic of the different secondary structures in the overall amide I spectral envelope is performed by band-narrowing resolution-enhancement techniques: either Fourier self-deconvolution or the second-derivative display. Quantitation is performed by fitting the individual components to the overall envelope of overlapping bands. Rates of (slow) exchange of the amide protons in the peptide backbone for deuterium can be determined, either from band shifts in the amide I region, or from the loss of intensity in the amide II region (35), to give information not only on secondary structure but also on tertiary assembly. If membranous samples can be oriented on a substrate, the orientation of the peptide backbone relative to the membrane surface can be determined from the dichroism of the amide I band with polarized infrared radiation (36).

3.1.1 FTIR spectrometer requirements

A Fourier-transform instrument is essential in order to be able to subtract the strong infrared absorption by water with a buffer blank sample by using the standard software supplied. A high-resolution instrument is not required because the bands are generally broad: 2–4 cm^{-1} resolution is adequate. Practically any instrument of the lower-cost range, with a dedicated PC-compatible computer for data collection and analysis, is suitable. A particular problem is the elimination of water vapour from the instrument

Table 1. Infrared absorption band positions in the amide I region of proteins in D_2O (34)

Structure	Mode	Frequency (cm^{-1})
β-sheet or	$\nu_{\perp}(\pi, 0)$	1624 ± 4, 1631 ± 3, 1637 ± 3
extended chain	$\nu_{\parallel}(0, \pi)$	1675 ± 4
α-helix	$\nu(0)$	1654 ± 3
Unordered (random)		1645 ± 4
Turns and bends [a]		1663 ± 4, 1670 ± 2, 1683 ± 2, 1688 ± 2
—		1694 ± 2

[a] These assignments are less certain.

because this gives sharp lines over a broad range of interest that can interfere both with quantitation and resolution enhancement. Care must be taken to purge the spectrometer with dry nitrogen, to minimize the length of time that the sample chamber is exposed to the atmosphere, and to allow adequate time for purging after changing samples. An automatic sample changer with which the sample chamber remains closed can be helpful in this respect. Vacuum instruments have improved performance with respect to contamination by water vapour but are much more expensive. Many interferograms (typically 100) are accumulated automatically to improve signal-to-noise ratio and these are then Fourier transformed to yield the whole spectrum from which the bands of interest may be selected. Triangular apodization prior to Fourier transformation, using the standard software, is recommended for improving signal-to-noise ratios without introducing negative 'side-lobes' in the spectrum and with least complications during resolution enhancement.

3.1.2 FTIR sample cells

As standard, a transmission sample cell with CaF_2 windows is used for aqueous samples. These are separated by a Teflon spacer-gasket of 25–50 μm thickness (for samples in D_2O), and this forms the chamber in which the sample is contained. For samples in H_2O, thinner spacers of 5 μm or less are required because of the stronger infrared absorption by H_2O. (Alternatively, the membranes may be partially dried or ATR methods—see below—may be used, for samples in H_2O.) The windows are clamped hermetically with Viton gaskets between metal holders through which thermostatted water may be circulated for temperature control. Liquid samples may be introduced via a small hole at the top of one of the windows, but thick membrane suspensions are better introduced by first opening the cell and then re-clamping it. Care should be taken not to introduce air bubbles into the sample. A system has been described to produce oriented membrane samples on an AgCl substrate by centrifugation combined with drying, for FTIR transmission studies with polarized infrared radiation (37).

For ATR (attenuated total reflection) measurements on oriented systems, the membrane must first be oriented on the germanium (or zinc selenide) ATR crystal. This can possibly be achieved by evaporation of a solution of lipid plus peptide on one of the reflecting surfaces of the ATR crystal and subsequent hydration with buffer. The hydrated sample must then be hermetically sealed to prevent drying out. This can best be achieved with a horizontal ATR cell. In certain cases, oriented membrane fragments may be produced by adsorption on an ATR crystal that is first coated with a single lipid bilayer by Langmuir–Blodgett techniques (38). For dichroism measurements an infrared polarizer is inserted in the infrared beam before the sample. ATR (liquid) cells may also be used for unoriented samples in H_2O, to circumvent the difficulty in handling extremely thin Teflon spacers.

3.1.3 FTIR sample requirements

Typically 2–3 mg (dry weight) of sample are required. This is less for signal-to-noise reasons than to ensure that the infrared spectrum from the peptide is sufficiently strong relative to the water absorption (which inevitably introduces baseline distortions after subtraction) and to the water-vapour lines. Samples are normally examined in D_2O buffer because D_2O both has a weaker infrared absorption than does H_2O and also, unlike H_2O, does not have an absorption in the amide I region. For membrane samples, repeated centrifugation and resuspension is required to replace the H_2O buffer by D_2O buffer. It should be noted that spectral shifts take place in the amide I region of the spectrum, depending on the degree of exchange of the amide hydrogens, i.e. on the duration and temperature of incubation in the D_2O buffer. Care should also be exercised both in storage of the D_2O buffer and in sample preparation, in order to minimize the exchange of the D_2O buffer with atmospheric H_2O. Compounds having carbonyl infrared bands in the region near the amide I band, e.g. carboxylic acids and urea, must be excluded from the buffer.

Because the sample volume is small (25 μm spacer and ca. 1×2.5 cm^2 area $\equiv$ 62.5 μl) concentrated samples are required. For membranous samples, concentration is best achieved by low-speed centrifugation in a bench-top centrifuge, or by dispersing a lipid–peptide mixture dried from organic solvent in a small volume of buffer, and then transferring the thick dispersion to the opened cell. For lipid–peptide mixtures, it should be noted that peptides can have different conformations in different organic solvents and this may affect the secondary structure in the resulting lipid dispersion. Also, drying from organic solvent may not produce samples of homogeneous lipid/peptide ratio (which should be checked by sucrose density-gradient centrifugation) and dialysis from an organic solvent miscible with water, e.g. 2-chloroethanol (39), may be preferable.

3.1.4 FTIR spectral analysis

The (primarily) carbonyl infrared bands corresponding to the different secondary structures of the peptide backbone are relatively closely spaced and all contribute to the overall spectral envelope in the amide I region at 1600–1700 cm^{-1}. Unless the peptide is essentially in a single conformational state (see *Figure 1*), the different components must be identified by using band-narrowing techniques and quantitated by band fitting. Of the resolution enhancement techniques, Fourier self-deconvolution (41), unlike the other popular second derivative display, has the advantage of preserving the relative band intensities and therefore may be used also for subsequent band-fitting (34). In all resolution enhancement techniques, care must be taken not to over-deconvolute, hence giving rise to spurious bands, e.g. from noise or water-vapour lines. With regard to both the latter, prior spectral smoothing may be

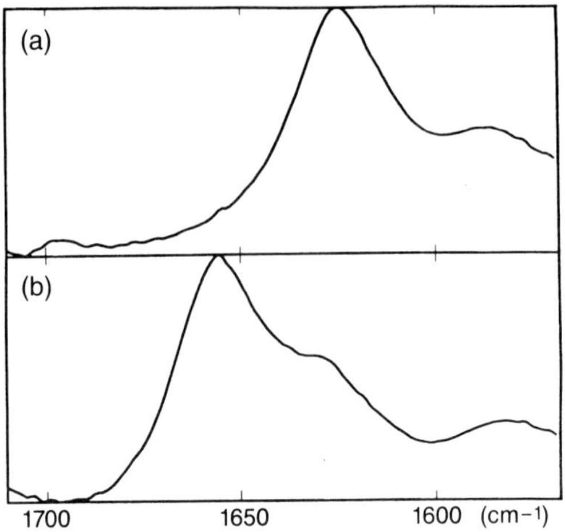

Figure 1. FTIR spectra of (a) the I_{sK} potassium channel-derived transmembrane peptide of sequence KEALYILMVLGFFGFFTLGIMLSYIR at a lipid/peptide ratio of 7.5 mol/mol and (b) a bipolar hydrophobic peptide of sequence $K_2GL_{20}K_2A$, both in dimyristoyl phosphatidylcholine bilayers dispersed in D_2O buffer at 30°C. Interferograms were collected on a Bruker IFS 25 spectrometer with DTGS detector, at a resolution of 4 cm^{-1}, apodized with a triangular function, and Fourier transformed after one level of zero-filling. The I_{sK} peptide (a) has a single band in the amide I region at 1626 cm^{-1} that is characteristic of a β-sheet structure. The $K_2GL_{20}K_2A$ peptide (b) has its major band at 1656 cm^{-1}, characteristic of α-helical structure, and a minor band at the position expected for β-sheet structure (40).

advantageous because both are sharper (and hopefully much less intense) than the spectral bands of interest.

Resolution enhancement by Fourier self-deconvolution (41) is best performed conservatively assuming an initial half-width at half-height (HWHH) of typically 17 cm^{-1} and taking a line-narrowing factor of K=2. The band-fitting of the resulting self-deconvoluted spectrum can then be performed with Gaussian bands under the reasonable constraint that all components have comparable widths. This is best done in stages by first holding the widths constant, then with the fixed-component frequencies so obtained, the band widths are varied, and finally the resulting parameter set is optimized by releasing both band frequencies and bandwidths. A typical example of a final fit is shown in *Figure 2*. Band fitting of the original spectrum is also possible.

3.2 Spin-label electron spin resonance (ESR) spectroscopy

Spin-label ESR spectroscopy is a method that is sensitive to the molecular mobility of membrane components and is applicable to opaque or highly

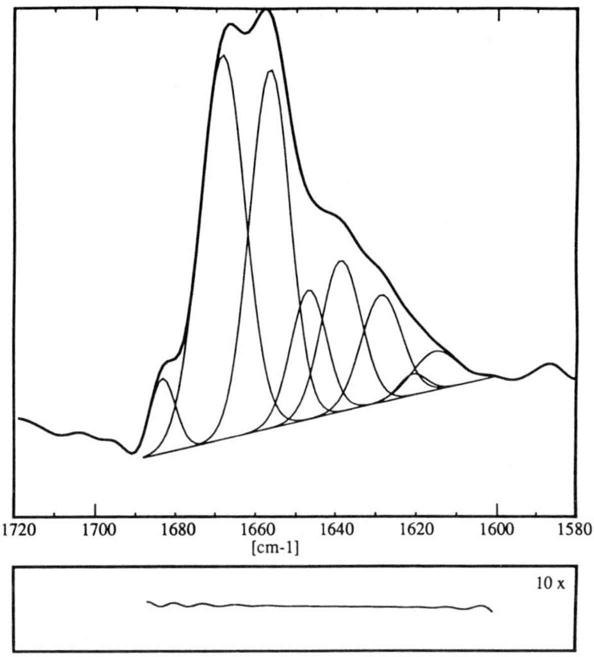

Figure 2. FTIR spectrum (amide I region) of 16 kDa channel-containing membranes from *Nephrops norvegicus* in D_2O buffer after Fourier self-deconvolution with a Lorentzian linewidth (HWHH) of 17 cm^{-1} and a band-narrowing factor of K=2. The spectrum was recorded at a temperature of 15°C. The deconvoluted spectra are fitted by single Gaussian bands as indicated, with the following frequencies and intensities: 1684 cm^{-1}, 4%; 1669 cm^{-1}, 34%; 1657 cm^{-1}, 28%; 1647 cm^{-1}, 9%; 1639 cm^{-1}, 12%; 1629 cm^{-1}, 8%; 1621 cm^{-1}, 1%; 1616 cm^{-1}, 3%. Component bandwidths are in the range of 7.5–13 cm^{-1}. The residuals from the fit in the regions of the individual band positions are <0.02–0.1% (42).

scattering membranous samples. Its principal advantage lies in the fact that the time-scale of sensitivity of conventional ESR spectra from nitroxide spin labels is optimally matched to the motions of the lipids in membranes, as is that of saturation transfer ESR (STESR) spectra to the overall motions of membrane proteins. In the field of membrane channel studies, generally useful applications are to the study of lipid–protein interactions by conventional ESR, which give a measure of the intramembranous perimeter of the channel assemblies, and to the determination of rotational rates of channel proteins, which are related directly to the cross-sectional area of the rotating species. Both are therefore applicable to determining the state of oligomeric or supramolecular organization of the channels in the membrane. Of at least equal importance are saturation studies, from which the accessibility of spin-labelled residues to paramagnetic ions and to molecular oxygen can be used to determine the location, orientation, and conformation of channel-peptide segments in the membrane.

3.2.1 ESR spectrometer requirements

Standard 9 GHz frequency (X-band) ESR spectrometers are suitable for spin-label studies. The spectra are centred in the normal g=2 region (at *c*. 3250–3400 Gauss), and do not require particularly high resolution, nor wide sweep-widths (normally 100 Gauss). Standard microwave cavities may be used, but for limited amounts of sample, loop-gap resonators (43) or dielectric resonators are advantageous because of higher filling factors, and also for saturation studies where higher microwave (H_1) fields are desirable. Most modern spectrometers are equipped with the second-harmonic detection and accurate phase-setting capability that is essential for saturation transfer ESR studies. Computer control of the spectrometer is not necessary, but a dedicated computer or PC is required in many cases for data collection and analysis. Accurate temperature control is necessary and best can be effected by using a gas-flow system with double-wall quartz Dewar in the microwave cavity. Experimental details and operating procedures can be found in ref. 44. Typically, modulation amplitudes in the region of 1.0 Gauss are used, microwave powers of 10 mW or less, field-scan time of 4 min with filter time constant of 0.1–0.3 sec, and spectral averaging of repeated scans, if necessary to improve signal-to-noise ratios.

3.2.2 ESR sample cells

The most useful sample cells for membrane samples are generally 1 mm i.d. glass capillaries such as are used for standard 100 μl positive-delivery pipettes. These can be flame-sealed readily. (Glass does have some microwave loss, but seldom gives particularly troublesome background signals. Ideally, ESR-grade quartz capillaries would be used, but these are expensive.) The membrane must be packed in the capillary—after transferring the homogenized suspension with a drawn-out Pasteur pipette—by slow-speed centrifugation with a bench-top centrifuge. Excess supernatant is then removed—leaving only a small amount to prevent the sample drying out—to minimize the microwave losses associated with liquid water. The sealed capillaries are accommodated in a standard 4 mm diameter quartz ESR tube (in the gas-flow Dewar) which contains dry, light silicone oil for thermal stability and a fine-wire thermocouple for temperature measurement. The membrane sample must be centred in the microwave cavity, so it will experience the maximum microwave and modulation magnetic fields.

Special quartz ESR flat cells are available for optimum sensitivity with aqueous solutions, but these are less suitable for particulate samples and are difficult to thermostat. Solutions in apolar (i.e. 'non-lossy') solvents can be examined in the standard 4 mm quartz ESR tubes.

3.2.3 ESR sample requirements

Typically 0.5 mg of sample is suitable; smaller amounts are possible, especially if loop-gap resonators together with a low-gain microwave preamplifier are used. For experiments with spin-labelled lipids, the membranes are doped at a level of 0.5–1 mol% spin label with respect to total lipid in the sample. For samples prepared by hydrating lipid–peptide mixtures dried from organic solvent, the lipid spin label can be added to the organic solvent. For membrane samples and often for samples reconstituted by dialysis, the spin-labelled lipid must be added exogeneously to the aqueous membrane dispersion as a small volume (*c.* 1 μl) of concentrated solution in ethanol (with vortexing). For spin-labelled phospholipids the sample must then be resolved from unincorporated spin label by repeated centrifugation and washing. Single-chain spin-labelled lipids, e.g. fatty acids, establish a partition equilibrium and the samples require only to be concentrated by a single centrifugation after labelling. Proteins are spin labelled by using spin-labelled derivatives of standard reagents for covalent modification of proteins, e.g. maleimide. Synthesis of spin-labelled fatty acids and phospholipids is described in ref. 45. (Certain spin-labelled stearic acids and spin-labelled protein reagents are available from Sigma.)

3.2.4 Lipid spin labels: lipid–protein interactions

Ideally, the spin label [4,4-dimethyloxazolidine-*N*-oxyl (DOXYL) derivative] is attached at the C14 atom of stearic acid, which is the *sn*-2 chain in the case of phospholipids. Of the commercially available (Sigma) stearic-acid derivatives, that labelled on the C16 atom (or possibly C12 atom) is most suitable. The reason for this is to obtain a narrow (i.e. mobile) spectrum from the fluid lipid regions of the membrane, in order to be able to resolve the motionally restricted lipids that are interacting directly with the protein, in the outer wings of the spectrum. The temperature at which the spectrum is recorded is adjusted to optimize the resolution of these two components, and this depends on the position of the spin label and the fluidity of the particular membrane system. Typical spectra are shown in *Figure 3*.

The two components must then be quantified by computer subtraction of the digitized spectra. The protocol is given in ref. 44. The single-component spectrum corresponding to the fluid lipid environment which is required for subtraction is obtained from a dispersion of the extracted membrane lipids (Bligh and Dyer extraction, see ref. 46), or from a lipid dispersion without peptide. These spectra are recorded at closely spaced (1°C) temperature intervals, in order to have a library from which to choose the one that best matches the fluid component in the two-component membrane spectrum. The single-component spectrum corresponding to the protein-interacting environment which is required for the complementary subtraction is obtained from sonicated, small unilamellar vesicles of dimyristoyl phosphatidylcholine

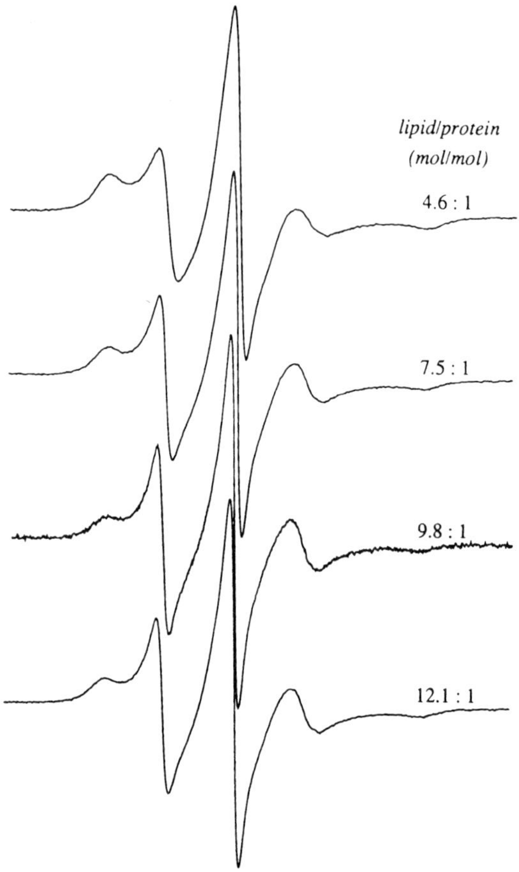

Figure 3. Conventional ESR spectra (first harmonic, in-phase) of the 14-(4,4-dimethyl-oxazolidine-*N*-oxyl) stearic acid (14-SASL) spin label in complexes of the I_{sK} potassium channel-derived transmembrane peptide (of sequence KEALYILMVLGFFGFFTLGIMLSYIR) with dimyristoyl phosphatidylcholine at lipid/peptide ratios of 4.6, 7.5, 9.8, and 12.1 mol/mol, as indicated. Spectra were recorded at 30°C on a Varian Century Line 9 GHz spectrometer with a modulation amplitude of 1.6 Gauss, modulation frequency of 100 kHz, microwave power of 10 mW, filter time constant of 0.128 sec, scan time of 4 min, field centre of 3250 Gauss, and scan width of 100 Gauss (40).

(DMPC) in the gel phase (i.e. below 20°C). Again a spectral library is recorded at closely spaced temperature intervals (from *c.* 0°C). For peptide–DMPC and similar mixtures, spectra recorded from sonicated DMPC above the phase transition may constitute a better library for matching the fluid component than do those from unsonicated DMPC. The advantage of the sonicated system is that it exhibits a wider range of linewidths. Subtraction end points are assessed by comparison of the difference spectrum with the complementary single-component spectrum. The best-matching single components are chosen in this way, by an iterative procedure.

The relative proportions of the two components are then obtained from the fraction, f, of the double-integrated intensity (ESR spectra are first-derivative displays) of the single-component relative to the two-component spectrum that has to be subtracted. Ideally the two subtraction factors from the complementary subtractions should add up to one, i.e. are f and $(1-f)$. If they are close, average the two f values. If the fluid component has narrow lines and dominates the two-component spectrum, it may prove difficult to match exactly and only the complementary subtraction of the motionally restricted component may be feasible.

The total phospholipid/protein molar ratio, n_t, of the sample must be determined by chemical analysis, i.e. protein (47) and phosphate (48) determinations. If the spin-labelled lipid does not display a preferential selectivity for the protein (this is usually the case for spin-labelled phosphatidylcholine, especially in peptide–phosphatidylcholine complexes), the number of lipids, N_b, interacting with the intramembranous perimeter of the protein is simply $N_b = n_t \times f$, where f is the fraction of the motionally restricted component. If the spin-labelled lipid exhibits a selectivity for the protein (this is quite frequently the case for fatty acids), the value of N_b is given (quite generally) by the intercept on the n_t axis in a plot of $(1-f)/f$ against n_t, for lipid–peptide complexes, where n_t can be varied experimentally (45).

The diameter, d, of a lipid chain is approximately 0.48 nm. Thus the effective intramembranous perimeter of the protein assembly is approximately $0.48 \times N_b$ nm (allowing for both chains of a diacyl phospholipid and both sides of the bilayer). This can then be compared with the dimensions of the protein monomer, in order to decide whether part of the monomer perimeter is not accessible to the lipid chains because of the oligomeric assembly of the channel proteins. Straightforward geometric considerations give an estimate of the degree of oligomerization. For n proteins of diameter D, $N_b \simeq \pi(D/d + 1) + nD/d$ for a sandwich arrangement ($n \geqslant 3$), and $N_b \simeq \pi(D/d + 1) + 2(n-1)D/d$ for a linear array. With $D \simeq 1$ nm, these expressions apply to bundles of α-helices (49).

3.2.5 STESR spectroscopy: protein rotation

The protein must be spin labelled with a nitroxide derivative of a protein-modification reagent such that there is little or no independent motion of the label relative to the protein. Of the commercially available reagents (Sigma) the maleimide derivative with the shortest link to a five-membered ring nitroxide (3-maleimido-2,2,5,5-tetramethyl-1-pyrrolidinyloxy) is probably the best. Labelling conditions should be optimized to produce a conventional ESR spectrum that consists wholly or predominantly of very strongly immobilized labels (i.e. maximum outer hyperfine splitting ⩾65 Gauss). With membrane proteins, a sharp spectral component is often found and can sometimes be masked by pre-labelling fast-reacting groups with *N*-ethylmaleimide.

After extensive washing to remove unreacted label, the sample is packed in a 1 mm diameter ESR capillary and excess sample and supernatant removed to produce a sample of 5 mm height. The sample is carefully centred in the microwave ESR cavity (in a position that minimizes microwave–microwave and modulation-field inhomogeneities).

Both sample configuration and instrumental settings are far more critical for saturation transfer ESR measurements than for conventional ESR. Standard operating conditions are given in ref. 50. The STESR spectrum is recorded in the second-harmonic mode with the phase of the detection set accurately 90° out of phase with respect to the field modulation. The detector phase is set such that a null signal is obtained from the sample at low microwave power (<1 mW) (44). The STESR spectrum is recorded at a microwave power such that the microwave field at the sample is $H_1 = 0.25$ Gauss. The exact microwave power required depends on the sample properties (e.g. dielectric loss) and the method of calibration and setting is given in ref. 50. Typically, for a 5 mm packed membrane pellet, the microwave power is in the region of 30–40 mW. The modulation amplitude is set to a fixed standard of 5.0 Gauss.

Typical STESR spectra are shown in *Figure 4*. The rotational correlation time, τ_R^{eff}, is estimated from the ratio of the spectral line heights, L'' and

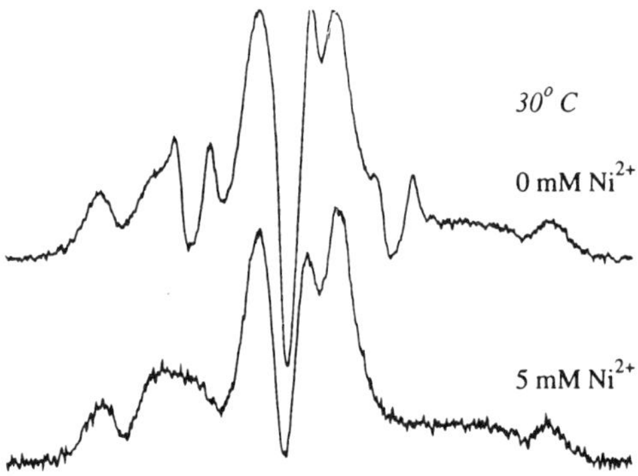

Figure 4. Saturation transfer ESR spectra (second harmonic, 90° out-of-phase) of the 2-[(1-oxyl-2,2,5,5-tetramethyl-3-pyrrolin-3-yl)methenyl]indane-1,3-dione spin label covalently bound to a cysteine mutant of the I_{sK} potassium channel-derived transmembrane peptide (of sequence KEALYILMVLGFFGFFTCGIMLSYIR) in a complex with dimyristoyl phosphatidylcholine, with a sample length of 5 mm. Spectra were recorded at 30°C on a Varian Century Line 9 GHz spectrometer with a modulation amplitude of 5.0 Gauss, modulation frequency of 50 kHz (detection at 100 kHz), microwave power of 35 mW, filter time constant of 0.25 sec, scan time of 4 min, field centre of 3250 Gauss, and scan width of 100 Gauss. The lower spectrum is in the presence of 5 mM Ni^{2+} to quench the sharp spectral component (40).

H'', at intermediate positions in the low- and high-field regions of the spectrum, respectively (which are sensitive to motion) to the line heights at the corresponding extrema, L and H (which are insensitive to motion). Calibrations for these line-height ratios are given by $\tau_R^{eff} = k/[(L''/L)_o - L''/L] - b$, where $k = 116$ μsec, $b = 67.9$ μsec, and $(L''/L)_o = 1.88$ for the low-field ratio, and correspondingly for the high-field ratio, H''/H: $k = 407$ μsec, $b = 210$ μsec, and $(H''/H)_o = 2.17$ (51).

The STESR correlation-time calibrations are for isotropic (i.e. random) rotation, but for a channel assembly in a membrane, the rotation is only about one axis, i.e. the membrane normal. In this case, the effective correlation time depends on the orientation, θ, of the spin-label axes relative to the rotation axis. The true correlation time is then given by ref. 50: $\tau_{R\|} = (\tau_R^{eff}/2) \sin^2 \theta$, which becomes $\tau_R^{eff}/2$ for $\theta = 90°$ which is the most probable orientation. The rotational correlation time is directly related to the volume, V, of the portion of the protein that is *within the membrane* (50): $\tau_{R\|} = 2\eta V/(3kT)$, where $\eta = 2.5$–5 poise is the membrane viscosity, k is Boltzmann's constant, and T is the absolute temperature. Clearly the rotational correlation time is related directly to the degree of aggregation, n, of the channel protein: $\tau_{R\|}$ ($n$mer) $= n \times \tau_{R\|}$ (monomer). Also, if values are assumed for the membrane viscosity and the spin-label orientation, the cross-sectional area, A, of the channel assembly can be estimated, remembering that the membrane thickness is $h \simeq 4.5$ nm and thus $A = V/h$.

3.2.6 ESR saturation studies: protein orientation and conformation

Measurement of the saturation of the ESR spectrum with increasing microwave power allows determination of the accessibility of a spin-labelled group on a channel protein to paramagnetic agents that enhance the spin-lattice relaxation and hence reduce the degree of saturation of the spin label. The principal relaxation agents are molecular oxygen that is preferentially concentrated in the lipid hydrocarbon regions of the membrane (relative to the aqueous phase and the protein interior) and paramagnetic transition-metal ions that are confined to the aqueous phase (52). By combining the results from both these reagents it is possible to say whether residues are exposed at the membrane surface, are exposed to the lipid, or are buried within the protein. This gives information on the transmembrane topology of the channel protein and on the orientation, relative to the intramembranous surface, of those sections of the protein that are in contact with lipid. By using site-specific substitution with cysteine residues (which are readily spin labelled), secondary structures may be distinguished (52) because a transmembrane α-helix will have a 3.6-residue periodicity in exposure to lipid, whereas a β-sheet will have a two-residue periodicity. Additionally, alleviation of saturation by paramagnetic ions may be used to determine the distance from the membrane surface of the spin-labelled residue (53).

The degree of saturation can be determined either from the variation in the intensity of the conventional ESR spectrum with progressively increasing microwave power, or from the integrated intensity of the STESR spectrum normalized to the double-integrated intensity of the conventional first-derivative ESR spectrum (i.e. to the spin-label concentration). The second method is unconventional, but is more directly sensitive to the degree of saturation. The effects of saturation are opposite for the two methods: the intensity of the conventional spectrum is reduced relative to that in the absence of saturation, but increased for the STESR spectrum.

For progressive saturation experiments on the conventional spectrum, requirements for sample size and calibration of the microwave field at the sample, etc., are the same as those given above for STESR spectroscopy. Either the spectrum height or its double-integrated intensity is plotted against the square root of the microwave power, P. In the absence of saturation (i.e. at low powers) the dependence is linear with $\sqrt{P}$, and this linear dependence is extrapolated to higher powers to determine the power $P_{½}$, at which the measured line height or intensity is half what it would be in the absence of saturation. [The double-integrated intensity saturates less rapidly than the line height and the exact power dependence, given by $\sqrt{P}/\sqrt{(1+3P_{½}/P)}$, should be used to fit the whole saturation curve (54).]

The quantity that is used to analyse the relaxation enhancement is $\Delta P_{½} = P_{½} - P^{o}_{½}$, where $P^{o}_{½}$ is the half-saturation power in the absence of the relaxation agent. This quantity is linearly proportional to the concentration of the relaxation agent and is used as the 'accessibility parameter' of the paramagnetic agent to the spin label. In comparing spin labels with different linewidths, this quantity must be normalized by dividing by the linewidth (55). Typically, for experiments with oxygen, a sample saturated with either oxygen or air is compared with a sample that has been carefully deoxygenated (or saturated with either argon or nitrogen). Commonly used paramagnetic ions are chromium oxalate [potassium tris(oxalato)chromate(III) trihydrate (56)] or aqueous $NiCl_2$. Typically, concentrations of 1–5 mM of paramagnetic ion may be used, but up to 30 mM might be appropriate in some cases, and the samples should be deoxygenated. In working with charged membranes, it should be remembered that chromium oxalate is negatively charged and the nickel ion is positively charged, and therefore electrostatics and/or ion binding can have a large effect on the local ion concentrations. Also it should be noted that aqueous nickel has a tendency to precipitate at high pH (above neutral) and that uncharged transition metal ion complexes tend to partition into membranes (55).

For the normalized integrated intensity, I_{ST}, of the STESR spectrum, the accessibility parameter is defined by: $\Delta(1/I_{ST}) = 1/I_{ST} - 1/I^{0}_{ST}$, where I^{0}_{ST} is the value of I_{ST} in the absence of relaxation enhancement. This again is linearly proportional to the concentration of the relaxation reagent, but requires no correction for the linewidth. Normally for either method, the

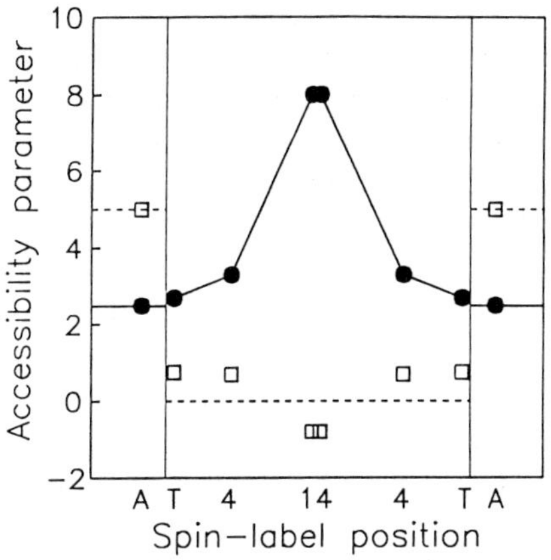

Figure 5. Dependence of the accessibility parameter to 100% oxygen (filled circles) and to deoxygenated 1 mM chromium oxalate (open squares) on the location of the spin label group in dimyristoyl phosphatidylcholine bilayers at 30°C. T is a spin label on the lipid headgroup, 4 and 14 are spin labels on the 4 and 14 carbon atoms (respectively) of the *sn*-2 chain of phosphatidylcholine, and A is a spin label in the aqueous phase. The vertical lines indicating the surface of the membrane have a separation of about 4 nm. Accessibility parameters are determined for saturation of the double-integrated intensity of the conventional ESR spectrum, are corrected for the linewidth, and are expressed in terms of the microwave field at the sample to give true relaxation times (units 10^{13} s^{-2} $Gauss^{-1}$) (55).

relative positional information is obtained by comparison of the accessibility parameters from spin labels at different locations (see, for example, *Figure 5*). In this connection, systematic site-directed spin-labelling is particularly useful for high-resolution studies (52). A direct distance calibration is available for STESR studies with aqueous Ni^{2+} ions: $d\{\Delta(1/I_{ST})\}/d[Ni^{2+}] = 14.8\ \{R^{-3} + (d_1-R)^{-3}\}\ mM^{-1}$, where R nm is the distance from the membrane surface and $d_1 \simeq 4.0$–4.5 nm is the membrane thickness (53). This latter allows positional information to be obtained for a single site of labelling.

3.3 Nuclear magnetic resonance (NMR) spectroscopy

High resolution NMR spectra are obtained only from relatively small molecules or aggregates for which there is rapid molecular rotation in solution. This limits studies of channel conformations by high-resolution techniques to (transmembrane) peptide sections either in organic solvents or bound to small micelles. Larger aggregates such as membranes or membrane fragments have NMR spectra with very broad lines and solid-state NMR methods are required, which at present are not capable of yielding all the conformational detail afforded by high-resolution NMR. In general, the amounts of sample required are very large in both cases.

3.3.1 High-resolution NMR

High-resolution, two-dimensional ^{1}H-NMR spectroscopy has been used to determine, for example, the structure of the 21-residue S4 segment of the sodium-channel protein (57) and of the c subunit (79 residues) of the ATP-synthase proton channel (58) in deuterated organic solvents as a membrane-mimetic environment. The structure of other transmembrane or surface active peptides (e.g. δ-haemolysin, melittin, glucagon, and gramicidin A (59)) have similarly been determined bound to micelles of perdeuterated detergents, which are likely to be a closer mimic of the membrane environment. In such studies, it must be borne in mind that the solvent used may influence the peptide conformation, for example certain solvents (trifluoroethanol, chloroethanol, etc.) are known to favour α-helix formation.

Peptide concentrations required are typically in the range of 5–10 mM. A high-field (300–500 MHz) machine is essential to obtain sufficient resolution, and for larger peptides isotopic substitution (e.g. by ^{15}N) is also needed (see, for example, ref. 58). Basic principles for both one-dimensional and two-dimensional Fourier-transform NMR spectroscopy can be found in, for example, ref. 60.

The first step is to obtain sequence-specific assignments of all the resonances, particularly those of the amide protons—the labile ones of which must be determined in protonated rather than in deuterated solvents, with appropriate solvent-suppression pulse sequences. The assignment is performed sequentially by using a variety of 2D-NMR techniques, mostly 2D correlation spectroscopy (COSY type) to determine through-bond connectivities (up to three bonds) and 2D nuclear Overhauser enhancement spectroscopy (NOESY type) to determine short-range, through-space connectivities (separations $\leqslant$ 2.5, 3.5 Å for strong- and medium-intensity cross peaks) (61). The structural information then comes from the nuclear Overhauser effects (NOE) which determine inter-proton distance constraints, including those of longer range (separations $\leqslant$ 5 Å for weak cross peaks). An important parameter in the NOESY experiment is the mixing time (typically 100–400 msec depending on protein size/mobility) which limits the range of distances over which NOEs may be obtained, by limiting spin diffusion to more remote nuclei.

Secondary structure can be identified readily from the inter-residue NOESY cross peaks (61). Those between residues three apart (i, i+3), and less frequently four apart (i, i+4), are characteristic of α-helical structures. Such cross peaks are absent for a β-sheet structure, which has strong intra-residue cross peaks between the α-carbon and amide protons that, on the other hand, are rather weak for an α-helical structure. When they can be measured, additional information comes from the spin–spin coupling constants ($^3J_{HN\alpha}$) between the α-carbon and amide protons, which are approximately 4 Hz for an α-helix and 9 Hz for a β-sheet structure (61).

The three-dimensional (i.e tertiary) structure comes from a consistent interpretation of all NOE distance constraints by using a distance geometry algorithm (61). For proteins of less than 100 residues, this usually results in a family of closely related structures that can be further refined by energy minimization and restrained molecular dynamics. Additional conformational constraints, e.g. dihedral angles from coupling constants, can be introduced into these calculations to improve precision.

3.3.2 Solid-state NMR

Two basically different types of solid-state NMR experiment are available, both of which require high radio-frequency powers (see ref. 62 for a review of techniques). These two categories of experiment are magic-angle spinning, which is used to produce high-resolution spectra from which chemical shifts may be determined, and broad-line methods, which yield angular-dependent information (i.e. powder patterns from unoriented samples). Both are specialized forms of experiment and require spectrometers specifically designed for solid-state applications.

Magic-angle spinning removes chemical-shift anisotropy and employs magnetically dilute nuclei, usually ^{13}C or ^{15}N, with cross polarization (CPMAS) from protons to improve sensitivity, combined with high-power proton decoupling to remove dipolar line broadening. A good example of the use of chemical shifts to give structural information is afforded by CPMAS studies on the ^{13}C-labelled retinal chromophore in bacteriorhodopsin (63).

Broad-line NMR requires, in addition to high powers, also high digitization rates that are available only in solid-state spectrometers. The angular-dependent quantities measured are either nuclear quadrupole splittings (principally for deuterium-labelled systems), chemical-shift anisotropies (for ^{13}C, ^{15}N, or ^{31}P), or heteronuclear dipolar couplings (^{15}N–^{1}H and ^{15}N–^{13}C). The principles for determination of the peptide backbone conformation by using the orientation information from these measurements has been established (64). An example is afforded by the high-resolution determination of the conformation of the gramicidin-A channel peptide in oriented lipid bilayers, by using combination of these angular constraints and subsequent energy refinement (65). Additionally, incorporation of ^{2}H-labelled amino acids can be used to study the orientation and mobility of the labelled side chains from the reduction in the quadrupole anisotropy by angular rotation (66, 67). The degree of mobility of the different residues is likely to reflect the secondary and tertiary folding and the environment of the channel protein. Such studies may be extended also to protein backbone motions by employing ^{13}C and ^{15}N isotopic labelling (66, 68). In addition, broad-line ^{31}P-NMR spectroscopy of the phospholipid component of membranes is, in many cases, diagnostic of the membrane morphology (see, for example, ref. 69). The lipid bilayer gives a characteristic powder pattern, arising from the chemical-shift anisotropy, that can be distinguished from the non-lamellar lipid arrangements which may be induced on incorporation of channel peptides (70).

3.4 Circular dichroism (CD) spectroscopy

Far ultra-violet (180–250 nm) circular dichroism is a sensitive method for investigating protein secondary structure in solution from the differential absorption (ellipticity) of right- and left-circularly polarized light by the peptide bonds (71). It is therefore widely applied to channel-derived peptides in organic solvents. The secondary structure content is determined by fitting the experimental CD spectrum with a weighted average of spectra either from standard reference systems that correspond to the different secondary structural elements or from a library of experimental spectra from proteins of known secondary structure. The ellipticity at a given wavelength is then given by, for example, $[\theta] = f_{\alpha}[\alpha] + f_{\beta}[\beta] + f_{R}[R]$, where f_i are the fractional contributions from the different secondary structural elements, α-helix, β-sheet, and random, etc. Precision is best for α-helical structures which have a characteristic CD spectrum with a pronounced negative peak at 222 nm, corresponding to a molar ellipticity of $[\theta] \simeq -30\,000°$ cm^2/decimol. This wavelength is frequently used for monitoring structural changes. Complications can arise because the CD spectra of the peptide bond may be affected by the interaction with neighbouring groups, particularly aromatic side chains which may have an induced optical activity. This can cause uncertainties in the determination of the secondary structure, but does confer sensitivity to other structural changes. Quite often, measurements are restricted to wavelengths greater than 200 nm, which reduces the degree of discrimination between the various secondary structures. Measurements below this wavelength (in the vacuum UV) require special precautions, since many substances including solvents, cuvette materials, buffer salts, and atmospheric oxygen have strong absorptions in this region. Experimental considerations can be found in ref. 72.

Membranes cause considerable problems for CD spectroscopy because of the differential scatter of circularly polarized light, and also because of the non-uniform distribution of chromophores in the light beam, which gives rise to absorption flattening. Experimental methods of correction for differential scatter are dealt with in ref. 72. These include positioning the photomultiplier detector close to the cuvette to increase the acceptance angle for light collection (limited to ±45° by internal reflection), and refractive index matching of the suspending buffer, e.g. by adding glycerol. Absorption flattening may be reduced by using high sample concentrations to increase homogeneity, but this requires ultra-short path lengths because of the high absorption, and refractive-index matching to compensate for the increased scatter (72). Otherwise, in both cases, it is necessary to reduce the particle size and improve homogeneity, either by sonication or solubilization, both of which may have attendant changes in the protein structure. Computational methods for correcting the distortions of CD spectra from membrane suspensions are given in refs 71, 72.

3.5 Fluorescence spectroscopy

The fluorescence emission either from intrinsic chromophores (e.g. tryptophan) or from extrinsic probe molecules is an extremely versatile reporter of membrane environments. This arises from several different properties of fluorescent molecules (73). The fluorescence spectrum and its quantum yield are sensitive both to environment and polarity (much more so than is the conventional absorption spectrum). The quantum yield is sensitive to the accessibility to fluorescence quenchers and to the distance from other fluorophores with which resonance energy transfer may take place. The polarization of the fluorescence spectrum is also sensitive to the rotational dynamics that take place within the fluorescence lifetime. Fluorescence spectroscopy is more sensitive than conventional absorption spectroscopy and far more sensitive than any of the other forms of spectroscopy discussed in this chapter. Submicromolar concentrations should be practicable in favourable circumstances. The main disadvantage in membrane studies is the interference of scattering by particulate suspensions, which requires the use either of small particles, refractive-index matching, front-face detection, or other techniques. Experimental aspects of fluorescence spectroscopy are treated in refs 74, 75.

Fluorescence can be quenched by paramagnetic and other species and this can be used to determine the localization of the fluorophore in the membrane. Quenchers that are located solely in the aqueous phase include paramagnetic transition-metal ions and acrylamide, and those that are located in the membrane include spin-labelled lipids that bear a nitroxide group at a given position in the fatty acid chain. If the quencher does not absorb appreciably at the excitation or emission wavelengths of the fluorophore (inner-filter effects), the extent of quenching of the fluorescence, F, is proportional to the concentration, [Q], of quencher: $F_0/F = 1 + K_Q[Q]$, where F_0 is the fluorescence intensity in the absence of quencher. For dynamic quenching, the quenching constant, K_Q, is the collisional rate constant times the intrinsic fluorescence lifetime, and for static quenching it is equal to the association constant for complex formation. An application to the quenching of the intrinsic (multi-tryptophan) fluorescence of the acetylcholine receptor by spin-labelled fatty acids is given in ref. 76.

Fluorescence energy transfer takes place between chromophores for which there is overlap between the emission spectrum of the donor and the absorption spectrum of the acceptor. The extent of energy transfer depends on the extent of spectral overlap which defines the characteristic distance, R_o, over which the transfer may take place (typically 30–50 Å). Experimentally, the energy transfer may be detected by the fluorescence quenching of the donor in the presence of the acceptor (or the fluorescence of the acceptor induced by excitation of the donor). Corrections, by subtraction, are necessary for any scattering and/or fluorescence of the acceptor induced by the exciting light. The distance dependence of the energy transfer efficiency is given by the Förster expression: $E = R^{-6}/(R^{-6}+R_o^{-6})$, where R is the distance between centres of the donor and acceptor. In using this method to detect specific association between membrane components it is necessary to establish, by using low acceptor concentrations, by calculation, or by control experiments, that the extent of energy transfer is appreciably greater than that expected for a two-dimensional random distribution of donor and acceptor (see ref. 77). An application to the study of the association and assembly of fluorescein donor- and rhodamine acceptor-labelled ($R_o \simeq 45$ Å) peptides derived from the putative pore region of the Shaker K^+ channel, in lipid vesicles, is given in ref. 78.

Polarized luminescence measurements are most usefully made with phosphorescent probes, e.g. eosin. This is because the longer lifetime of the triplet state (compared with that of prompt fluorescence) allows measurement of the rotational rates of proteins and peptide aggregates, hence yielding information similar to that from spin-label STESR spectroscopy (cf. above). Time-resolved experiments must be performed with polarizers in both the incident and detected light beams; details of methods and applications to membranes may be found in ref. 79.

Further details of fluorescence measurements on membranes may be found in refs 73, 74.

References

1. Eisenberg, D., Weiss, R. M., and Terwilliger, T. C. (1984). *Proc. Natl Acad. Sci. USA* **81**, 1401.
2. Kyte, J. and Doolittle, R. F. (1982). *J. Mol. Biol.* **157**, 105.
3. Cornette, J. L., Crease, K. B., Margalit, H., Sponge, J. L., Beozofsky, J. A., and DeLisi, C. (1987). *J. Mol. Biol.* **195**, 659.
4. Rees, D. C., DeAntonio, L., and Eisenberg, D. (1989). *Science* **245**, 510.
5. Attwood, T. K. and Findlay, J. B. C. (1994). *Protein Engineering* **7**, 195.
6. Donnelly, D., Overington, J. P., Ruffle, S. V., Nugent, J. H. A., and Blundell, T. L. (1993). *Protein Sci.* **2**, 55.
7. Baldwin, J. M. (1993). *EMBO J.* **12**, 1693.
8. Pongs, O. (1993). *J. Membrane Biol.* **136**, 1.
9. Catterall, W. A. (1993). *Trends Neurosci.* **16**, 500.
10. Thompson, P. and Findlay, J. B. C. (1984). *Biochem. J.* **220**, 773.
11. Ohguro, H., Johnson, R. S., Ericsson, L. H., Walsh, K. A., and Palczewski, K. (1994). *Biochemistry* **33**, 1023.
12. Li, B., Langer, J. A., Schwartz, B., and Pestka, S. (1989). *Proc. Natl Acad. Sci. USA* **86**, 558.
13. von Heijne, G. (1992). *J. Mol. Biol.* **225**, 487.
14. Weiss, M. S., Kreusch, A., Chiltz, E., Nestel, U., Welte, W., Weckesser, J., and Schultz, G. E. (1990). *FEBS Lett.* **267**, 268.

15. Findlay, J. B. C. (1987). In *Biological membranes: a practical approach* (eds J. B. C. Findlay and W. H. Evans), pp. 179–217. IRL Press, Oxford.
16. Falke, J. J., Dernburg, A. F., Sternburg, D. A., Zalkin, N., Milligan, D. L., and Koshland, D. E. Jr. (1988). *J. Biol. Chem.* **263**, 14850.
17. Patula, A. A. and Simon, M. I. (1992). *Proc. Natl Acad. Sci. USA* **89**, 4144.
18. Taddayon, M., Zhang, Y., Gnaueshan, S., Hunt, L., Mehraein-Ghomi, F., and Broome-Smith, J. K. (1992). *Biochem. Soc. Trans.* **20**, 598.
19. Boyd, D. (1993). The use of gene fusions to study membrane protein topology. In *Membrane protein structure: experimental approaches* (ed. S. H. White). Oxford University Press, Oxford.
20. Wan, H.-Y., Lipfort, L., Malbon, C. C., and Bahouth, S. (1989). *J. Biol. Chem.* **264**, 14424.
21. Unwin, N. (1993). *J. Mol. Biol.* **229**, 1101.
22. Eliopoulos, E. E., Geddes, A. J., Brett, M., Pappin, D. J. C., and Findlay, J. B. C. (1982). *Int. J. Biol. Macromol.* **4**, 263.
23. Earnest, T. N., Herzfeld, J., and Rothschild, K. J. (1990). *Biophys. J.* **58**, 1539.
24. Holtzenburg, A., Jones, P. C., Franklin, T., *et al.* (1993). *Eur. J. Biochem.* **213**, 21.
25. Altenbach, C., Marti, T., Khorana, H. G., and Hubbell, W. L. (1990). *Science* **248**, 1088.
26. Davison, M. and Findlay, J. B. C. (1986). *Biochem. J.* **234**, 413.
27. Blanton, M. P. and Cohen, J. B. (1992). *Biochemistry* **31**, 3738.
28. Donnelly, D., Johnson, M. S., Blundell, T. L., and Saunders, J. (1989). *FEBS Lett.* **251**, 109.
29. Donnelly, D. and Cogdell, R. J. (1993). *Protein Eng.* **6**, 629.
30. Brett, M. and Findlay, J. B. C. (1979). *Biochem. J.* **177**, 215.
31. Diesenhofer, J. and Michel, H. (1991). *Ann. Rev. Biophys. Biophys. Chem.* **20**, 247.
32. Kühlbrandt, W. and Wang, D. N. (1991). *Nature* **350**, 130.
33. Weiss, M. S., Kreusch, A., Chiltz, E., *et al.* (1990). *FEBS Lett.* **267**, 268.
34. Byler, D. M. and Susi, H. (1986). *Biopolymers* **25**, 469.
35. Heimburg, T. and Marsh, D. (1993). *Biophys. J.* **65**, 2408.
36. Rothschild, K. J., Sanches, R., and Clark, N. A. (1982). In *Methods in enzymology* (ed. L. Packer), Vol. **88**, 696. Academic Press, New York.
37. Clark, N. A. and Rothschild, K. J. (1982). In *Methods in enzymology* (ed. L. Packer), Vol. **88**, p. 326. Academic Press, New York.
38. Fringeli, U. P., Apell, H.-J., Fringeli, M., and Läuger, P. (1989). *Biochim. Biophys. Acta* **984**, 301.
39. Boggs, J. M., Vail, W. J., and Moscarello, M. A. (1976). *Biochim. Biophys. Acta* **448**, 517.
40. Horváth, L. I., Heimburg, T., Kovachev, P., Findlay, J. B. C., Hideg, K. and Marsh, D. (1995), *Biochemistry* **34**, 3893.
41. Kauppinen, J. K., Moffatt, D. J., Mantsch, H. H., and Cameron, D. G. (1981). *Appl. Spectrosc.* **35**, 271.
42. Holzenburg, A., Jones, P. C., Franklin, T., *et al.* (1993). *Eur. J. Biochem.* **213**, 21.
43. Hyde, J. S. and Fronciscz, W. (1989). In *Advanced EPR. Applications in biology and biochemistry* (ed. A. J. Hoff), pp. 277–305. Elsevier, Amsterdam.

44. Marsh, D. (1982). *Tech. Life Sci: Biochem.* **B4/II**, B426/1.
45. Marsh, D. and Watts, A. (1982). In *Lipid–protein interactions* (ed. P. C. Jost and O. H. Griffith), Vol. 2, pp. 53–126. Wiley-Interscience, New York.
46. Bligh, E. G. and Dyer, W. J. (1959). *Can. J. Biochem. Physiol.* **37**, 911.
47. Lowry, O. H., Rosebrough, N. J., Farr, L., and Randall, R. J. (1951). *J. Biol. Chem.* **193**, 265.
48. Eibl, H. and Lands, W. E. M. (1969). *Anal. Biochem.* **30**, 51.
49. Marsh, D. (1993). In *Protein–lipid interactions* (ed. A. Watts), pp. 41–66. Elsevier, Amsterdam.
50. Marsh, D. and Horváth, L. I. (1989). In *Advanced EPR. Applications in biology and biochemistry* (ed. A. J. Hoff), pp. 707–52. Elsevier, Amsterdam.
51. Marsh, D. (1992). *Appl. Magn. Reson.* **3**, 53.
52. Altenbach, C., Marti, T., Khorana, H. G., and Hubbell, W. L. (1990). *Science* **248**, 1088.
53. Páli, T., Bartucci, R., Horváth, L. I., and Marsh, D. (1992). *Biophys. J.* **61**, 1595.
54. Páli, T., Horváth, L. I., and Marsh, D. (1993). *J. Magn. Reson.* **A101**, 215.
55. Snel, M. M. E. and Marsh, D. (1993). *Biochim. Biophys. Acta* **1150**, 155.
56. Bailar, J. C. Jr. and Jones, E. M. (1939). *Inorg. Synth.* **1**, 35.
57. Mulvey, D., King, G. F., Cooke, R. M., Doak, D. G., Harvey, T. S., and Campbell, I. D. (1989). *FEBS Lett.* **257**, 113.
58. Norwood, T. J., Crawford, D. A., Steventon, M. E., Driscoll, P. C., and Campbell, I. D. (1992). *Biochemistry* **31**, 6285.
59. Lomize, A. L., Orechov, V. Yu., and Arseniev, A. S. (1992). *Bioorg. Khim.* **18**, 182.
60. Sanders, J. K. M. and Hunter, B. K. (1987). *Modern NMR spectroscopy. A guide for chemists.* Oxford University Press, Oxford.
61. Wüthrich, K. (1986). *NMR of Proteins and Nucleic Acids.* Wiley, New York.
62. Griffin, R. G. (1981). In *Methods in enzymology* (ed. J. M. Lowenstein), Vol. **72**, 108. Academic Press, New York.
63. Smith, S. O. and Griffin, R. G. (1988). *Ann. Rev. Phys. Chem.* **39**, 511.
64. Opella, S. J., Stewart, P. L., and Valentine, K. G. (1987). *Quart. Rev. Biophys.* **19**, 7.
65. Ketchem, R. R., Hu, W., and Cross, T. A. (1993). *Science* **261**, 1457.
66. Leo, G. C., Colnago, L. A., Valentine, K. G., and Opella, S. J. (1987). *Biochemistry* **26**, 854.
67. Hu, W., Lee, K.-C., and Cross, T. A. (1993). *Biochemistry* **32**, 7035.
68. Nicholson, L. K., Teng, Q., and Cross, T. A. (1991). *J. Mol. Biol.* **218**, 621.
69. Knowles, P. F. and Marsh, D. (1991). *Biochem. J.* **274**, 625.
70. Killian, J. A. (1992). *Biochim. Biophys. Acta* **1113**, 391.
71. Long, M. M. and Urry, D. W. (1981). In *Membrane spectroscopy* (ed. E. Grell), pp. 143–71. Springer, Berlin.
72. Schneider, A. S. and Rosenheck, K. (1982). *Tech. Life Sci: Biochem.* **B4/II**, B424/1.
73. Yguerabide, J. and Foster, M. C. (1981). In *Membrane spectroscopy* (ed. E. Grell), pp. 199–269. Springer, Berlin.
74. Lee, A. G. (1982). *Tech. Life Sci: Biochem.* **B4/II**, B422/1.
75. Harris, D. A. and Bashford, C. L. (eds) (1987). *Spectrophotometry and spectrofluorimetry: a practical approach.* IRL Press, Oxford.

76. Marsh, D. and Barrantes, F. J. (1978). *Proc. Natl Acad. Sci. USA* **75**, 4329.
77. Fung, B. K. K. and Stryer, L. (1978). *Biochemistry* **17**, 5241.
78. Peled, H. and Shai, Y. (1993). *Biochemistry* **32**, 7879.
79. Cherry, R. J. (1979). *Biochim. Biophys. Acta* **559**, 289.

10

Electron microscopical analysis of ion channels

ANDREAS HOLZENBURG

1. Introduction

The initial steps in approaching the structural determination of ion channels should always include electron microscopical studies, which

- are easy to perform;
- reveal a lot of useful information about the protein within a short period of time;
- require only minute amounts of sample.

The latter criterion makes electron microscopy especially suitable for structural studies of proteins which are only available in small amounts (< 1 mg). 50 μl of a sample containing a total of 5 μg protein is, in principle, sufficient to elucidate the 3D structure of a protein! The instruments with the widest range of applications for structural studies of integral membrane proteins are the transmission electron microscopes (TEM). While the scanning TEM (STEM) is mainly used to determine the mass of biological macromolecules and/or macromolecular complexes by recording elastic annular dark-field images (an angular detector specifically retrieves elastically scattered electrons at large angles, i.e. dark-field images) which represent two-dimensional distributions of the projected mass (1, 2), the conventional TEM (CTEM) can be considered the main workhorse in structural molecular biology. Its cryo-version (cryo-TEM) is used to retrieve high-resolution structural information (< 1 nm) from ordered (crystalline) specimens and to gain structural information devoid of possible staining artefacts. In cryo-electron microscopy one often takes advantage of the fact that high-resolution amplitude information can be more reliably obtained from electron diffraction patterns, which can then be combined with the phase information from the image. This approach to obtaining high-resolution structural information falls within the remit of electron crystallography (3).

The value of electron crystallography for the structural elucidation of membrane proteins at near-atomic resolution cannot be overestimated,

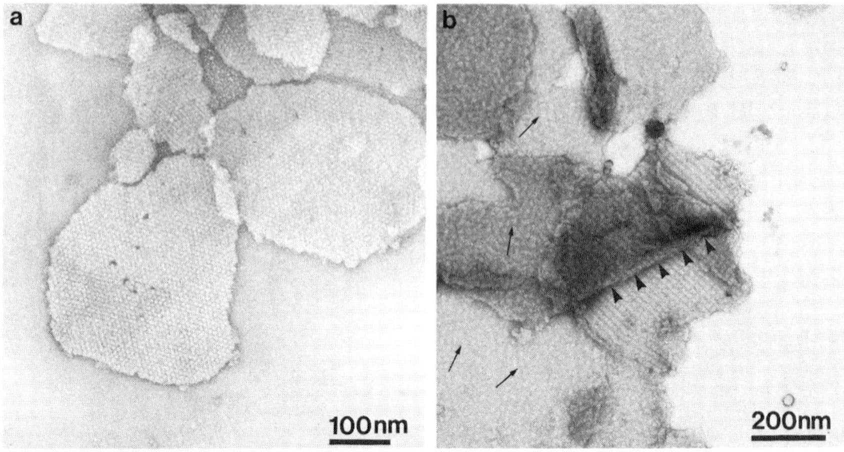

Figure 1. Electron micrographs of negatively stained 2D crystals of (a) hexameric complexes of the 16 kDa protein (major protein of the membrane sector of the vacuolar H^+ATPase) isolated from the American lobster (*Homarus americanus*) and (b) photosystem II complex from spinach (*Spinacia oleracea*). Although photosystem II is not an ion channel, it was chosen since it represents a good example of *in-situ* crystallization (9). In both micrographs, periodicity and unit cells are clearly discernible. The difference in dimensions of the proteins involved should be noted: (a) shows a small channel-forming protein and (b) a large protein (including chlorophyll and carotenoids) complex of about 850 kDa. Beside the crystalline array, parts of which are folded over (indicated by arrowheads), membranes are observed which contain protein complexes (some of which are marked by arrows) randomly arranged in the membrane plane.

as recent progress in this field demonstrates that electron crystallography is making substantial inroads into realms once dominated by X-ray crystallography (4–6). A comprehensive treatise of this rather complex subject cannot fall within the scope of a chapter of this size. Instead, possible strategies will be presented for arriving at specimens suitable for electron crystallographical studies, i.e. ordered two-dimensional protein–lipid arrays which, in the following, are referred to as 2D crystals (*Figure 1*). The chapter also includes hints for overcoming some cryo-TEM-specific problems and reference is made to image analysis and processing. It is also a major aim of this section to provide sufficient information for those individuals only slightly familiar with the field of electron microscopy, concerning the preparation of an ion channel protein for imaging in the TEM.

2. Reconstitution of ion channels into 2D crystals

The idea is to incubate detergent–lipid mixed micelles with detergent-solubilized protein, remove the detergent (see *Protocol 1*), and end up with 2D crystals, i.e. ordered 2D protein–lipid arrays. Two examples of how 2D crystals may appear in the TEM are shown in *Figure 1*. Before advancing to any reconstitution procedures, one should be aware of the fact that 2D crystals may occur naturally and can, therefore, sometimes be observed after simple isolation procedures (7, 8), or can be obtained by *in-situ* crystallization induced by detergent (or other) treatments (8, 9). If it is possible to obtain membranes which are enriched in one particular protein, then an *in-situ* approach should certainly be considered. In the following, a comprehensive description of reconstitution experiments is given, preceded by introductions to the different parameters involved.

2.1 Reconstitution parameters

2.1.1 Lipid component

It is crucial for a successful reconstitution to find the 'correct' lipid. To this end, there are two important considerations, the shape of the lipid and the length of its alkyl chain (e.g. refs 10 and 11).

The shape of a lipid is, among other things, defined by the cross-sectional area of the alkyl chains (larger for unsaturated and smaller for saturated alkyl chains), the cross-sectional head-group area, head group hydration and charge, hydrogen bonding, and counterions like Ca^{2+}. Phosphatidylcholine, for instance, assumes a cylindrical molecular shape and would, therefore, favour the formation of planar bilayers while phosphatidylserine in water would assume the shape of a truncated cone and support the formation of flexible bilayers or vesicles. The addition of Ca^{2+} to phosphatidylserine renders its shape cylindrical due to a decreased head-group area. Saturated phosphatidylethanolamines also exhibit a cylindrical shape, even though, compared to phosphatidylcholine, the head groups are smaller and less hydrated, and additional intermolecular hydrogen bonds may decrease the head-group area still further. However, an increase in the cross-sectional area of the alkyl chains introduced by double bonds will give unsaturated phosphatidylethanolamines the shape of inverted truncated cones and result in hexagonal H_{II} phase structures (inverted micelles). Lastly, the pH determines the state of dissociation and hence the head-group area. For example with phosphatidylserine, a smaller head-group area will be present at $pH < 4$ than at neutral pH, since the more dissociated a head group the larger its area.

The alkyl chain length can be optimized using the following rationale. In an ordered 2D lattice, the protein molecules interact specifically with each other. These specific contacts are of hydrophilic as well as hydrophobic nature. While the hydrophobic regions should always be encased by the lipid (or detergent), the hydrophilic regions should remain accessible. Hence, a lipid should be found which, because of its alkyl chain length, encases the hydrophobic centre region but does not cover up any specific hydrophilic contact areas. Hydrophobic interactions, it seems, can also be mediated through the hydrophobic tails of the lipids, so that for the latter interactions there is less of a demand for the 'correct' lipid, but the lipids do need to be sufficiently long to shield the hydrophobic area of the protein. A sensible range for the variation of the chain length is 12–18 carbon atoms.

Furthermore, a lipid should be chosen which exhibits not too high a gel-to-fluid phase transition temperature which could interfere with the stability of the protein (see also Section 2.1.4). The transition temperature is strongly dependent on the length and degree of saturation of the alkyl chain: the shorter the chain and the more unsaturated, the lower the transition temperature. For information on the phase transition temperatures of different lipids the reader is referred to ref. 12.

The assembly of naturally occurring or *in-situ* formed crystals, as mentioned above, takes place in native membranes, i.e. in an endogenous lipid environment. These endogenous lipids may prove vital for the formation of crystals and should, therefore, also be included in the list of lipids to be tested in reconstitution experiments.

2.1.2 Detergents

Detergents interact with the accessible hydrophobic regions of a protein. Consequently, the hydrophilic head of a detergent may camouflage some of the protein's hydrophilic regions. This, in turn, may interfere with protein–lipid and protein–protein recognition processes, of which the former are required for the formation of a protein–lipid bilayer, and the latter are almost certainly a prerequisite for the formation of 2D crystals. The optimal detergent should be compatible with the formation of 2D crystals and conserve the activity of the protein. As a general rule: non-ionic detergents with large head groups and long alkyl chains are milder than zwitterionic detergents with smaller head groups and shorter alkyl chains, e.g. Na^+K^+ATPase is active in $C_{12}E_8$ but not in 8-Gluc, and the half-life of rhodopsin at 20°C ranges from < 15 sec (in Triton X-100), 15 min (in 8-Gluc), to 24 h (in cholate), and even > 40 h in 12-Malt (8, and refs therein). If an activity-maintaining detergent does not prove suitable for 2D crystallization, the addition of small amphiphiles might provide a solution. Welte and his colleagues (13, 14) have shown that the addition of heptyl- or octyldimethylamineoxide (at 3% (w/v) or 5% (w/v), respectively), or heptane-triol (at 3% (w/v)) to the primary detergent improves protein–protein interactions, presumably via a decrease in micelle size. The addition of taurine as a 2D crystal-inducing amphiphile (15) might also be considered.

If no single detergent seems suitable, one should not hesitate to try binary mixtures of detergents. The c.m.c. (critical micellar concentration) of such mixtures is given by the expression $1/\text{c.m.c.}_{\text{mixture}} = \alpha/\text{c.m.c.}_1 + (1-\alpha)/\text{c.m.c.}_2$ with α being the mole fraction of detergent 1 in the mixture (16). Alternatively, the detergent used for solubilization can differ from the one used for reconstitution. To achieve this, the detergent can either be changed prior to or during reconstitution. The efficiency of the latter strategy needs to be established for each individual system and is based on the utilization of a

different detergent for the preparation of the mixed micelles (see *Protocol 1*).
Finally, one may also wish to consider the influence of the overall shape of the detergent (17) on the reconstitution experiment.

2.1.3 Lipid-to-protein ratio

The lipid-to-protein ratio (LPR) is another parameter which needs to be optimized, since it controls, during the course of dialysis, the balance between protein–protein and protein–lipid interactions as well as their absolute magnitudes, factors which crucially influence the formation of 2D crystals and the final lattice parameters (packing density and symmetry). Optimum LPR have been reported to be in the range 0.1 to 10.0 (w/w). It seems that membrane proteins with bulky extramembranous domains require a higher LPR than 'lean' ion channels. In initial trials, one should include at least three different LPRs (such as 0.3, 1.0, and 3.0) with a slight bias towards lower LPRs, since one would expect to find better ordered crystals when the lipid content is low. Additionally, the slow removal of phospholipids from reconstituted membranes can induce crystal formation. This strategy involves the use of phospholipase A_2 from bee venom. The protein molecules come gradually closer together as a result of the slow removal of phospholipids by hydrolysis and crystals eventually form, as demonstrated for mitochondrial outer membrane channels (18). The enzyme should be used at 4°C together with dialysis, as the removal of the products of phospholipid hydrolysis has been found to be crucial. A detailed account on the use of this enzyme is given by Mannella (18).

2.1.4 Temperature

The temperature should be chosen so that it is above the gel-to-fluid phase transition temperature of the lipid in use because this facilitates the incorporation of protein in the bilayer. A second temperature-governing factor is the stability of the protein, keeping in mind that the protein might even be stable at unexpectedly high temperatures once incorporated into the bilayer. Having found a temperature at which protein builds into the bilayer, but where no or only poorly ordered 2D crystals are repeatedly obtained, increasing and decreasing the temperature through the phase transition temperature (tempering) might help to increase the order or even induce crystal formation. By continually inducing (partial) expulsion and reintegration of the protein in and out of the bilayer, isotropically distributed protein–protein interactions are more likely to be established, and the formation of crystals is favoured. One should also consider employing a temperature ramp following (or during) the aforementioned phospholipase treatment, i.e. incubating the membranes overnight at 37°C and then letting the membranes cool slowly.

2.1.5 Ionic milieu

When choosing the ionic milieu for a reconstitution experiment, two primary constraints need to be considered:

- Compatibility with specimen preparation procedures, e.g. negative staining (see Section 3.2).
- Compatibility with protein activity, i.e. the protein ought to be stable (active) in the buffer solution employed.

A third constraint is the surface charges which are believed to be important for those regions of the protein which protrude from the membrane surface. These charges seem to influence crystal formation and crystalline order, and can be adjusted experimentally by adding NaCl or $MgCl_2$ to the buffer system (19) and changing the pH (20). The quantities to be used should resemble the concentrations required to keep a soluble protein in solution (20–200 mM). Higher concentrations may result in the loss of long-range order.

2.1.6 Lipid-to-detergent ratio

Upon detergent removal, spherical mixed micelles grow by fusion processes into discoidal or cylindrical micelles. These processes are driven by a decrease in edge energy, which is associated with the exposure of hydrophobic regions to the aqueous environment (bearing in mind that detergent molecules will cover the rims of the micelles where hydrophobic areas are particularly exposed). At a certain point, the size of a discoidal/cylindrical micelle is such that closure to form a vesicle is the most energetically favourable option, due to the increase in curvature energy (energy associated with the stretching/compression of opposing leaflets) on vesicle closure being less than the resultant decrease in edge energy. According to these considerations, the formation of planar bilayers is favoured over the formation of vesicles when the increase in curvature energy is greater than the drop in edge energy. Given lipids which would, as a result of their shape, favour the formation of planar bilayers, the curvature energy required for vesicle formation would be much greater than the edge energy required for the maintenance of planar bilayers.

The size of the detergent–lipid mixed micelles does not depend on the detergent used while the ultimate size of the vesicles does depend on the detergent. This dependence is given through detergent-induced post-vesiculation size growth (PVSG) processes (21). Vesicle growth by fusion is much more likely to occur within a critical range of lipid-to-detergent ratios (LDRs), regardless of whether this situation is arrived at by reduction of the detergent concentration via dialysis, or addition of detergent to a sample containing none (integrating detergent into lipid bilayers also causes an increase in bilayer fluidity and this also favours fusion). In order to obtain large vesicles, one needs to make sure that the PVSG processes are rapid with respect to the rate of detergent removal. This can be achieved by (i) slow removal of detergent, and (ii) fast PVSG processes. Although in some cases fast detergent removal (e.g. by polystyrene beads) may yield excellent results (22), slow removal of detergent enables the sample to remain at an optimum LDR

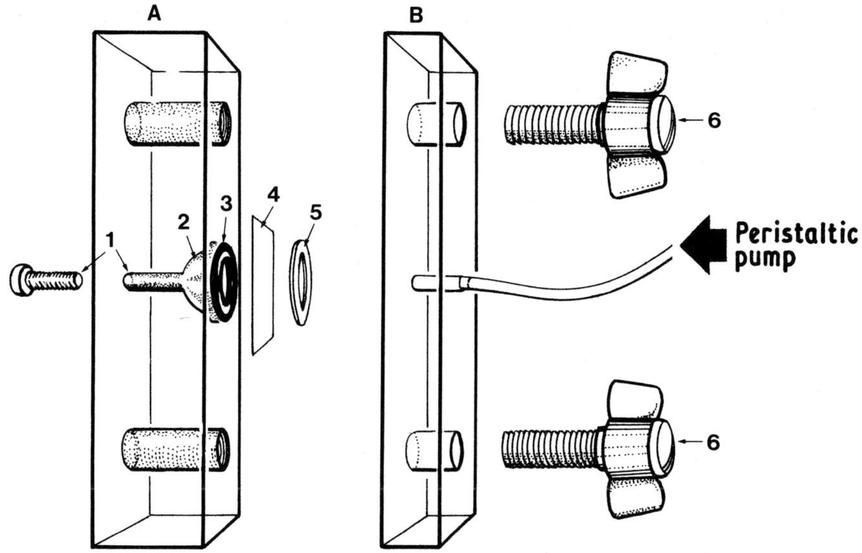

Figure 2. Exploded view of a flow-dialysis chamber. This chamber is a somewhat simpler version of a device described earlier (8, 9, 23) and can be built easily by most departmental workshops. The major parts are as follows: A and B constitute the Perspex lower and upper chamber housing, respectively; 1, sample port with screw; 2, hemispherical sample chamber with a capacity, generally, between 0.1 and 1 ml; 3, rubber O-ring; 4, dialysis membrane; 5, porous polymer washer (plastic frit); 6, PVC thumbscrews. The rate of detergent removal is adjusted by the flow rate of the dialysis buffer which is controlled via a peristaltic pump. For even finer tuning of the rate of detergent removal, the detergent concentration in the dialysis buffer can be varied with time using a gradient mixer. Temperature is controlled by leading the tubing delivering the dialysis buffer through a thermostatted water bath. Temperature profiles as a function of time can be generated using a microprocessor-controlled water bath. (*Assembly*: After filling the sample chamber, with the sample port screw and the O-ring in place, the dialysis membrane and, subsequently, the porous polymer washer are carefully placed on top of the O-ring, before the upper chamber housing is gently lowered on to the washer and fastened using the two thumbscrews until finger-tight. Too much pressure may cause the upper housing to curve at both ends towards the lower housing, thus leading to possible leakage. Before starting the experiment, the chamber should be set up in such a way that the dialysis buffer is supplied from below the dialysis membrane.)

for longer periods so that fusion events are much more likely to take place and PVSG processes are favoured. This means in practice that flow-dialysis (23, 8) rather than conventional equilibrium dialysis should be used to allow precise adjustment of the detergent removal parameters (*Figure 2*). The speed of detergent removal depends, however, not only on the dialysis parameters but also on the detergent used, e.g. cholate removal proceeds faster than the removal of CHAPS. To gauge speed of removal, the notion that higher c.m.c. detergents can be more rapidly removed than lower c.m.c. detergents, and that biphenyl detergents (24) can hardly be removed at all from the lipid bilayer may be used as rough guidelines. Furthermore, the speed of PVSG processes is detergent dependent, e.g. CHAPS causes a faster size growth than cholate. Fast PVSG processes have also been reported for 8-Gluc, which, interestingly enough, is also a detergent for which the micelle-to-vesicle transition occurs at a lower LDR than for cholates. The increase in hydrodynamic radius (R_h) of aggregates formed by lipid–8-Gluc mixtures beyond the transition region (R_h for mixed micelles < 10 nm, R_h for vesicles > 10 nm) is orders of magnitude larger than with cholates. The detergents C_8E_5 and 8-POE fall into the same category as 8-Gluc, and the resulting vesicles are much larger than with cholates. Assuming that the above considerations for binary lipid–detergent systems are, within reason, transferable to ternary lipid–detergent–protein systems, screening for an optimum detergent for reconstitution experiments should ideally include each of the above three detergents (C_8E_5, 8-POE, and 8-Gluc).

During the course of dialysis, as the LDR increases, vesicles will form and the protein may or may not be incorporated into the bilayer, depending on how the chosen parameters govern the interactions between protein, lipid, and detergent. The exact sequence and nature of events regarding the combination of protein and lipid to form a bilayer is unclear and probably variable between systems. It has, however, been proposed that vesicle formation may precede incorporation of protein into the lipid bilayer when detergent is removed slowly, and that proteins are incorporated during micelle growth when detergent is removed quickly; this phenomenon is modulated according to the nature of the detergent used (25). In any case, the incorporation seems to take place long before the c.m.c. of the detergent is reached, and subsequently the protein molecules may assemble into ordered lattices. That the lattice assembly is initiated shortly before the c.m.c. is reached has been suggested for the OmpF porin from *E. coli* (26, 28; *Figure 3*). Some time between the initial conditions and reaching the c.m.c., LDR becomes critical and vesicle growth takes place. Once the detergent concentration is close to the c.m.c., the vesicles start to grow more rapidly. This rapid growth may be concomitant with an execution of lattice assembly. Later during dialysis, after the vesicle growth processes have ceased, further removal of detergent can improve the crystal order. While in the earlier stages during dialysis some detergent is needed for size growth processes, complete removal of the detergent should be striven for towards the end of the reconstitution process to obtain the best crystalline order possible. For further reading the reader is referred to some recent articles (8, 27, 28).

2.2 Monitoring progress

The beauty of electron microscopical studies is that the effect of many parameter changes can be almost immediately monitored. The most straightforward way to prepare specimens for TEM is by negative staining (see

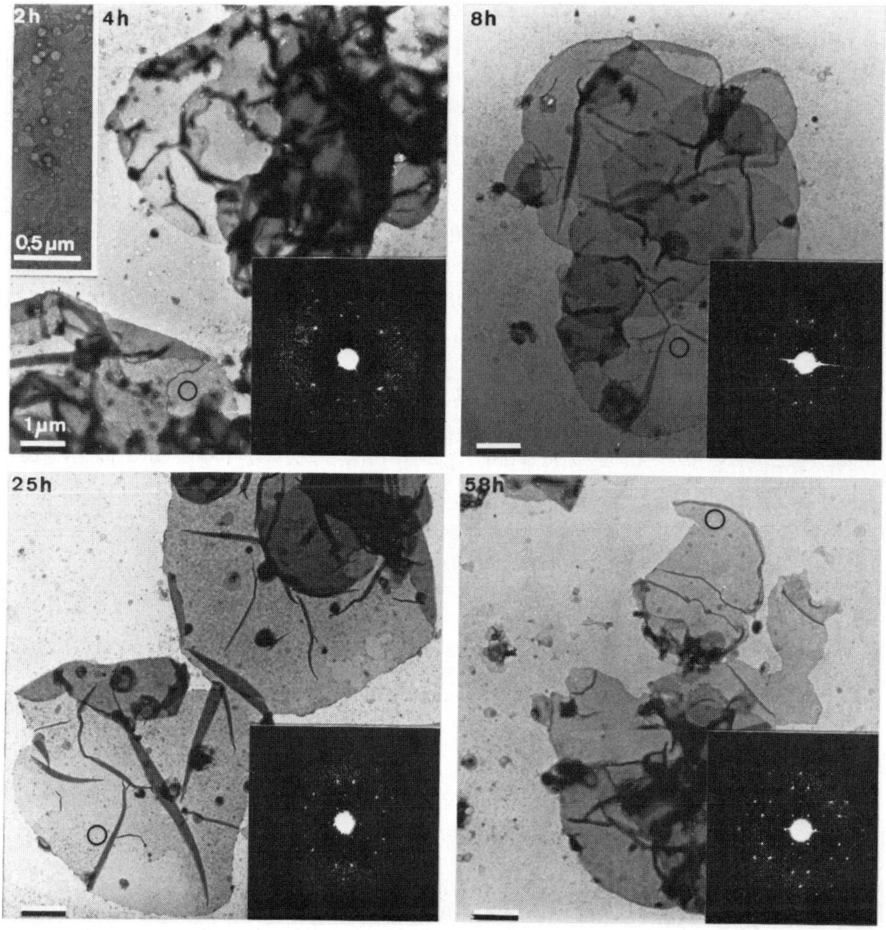

Figure 3. Time course of the reconstitution of OmpF porin as monitored by electron microscopy of negatively stained specimens (from ref. 26). The times at which the samples were taken are given. After 2 h mixed micelles as well as small vesicles with a diameter of 50 nm were visible. After 4 h membranes were observed which already showed a considerable degree of order as revealed by optical diffraction (diffractograms are depicted in the insets). The lattice constant of the 2D crystals is 9.2 nm. Owing to the low primary magnification (×5000) of the electron micrographs shown, ordered structures are not visible. Diffraction patterns of the circled areas were obtained from electron micrographs taken at a much higher primary magnification (×50 000). The membranes do not undergo any further growth processes between 4 and 58 h but the crystalline order improves continuously. The latter phenomenon may be due to the gradual removal of detergent. Since the c.m.c. is reached after about 2.2 h and the first ordered arrays are observed within less than 2 h thereafter, it is likely that lattice assembly is initiated shortly before the c.m.c. is reached.

Protocols 2 and *3*) which could be viewed as the method of choice for monitoring reconstitution experiments. It is easy to judge from negatively stained specimens whether vesicles have formed or not, whether formed vesicles are devoid of protein molecules, and whether 2D crystals have formed.

Protocol 1. Reconstitution of ion channels for formation of 2D crystals

Before attempting any reconstitution experiment, one should ensure that all protein, detergent and lipid concentrations can be relied upon. Note that lipids and detergents can generally be used without further purification. Be aware of the possibility that the mixed micelle solutions once prepared can have properties which are never again reproduced. For this reason, one should prepare a batch of sufficient size (e.g. 25 ml of a solution containing 2 mg DMPC/ml) to last for a large number of experiments, and freeze the whole batch in small aliquots (~250 μl). In this way the starting conditions can always be standardized.

Equipment and reagents

- Flow-dialysis apparatus including programmable water bath, peristaltic pump, etc. (see *Figure 2*)
- Dialysis membranes (boiled in 0.35 M $NaHCO_3$ and 1 mM EDTA prior to use)
- Suitable lipid
- Chloroform–methanol (2:1) or chloroform alone
- Suitable detergent dissolved in buffer (detergent–buffer; variation 1) or chloroform–methanol–(lipid) (variation 2)
- Detergent-solubilized protein solution

Method

1. Dissolve 50 mg of lipid in 1 ml chloroform–methanol (2:1) or chloroform alone in a small round-bottomed flask, preferably under nitrogen or argon to avoid oxidation.
2. *Either* (variation 1): evaporate the solvent in a rotary evaporator to produce a thin film of lipid and add detergent–buffer to resuspend and solubilize the lipid, then adjust the detergent concentration as required; *or* (variation 2): add the detergent to the resuspended lipid or dissolve it in chloroform–methanol prior to its addition to the lipid, then evaporate the solvent, vacuum dry, and subsequently add buffer.[a]
3. Add detergent-solubilized protein solution to the mixed micelles. Adjust the temperature to the value at which the dialysis will be carried out prior to the addition of the protein. The detergent concentration in the detergent-solubilized protein solution should be about 1.5 × c.m.c. or whatever is required to maintain protein solubilization. Check the sample in the CTEM for homogeneity, precipitates, etc., before proceeding.

4. Mix the detergent-solubilized protein with the detergent–lipid mixed micelles—the final detergent concentration should be about 3 × c.m.c.
5. Load the flow-dialysis chamber (see *Figure 2*) and make sure that the volume used totally fills the chamber.
6. Start the dialysis and note down all relevant parameters and changes of them (e.g. temperature, flow rate, composition of dialysis buffer used).
7. Take samples after, say, 0.5, 1, 2, 3, 4, 6, 12, 24, and 48 h, and prepare specimens for CTEM (see *Protocols 2* and *3*). Document the results by taking micrographs or via on-line digital image acquisition.
8. Assess the specimens for incorporation of protein into bilayers and the formation of ordered patches.
9. Check the quality of any ordered arrays using a laser diffractometer or on-line fast Fourier transforms. It is imperative to note down every detail, from the specimen preparation method and support films used, to the indices of the highest reflections observed.

[a] Note that in reconstitution experiments starting from mixed micelles prepared according to variation 2, a longer period of dialysis is required for vesicles to form than in experiments starting from variation 1 micelles (S. Renfrey, personal communication).

2.3 Trouble-shooting

If non-satisfactory results have been obtained using *Protocol 1*, attempt to optimize the parameter configuration using the foregoing rationales, and bearing in mind that parameters are easily changed—but maintained with difficulty. Therefore, every time one thinks about changing a parameter, one should also check whether it will be possible to keep the changed parameter at a constant value. Because of the complexity of reconstitution experiments, all parameters need to be controlled most thoroughly, and it is highly advisable to formulate a record sheet to keep all data in a standardized, easily reviewable format. Furthermore, always use the electron microscope to analyse what happened during the course of dialysis. Some of the principal indicators for a lack of reconstitution and/or formation of 2D crystals are given below, with possible explanations:

- No vesicle formation—detergent not dialysable, defective dialysis apparatus, too small an amount of lipid present.
- No protein inside the vesicles—wrong lipid environment, protein clumped together prior to insertion, dialysis membrane exclusion limit too large, too rapid a removal of detergent, unsuitable temperature.
- Protein inserted but no formation of 2D crystals—LPR too high, unsuitable detergent/lipid, inappropriate flow rate, protein denatured, protein of low

intrinsic symmetry (higher intrinsic symmetry may be obtainable by oligomerization or by limited digestion).

- Formation of poorly ordered arrays—incomplete removal of detergent, LPR too high, sub-optimal detergent/lipid, non-ideal temperature or temperature ramps chosen (tempering), inadequate phospholipase treatment procedures.

3. Conventional electron microscopy

3.1 Microscope requirements

A state-of-the-art CTEM should have acceleration voltages of at least 100 keV, be equipped with a low-dose kit (including electron-dose measuring facilities and beam blank), an eucentric goniometer (for 3D data acquisition), spot size adjustment, motor-driven translation controls (including a memory function), and be upgradable for cryo-work. Regarding the latter, it is advantageous if the microscope is supplied with an ion getter pump and a standard anti-contaminator for a clean and dry ultra-high vacuum ($< 10^{-7}$ Torr), and possesses easily expandable/upgradable microscope control software in order to be able, for example, to include cryo-specific pumping sequences.

3.2 Specimen preparation

In order to gain sufficient contrast to visualize proteins (composed of atoms of low atomic number) in the CTEM, the protein molecules need to be stained with chemicals which contain high atomic number atoms (e.g. uranyl salts) and hence interact significantly with the electron beam by, for example, scattering. In contrast to that, the specimen-support film needs to be as 'electron-neutral' as possible in order not to introduce significant levels of background noise to the image. Suitable support films are (i) simple carbon films (10–20 nm thick) or (ii) carbon-stabilized collodion films (i.e. carbon–collodion-coated grids). Carbon films should be produced by indirect (e.g. ref. 29) resistance evaporation of carbon on to freshly cleaved mica as proposed by Bradley (30). The indirect coating method (allowing only those carbon particles which have been reflected from a surface after evaporation from the carbon rods to sublime on to the mica surface) has the advantage of producing a carbon film of finer granularity and a higher degree of homogeneity.

Collodion films of 30–50 nm thickness (as judged from their grey–silver/grey colour; see, for example, ref. 31) can easily be prepared by the 'falling drop' method (32) using an approximately 8 μl droplet of a 1% collodion solution (w/v in isoamyl acetate) and a Büchner funnel with an internal diameter of 80 mm. The dried collodion-coated grids are subsequently coated with carbon by carbon resistance evaporation.

The procedure most commonly used to obtain a first impression of a sample is negative staining. Negative staining is one of the most convenient and easily applicable methods of contrast enhancement, suitable for objects ranging in size from biological macromolecules to whole organisms like bacteria. Staining is achieved by using aqueous solutions of heavy metal salts, which, upon drying, form a glassy cast around the protein. Ideally, no positive staining should occur and the stain distribution should be even throughout the entire structure. Assuming ideal behaviour, the stain-excluding areas correspond to protein, i.e. protein appears white. However, one cannot necessarily rely on ideal staining behaviour as shown in several cases (33–37). A procedure is given in *Protocol 2* which works well for most membrane proteins.

Protocol 2. Preparation of negatively stained specimens: droplet method

All work should be carried out in a clean environment with minimum haste. Care should be taken when handling uranyl salts since they are toxic and radioactive.

Equipment and reagents

- Collodion–carbon-coated 300 or 400 mesh/inch copper grids (freshly glow-discharged for 30 sec under a reduced atmospheric pressure of 0.1 Torr)
- Good illumination source over the working area
- Parafilm, Pasteur pipettes, filter paper, fine curved forceps
- Double-distilled water
- Staining solution, e.g. aqueous solution of uranyl acetate (4% (w/v), pH 4.25)

Method

1. Place 5–25 μl sample droplets[a] (i), each followed by two droplets of double-distilled water (ii, iii) and one droplet of the staining solution (iv) on to the clean inner surface (facing the backing paper) of a strip of Parafilm.
2. Place grids with the support film towards the droplet on to (i) for 10–30 sec, quickly remove excess liquid by blotting on filter paper (do not blot dry!), then place the grid on to (ii) and (iii) for about 1 sec each (each followed by a blotting step), stain for 30 sec on (iv) and blot dry with the grid perpendicular to the filter paper.

[a] Dependent on the amount of detergent present in the sample solution, the droplets may or may not spread out. If the droplets are quite flat, the grids should be placed on top of them whilst being held holding between the tips of a pair of forceps. Due to capillary forces, the liquid can easily be displaced from the grid surface towards the tips of the forceps. This phenomenon can be avoided by purchasing special anti-capillary forceps, or by Teflon-coating the tips of the forceps.

As far as the preparation procedure parameters are concerned, the time the grid is in contact with the sample droplet as well as the protein concentration in the sample will ultimately determine the amount of sample which adsorbs to the support film. The protein concentration required depends on the molecular mass of the protein: the higher its mass the greater the required concentration. As a general rule of thumb, a 250 kDa protein at 50 μg protein/ml should give a suitable density on the support film when adsorbed for about 15 sec. Obviously the hydrophilicity/-phobicity of the support film and the protein also play a major role in determining the degree of adsorption. As simple as the procedure is, minor variations are possible in the numerous fine details of technique. Trial and error remains the best for finding the optimum preparation conditions. This applies, for example, to the blotting times used or the position of the grid relative to the filter paper during blotting. For instance, if one was to aim for a deeper stain, then one would, for example, allow the stain to 'dry' on to the specimen a little prior to the final blotting step, and then slowly blot the solution off with pieces of filter paper, approaching the grid from different positions from around its margin, keeping the filter paper parallel to the grid.

From the point of view of compatibility of buffer solutions, the least problematic solutions that can be used in the procedure, apart from double-distilled water, are slightly alkaline solutions containing an organic buffer system (e.g. 20 mM Tris) and low salt concentrations (avoid phosphates when using uranyl salts!). Many other aqueous heavy metal salt solutions can be used in the place of uranyl acetate (e.g. 0.75% (w/v) uranyl formate, 2% (w/v) phosphotungstic acid which can be titrated to pH 7 with NaOH or solutions of methylamine tungstate (38), ammonium molybdate etc.; see ref. 39 for further details). Uranyl formate has the advantage of a finer grain size than uranyl acetate, and, in some cases, it helps avoid protein aggregation. However, aqueous solutions of uranyl formate are very unstable and need to be used immediately.

Should the specimen preparation procedure following *Protocol 2* fail to yield satisfactory results, try *Protocol 3* (after ref. 40).

Protocol 3. Preparation of negatively stained specimens: flotation method

This protocol should only be used for sample solutions which do not contain a detergent above its c.m.c. All work should be carried out in a clean environment with minimum haste. Care should be taken when handling uranyl salts since they are toxic and radioactive.

Equipment and reagents

- Mica, which has been indirectly coated with carbon (see, for example, ref. 29 for further details)
- Good illumination source over the working area
- Parafilm *or* tops of 1.5 ml Eppendorf cups
- Pasteur pipettes, filter paper, fine curved forceps
- Double-distilled water
- Staining solution, e.g. aqueous solution of uranyl acetate (4% (w/v), pH 4.25)

Methods

1. Pipette 50–100 μl of sample (i), washing (ii), and staining (iii) solutions on to a clean strip of Parafilm *or* pipette the solutions into three small wells (e.g. the tops of Eppendorf cups). When filling the wells, make sure a pronounced convex meniscus is obtained.
2. Using sharp, pointed scissors cut a 3 × 3 mm square from the carbon-coated mica.[a]
3. Holding the mica at one end with forceps, introduce it into the sample solution at an angle of 30–45°. The carbon film will start to float off but is held in place by the forceps.
4. Allow the sample material to adsorb on to the carbon film for 10–30 sec depending on the sample concentration (see also above). The adsorption time determines the final particle density on the film.
5. Remove the mica slowly. The carbon will fall back to its original position. Drain off excess fluid with filter paper, but do not blot dry.
6. Using the procedure described in steps 3–5, introduce the mica into the washing solution (double-distilled water or a suitable buffer). The 'washing' time should only be 0.5–3 sec.
7. Immediately after draining off excess washing solution, completely float off the carbon from the mica on the staining solution (e.g. aqueous solution of uranyl acetate; see also above) and let the carbon float on the droplet for 30 sec.
8. Using forceps, pick up a grid and touch the surface of the carbon film with the shiny grid surface facing the carbon. Allow the grid to make contact with the carbon for a few seconds—pressing it gently against the carbon—before lifting the grid tangentially away from the droplet, and turning it upside down.
9. Blot off the staining solution completely by pressing the grid–carbon–sample sandwich against filter paper.

[a] A larger piece of indirectly carbon-coated mica can also be used for the preparation of carbon-coated grids by directly floating off the carbon on double-distilled water, and then decorating the carbon film with grids. Before lifting the grids off the water surface, they need to be separated from each other, and any carbon extending over the grid margin needs to be removed.

4. Image analysis and processing

The analysis of electron micrographs in its most simplistic form can be carried out by eye. The experimenter's visual integrating capacity can be trained to an astonishing extent by looking at similar specimens over and over again. After a while it is possible to recognize typical projections of the protein and judge the behaviour of the protein in solution. From simple measurements (using a ruler or an eyepiece graticule) on different projections of single particles, one can arrive at an estimate of the overall volume of a protein and, consequently, its mass according the the formula m (Da) = volume of the protein (ml) × Avogadro's number × 1/υ (ml/g) with υ being the partial specific volume. The partial specific volume varies between 0.73 and 0.77 ml/g for dehydrated and partially hydrated proteins, respectively (41). Furthermore, it is possible to retrieve information about the quaternary structure of a protein by visual analysis of electron micrographs and simple procedures like Markham rotation (42). The latter procedure can be carried out in any photographic laboratory, while the more advanced analogue techniques require a laser-powered optical diffractometer equipped with a lens system for image reconstruction (43). In order to render even the simplest measurements more trustworthy, it is essential to get into the habit of working with calibrated primary magnifications rather than relying on the nominal magnifications displayed on the microscope's panel. Routines for the swift and reproducible operation of an electron microscope have been described (44, 45).

In order to reliably retrieve structural information about a membrane protein in any more detail than that described above, the signal-to-noise ratio needs to be significantly improved. This can only be achieved by averaging over many motifs. A prerequisite for averaging is that each motif represents the same species of molecule in an identical projection, or, for three-dimensional data sets, in different known projections. For large macromolecules (≥ 500 kDa), single-particle averaging is a viable approach. Several well documented image processing software packages specifically designed to carry out averaging of single molecules exist (see below). The biggest difficulty with single-particle averaging is the analysis of the individual projections the molecules have assumed; multivariate statistical analysis in conjunction with classification techniques need to be employed to analyse the projections of randomly oriented particles and prepare them for the final averaging and 3D structure calculation. If an averaged projection is all that is required, then preference should be given to the imaging of the protein while in the membrane. Although the molecules may be randomly distributed in the membrane plane, they do not assume random projections, nor do they reveal strong variations in stain distribution. Due to the confinement of the lipid environment, the protein molecules are aligned along an axis perpendicular to the membrane plane which aids the alignment procedures required for single-particle averaging. In contrast, detergent-solubilized molecules are more likely to assume a whole range of projections due to non-specific adsorption to the support film. However, using single particles, the resolution of final averaged projections and 3D maps will rarely be better than 2 nm. In the case of ion channels (due to the small masses of most of these proteins) it is therefore highly desirable for CTEM studies (~ 2 nm resolution), and an absolute requirement for high-resolution (< 1 nm) cryo-

TEM studies, to grow 2D crystals. The major advantage of 2D crystals over single particles is due to the fact that all structural motifs are already precisely aligned.

In order to access the information contained within 2D crystals, digital image processing is required. Regarding the analysis of electron micrographs and image processing procedures, the reader is referred to some excellent books and articles (see, for example, refs 46–56). In most instances it will be mandatory to have at least some understanding of the underlying principles before attempting any type of image processing, although some software packages allow even absolute beginners to undertake complex processing steps. A very useful survey of the different image processing systems available, including references for more detailed information, is given in ref. 57. Two recent software packages not included in ref. 57 are Spectra (58) and CRISP/ELD (59, 60). SPECTRA is a colour X-Window Application running under UNIX designed for fast refinements of lattice parameters and correction of lattice distortions. CRISP/ELD is a comprehensive PC-based package for electron crystallographic image processing which was recently adapted to run under Microsoft Windows. Both packages are extremely user-friendly and are based on many of the MRC-LMB (Medical Research Council, Laboratory of Molecular Biology, Cambridge) routines. An account of another software library for 3D reconstructions which can be run on small workstations under the operating system AIX has recently been given (61). It should be noted that there is nowadays a general move to substitute the programs of some packages or program collections (e.g. SPIDER, MRC-LMB programs) formerly run under the VMS operating system with their corresponding UNIX versions.

5. High-resolution electron microscopy

High-resolution electron microscopy of biological macromolecules relies on the preservation of high-resolution structural details in the specimen, and, of course, on compatible imaging conditions. The former prerequisite can be fulfilled by observing specimens unstained and in the hydrated state. The hydration of the specimen, however, is incompatible with the vacuum and also leads to increased radiation sensitivity which can only be overcome by minimizing the electron dose and lowering the temperature, hence the need for cryo-electron microscopy. By lowering the temperature of the specimen while in the microscope to less than −160°C, a significant decrease in the propagation of free radicals, a better energy dissipation, and increased conductance are achieved, leading to increased radiation resistance. To this end 2D crystals are particularly useful since the electron dose can be distributed over a large number of identical motifs. The preparation of high-resolution specimens (see ref. 62 for further details) from 2D crystals typically entails adsorbing the sample to a carbon-coated grid (see footnote *a*, *Protocol 3*) or a grid coated with a perforated carbon film (63) and embedding it, for example, in vitreous water (frozen-hydrated), tannic acid/tannin, or glucose (62) before plunging into the cryogen (ethane or propane should be preferred over liquid nitrogen; see also 64 and refs therein). In many cases, the cooling rate needs to be sufficiently high to allow the vitrification of the specimen to take place (see, for example, ref. 62), which is required for the preservation/ observation of high-resolution structural detail. It is also possible to quick-freeze samples which have been embedded in aqueous solutions of uranyl salts (e.g. 0.75% (w/v) uranyl acetate) or aurothioglucose (65) in order to enhance the contrast. The specimen is then transferred into the cryo-holder and subsequently into the microscope. It is crucial that all of these transfers are carried out under frost-free conditions. A pre-cooled cryo-transfer chamber flushed with dried N_2 as well as dried, N_2-flushed protective bellows surrounding the cryo-holder greatly assist in this exercise. While the preparation procedures may appear trivial, considerable experience is required to obtain reproducible results. Almost the same is true regarding the high-resolution imaging in cryo-electron microscopy:

(a) The low contrast inherent to unstained specimens makes it necessary to record images considerably underfocus. While this aids the retrieval of high-resolution information, it also causes the phase contrast transfer function to oscillate and hence represent structural details corresponding to different spatial frequencies with reversed contrast and differentially attenuated amplitudes. This requires the phase and amplitude information to be corrected using numerical methods. Alternatively, the correct amplitude information can be obtained from electron diffraction patterns which are not influenced by the transfer function, provided that one is dealing with large (a few thousand unit cells; see ref. 54), highly ordered 2D crystals.

(b) Another problem concerns the mechanical stability of a cryo-stage and beam-induced movements of the specimen, both leading to drift. To this end, one should seek a cryo-stage which provides the least thermal gradients along the specimen holder rod and between the specimen and its immediate surroundings. Differential deposition of crystalline water (contamination) obstructs the observation of the specimen and also adds to the drift, hence good anti-contamination is required. The design (double blades, tube) should be such that adequate anti-contamination is guaranteed even when the specimen holder is in a tilted position. Ideally, the only thing one should see, if one was able to view in a direction perpendicular to the grid surface and facing away from it, should be the anti-contaminator. With respect to the amplitudes, this drift problem can be overcome by acquiring the amplitudes from the electron diffraction patterns which are translationally invariant. The phases, however, have to come from the image itself, and there are two ways to attenuate residual drift problems: firstly, by a faster acquisition time using a more sensitive recorder (e.g.

a slow-scan CCD camera rather than photographic emulsion), and, secondly, by spot-scan imaging (66). In the latter case, the area of interest is scanned with an electron beam of small spot size. This has two advantages: the acquisition of each spot only takes a fraction of a second, so that stage drift problems can be reliably overcome; and the specimen area illuminated at any given time is decreased, reducing beam-induced movements caused by specimen charging and heating. Spot-scan imaging software is available commercially for different makes of microprocessor-controlled microscopes. However, in order to arrive at short acquisition times, electron sources of high intensity, like LaB_6 cathodes or field emission guns, are essential. The latter types of electron source require a very good vacuum ($< 10^{-6}$ and $< 10^{-9}$ Torr, respectively), such as provided by ion getter pumps.

(c) The final problem constitutes a general hurdle associated with the acquisition of images containing high-resolution information: that is, radiation sensitivity. The increase in radiation resistance the specimen acquires by observation under cryo-conditions, i.e. with the temperature of the specimen at, for example, $-120°C$ is, taking a conservative estimate, in the region of three- to five-fold. While the 3D structure will be affected with doses between 50 and 100 electrons/nm^2 at room temperature, the critical dose under cryo-conditions is in the range between 150 and 500 electrons/nm^2. In practice, this means that one should definitely try to work at doses well below the 500 electrons/nm^2 threshold. In keeping with these electron-dose requirements, one also circumnavigates any problems arising from pre-sublimation processes (affecting the amorphous ice as well as the protein), which not only result in a loss of the protein's structural integrity but also in increased specimen movements. For ideal specimens (i.e. where most membrane patches show crystallinity), such low doses are not necessarily difficult to achieve, but for non-ideal specimens (where only the minority, perhaps $\leqslant 1\%$, of the membranes are crystalline) many grid windows have to be screened thoroughly (possibly at higher magnifications) for suitable areas, resulting in doses which may be incompatible with high-resolution data acquisition. In the latter case, a sensitive on-line image acquisition and analysis package (slow-scan CCD camera, frame grabber, workstation with array processor and high-resolution monitor, image analysis software for the on-line calculation of power spectra, Fourier peak filtered images, band-pass filtered images, etc.) is required to enable rapid acquisition of images and analysis of the quality of the corresponding specimen area at sufficiently low electron doses to avoid radiation damage and/or sublimation of water.

From the instrumentational point of view, the ultimate electron microscope for imaging biological macromolecules at near atomic resolution is an intermediate voltage instrument (300 keV) fitted with a field emission gun (67) and, possibly, equipped for imaging the specimen at liquid helium temperature.

The higher accelerating voltage gives a greater depth of field (not so crucial for thin 2D crystals), and leads to a decrease of the Ewald sphere curvature, thus safeguarding a more reliable determination of diffraction spots at high tilt angles. The field emission gun improves the spatial and temporal coherence (angular and energy spread of the incident beam, respectively), leading to a decrease in the attenuation of high-resolution information. It should, however, be mentioned that such dedicated high-resolution electron microscopes will have lost the ease of operation of a 100 or 120 keV instrument with an ordinary tungsten filament as electron source, which should be considered particularly if the screening of many specimens is important.

In conclusion, the most limiting parameter in high-resolution electron microscopy is the quality of the specimen. Once 2D crystals which diffract to less than 2 nm resolution in negative stain can be obtained reproducibly, the specimen-relevant resolution should be re-assessed under cryo-conditions. If the latter assessment confirms that high-resolution information can indeed be obtained, then the way is opened for a more comprehensive study of the sample in one of the few worldwide existing centres of excellence specialized in high-resolution electron crystallography.

Acknowledgements

I would like to thank Drs F. H. Shepherd (University of Leeds), S. Renfrey (Department of Pathology, University of Cambridge), K. H. Downing (LBL, UC Berkeley, USA), and W. Welte (Institute for Biophysics, University of Freiburg, Germany) for helpful discussions and suggestions, Dr A. Hoenger (Scripps Research Institute, La Jolla, CA, USA) for permission to use an illustration from his Thesis, and Miss A. Durbin (University of Leeds) for producing the drawing shown in *Figure 2*.

References

1. Engel, A. and Reichelt, R. (1984). *J. Ultrastruct. Res.* **88**, 105.
2. Reichelt, R., Holzenburg, A., Buhle Jr., E. L., Jarnik, M., Engel, A., and Aebi, U. (1990). *J. Cell Biol.* **110**, 883.
3. Glaeser, R. M. (1985). *Ann. Rev. Phys. Chem.* **36**, 243.
4. Henderson, R., Baldwin, J. M., Ceska, T. A., Zemlin, F., Beckman, E., and Downing, K. H. (1990). *J. Mol. Biol.* **213**, 899.
5. Kühlbrandt, W., Wang, D. N., and Fujiyoshi, Y. (1994). *Nature* **367**, 614.
6. Jap, B. K., Walian, P. J., and Gering, K. (1991). *Nature* **350**, 167.
7. Finbow, M. E., Buultjens, T. E. J., Lane, N. J., Shuttleworth, J., and Pitts, J. D. (1984). *EMBO J.* **3**, 2271.
8. Jap, B. K., Zulauf, M., Scheybani, T., Hefti, A., Baumeister, W., Aebi, U., and Engel, A. (1992). *Ultramicroscopy* **46**, 45.

9. Holzenburg, A., Wilson, F. H., Finbow, M. E., and Ford, R. C. (1992). *Biochem. Soc. Trans.* **20**, 591.
10. Cullis, P. R. and Hope, M. J. (1985). In *Biochemistry of lipids and membranes* (ed. D. E. Vance and J. E. Vance). Benjamin/Cummings, Menlo Park.
11. Israelachvilli, J. N., Marcelja, S., and Horn, R. G. (1980). *Quart. Rev. Biophys.*, **13**, 121.
12. Cevc, G. and Marsh, D. (1987). *Phospholipid bilayers*. Wiley, New York.
13. Welte, W. and Wacker, T. (1991). In *Crystallization of membrane proteins* (ed. H. Michel), p. 107. CRC Press, Boca Raton.
14. Timmins, P., Hauk, J., Wacker, T., and Welte, W. (1991). *FEBS Lett.* **280**, 115.
15. Boekema, E. J. (1990). *Electron Microsc. Rev.* **3**, 87.
16. Jost, F., Leiter, H., and Schwuger, M. J. (1988). *Coll. Polym. Sci.* **266**, 554.
17. Israelachvilli, J. N., Mitchell, D. J., and Ninham, B. W. (1976). *J. Chem. Soc. Faraday Trans. II*, **72**, 1525.
18. Mannella, C. A. (1985). In *Methods in enzymology* (ed. S. Fleischer and B. Fleischer), Vol. 125, p. 595. Academic Press, Orlando.
19. Ford, R. C., Hefti, A., and Engel, A. (1990). *EMBO J.* **9**, 3067.
20. Hovmöller, S., Slaughter, M., Berriman, J., Karlsson, B., Weiss, H., and Leonard, K. (1983). *J. Mol. Biol.* **165**, 529.
21. Rotenberg, M. and Lichtenberg, D. (1991). *J. Colloid Interface Sci.* **144**, 591.
22. Wingfield, P., Arad, T., Leonard, K., and Weiss, H. (1979). *Nature* **280**, 696.
23. Holzenburg, A., Engel, A., and Aebi, U. (1988). In *Proceedings of the 9th European Congress on Electron Microscopy*, Inst. Phys. Conf. Ser. No. 93, Vol. 3 (ed. H. G. Dickinson and P. J. Goodhew), p. 453. Institute of Physics, Bristol.
24. Schleicher, A., Franke, R., Hofmann, K. P., Finkelmann, H., and Welte, W. (1987). *Biochemistry* **26**, 5908.
25. Lévy, D., Gulik, A., Bluzat, A., and Rigaud, J.-L. (1992). *Biochim. Biophys. Acta* **1107**, 283.
26. Hönger, A. (1989). Diploma Thesis, Biocentre at the University of Basel, Switzerland.
27. Kühlbrandt, W. (1992). *Quart. Rev. Biophys.* **25**, 1.
28. Engel, A., Hoenger, A., Hefti, A., *et al.* (1992). *J. Struct. Biol.* **109**, 219.
29. Robinson, D. G., Ehlers, U., Herken, R., Herrmann, B., Mayer, F., and Schürmann, F.-W. (1987). *Methods of preparation for electron microscopy*. Springer, Berlin.
30. Bradley, D. E. (1965). In *Techniques for electron microscopy* (ed. D. Kay), p. 58. Blackwell, Oxford.
31. Hayat, M. (1981). *Principles and techniques of electron microscopy–biological applications*. Arnold, London.
32. Drummond, D. G. (1950). *J. Roy. Microsc. Soc.* **70**, 17.
33. Baker, T. S. and Caspar, D. L. D. (1984). *Ultramicroscopy* **13**, 137.
34. Harris, J. R. and Holzenburg, A. (1989). *Micron Mirosc. Acta* **20**, 223.
35. Sikerwar, S. S., Downing, K. H., and Glaeser, R. M. (1991). *J. Struct. Biol.*, **106**, 255.
36. Mosser, G. and Brisson, A. (1991). *J. Electron Microsc. Technique* **18**, 387.
37. Holzenburg, A., Bewley, M. C., Wilson, F. H., Nicholson, W. V., and Ford, R. C. (1993). *Nature* **363**, 470.
38. Faberge, A. C. and Olivier, R. M. (1974). *Microscopie* **20**, 242.

39. Spiess, E., Zimmermann, H.-P., and Lünsdorf, H. (1987). In *Electron microscopy in molecular biology: a practical approach* (ed. J. Sommerville and U. Scheer), p. 147. IRL Press, Oxford.
40. Valentine, R. C., Shapiro, B. M., and Stadtman, E. R. (1968). *Biochemistry* **7**, 2143.
41. Richards, F. M. (1977). *Ann. Rev. Biophys. Bioeng.* **6**, 151.
42. Markham, R., Frey, S., and Hills, G. J. (1963). *Virology* **20**, 88.
43. Beeston, B. E. P., Horne, R. W., and Markham, R. (1972). In *Practical methods in electron microscopy*, Vol. 1 (ed. A. M. Glauert), p. 185. North-Holland, Amsterdam.
44. Chescoe, D. and Goodhew, P. J. (1984). *The operation of the transmission electron microscope*, Microscopy Handbooks No. 2 of the Royal Microscopical Society. Oxford University Press, Oxford.
45. Agar, A. W., Alderson, R. H., and Chescoe, D. (1974). In *Practical methods in electron microscopy*, Vol. 2 (ed. A. M. Glauert). North-Holland, Amsterdam.
46. Engelhardt, H. (1988). *Methods Microbiol.* **20**, 357.
47. Frank, J. (1979). *J. Microsc.* **117**, 25.
48. Misell, D. L. (1978). In *Practical methods in electron microscopy*, Vol. 7 (ed. A. M. Glauert). North-Holland, Amsterdam.
49. Saxton, W. O. (1978). *Computer techniques for image processing in electron microscopy*. Academic Press, London.
50. Misell, D. L. and Brown, E. B. (1987). In *Practical methods in electron microscopy*, Vol. 12 (ed. A. M. Glauert). Elsevier, Amsterdam.
51. Chiu, W., Schmid, M. F., and Venkataram Prasad, B. V. (1993). *Biophys. J.*, **64**, 1610.
52. Hawkes, P. W. and Valdrè, U. (1990). *Biophysical electron microscopy*. Academic Press, London.
53. Amos, L. A., Henderson, R., and Unwin, P. N. T. (1982). *Prog. Biophys. Molec. Biol.* **39**, 183.
54. Unwin, P. N. T. and Henderson, R. (1975). *J. Mol. Biol.* **94**, 425.
55. Baldwin, J. and Henderson, R. (1984). *Ultramicroscopy* **14**, 319.
56. Henderson, R., Baldwin, J. M., Downing, K. H., Lepault, J., and Zemlin, F. (1986). *Ultramicroscopy* **19**, 147.
57. Hegerl, R. (1992). *Ultramicroscopy* **46**, 417.
58. Schmid, M. F., Dargahi, R., and Tam, M. W. (1993). *Ultramicroscopy* **48**, 251.
59. Hovmöller, S. (1992). *Ultramicroscopy* **41**, 121.
60. Zou, X., Sukharev, Y., and Hovmöller, S. (1993). *Ultramicroscopy* **49**, 147.
61. Lanzavecchia, S., Bellon, P. L., and Scatturin, V. (1993). *J. Microsc.* **171**, 255.
62. Wang, D. N. and Kühlbrandt, W. (1991). *J. Mol. Biol.* **217**, 691.
63. Toyoshima, C. (1989). *Ultramicroscopy* **30**, 439.
64. Stewart, M. (1991). In: *Celecton Microscopy in biology—a practical approach* (ed. J. R. Harris), p. 229. IRL Press, Oxford.
65. Kühlbrandt, W. (1982). *Ultramicroscopy* **7**, 221.
66. Downing, K. H. (1991). *Science* **251**, 53.
67. Zhou, Z. H. and Chiu, W. (1993). *Ultramicroscopy* **49**, 407.

A1

Addresses of suppliers

Aldrich

Aldrich Chemical Company Ltd, The Old Brickyard, New Road, Gillingham, Dorset, SP5 4BR, UK.

Aldrich Chemical Company Inc., 940 West St. Paul Avenue, Millwaukee, WI 53233, USA.

Amari Plastics

Amari Plastics, London, UK.

Amersham International

Amersham International PLC, Lincoln Place, Green End, Aylesbury, Bucks., HP20 2TP, UK.

Amersham Corporation, 2636 South Clearbrook Drive, Arlington Heights, IL 60005, USA.

Avanti Polar Lipids

Avanti Polar Lipids, Alabaster, AL, USA.

Axon Instruments

Axon Instruments, Foster City, CA, USA.

BASF

BASF, Mannheim, Germany.

BDH Chemicals

BDH Chemicals, Poole, Dorset, UK.

Beckman Instruments

Beckman Instruments UK Ltd., Progress Road, Sands Industrial Estate, High Wycombe, Bucks., HP12 4JL, UK.

Beckman Instruments Inc., PO Box 3100, 2500 Harbor Boulevard, Fullerton, CA 92634, USA.

Behring Diagnostics

Behring Diagnostics, La Jolla, CA, USA.

Biologic

Biologic, Claix, France.

Bio-Rad Laboratories

Bio-Rad Laboratories Ltd., Maylands Avenue, Hemel Hempstead, Herts HP2 7TD, UK.

Bio-Rad Laboratories, Division Headquarters, 3300 Regatta Boulevard, Richmond, CA 94804, USA.

Boehringer-Mannheim

Boehringer-Mannheim UK (Diagnostics/Biochemicals) Ltd., Bell Lane, Lewes, East Sussex, BN7 1LG, UK.

Boehringer-Mannheim Corporation, Biochemical Products, PO Box 50414, Indianapolis, IN 46250-0414, USA.

Boehringer-Mannheim GmbH Biochemica, PO Box 31 01 20, D-6800 Mannheim, Germany

Braun

Braun (B. Braun Medical Ltd.), 13–14 Farmborough Close, Aylesbury Vale Industrial Park, Bucks. HP20 1DQ, UK.

Braun (B. Braun Melsungen AG), PO Box 120, D-3508 Melsungen, Germany.

Calbiochem Novabiochem

Calbiochem-Novabiochem (UK) Ltd., 3 Heathcoat Building, Highfields Science Park, University Boulevard, Nottingham NG7 2QJ, UK.

Calbiochem, PO Box 12087, San Diego, CA 92112-4180, USA.

Cambridge Electronic Design

Cambridge Electronic Design, Cambridge, Camb., UK.

Camden Instruments

Camden Instruments, Loughborough, Leics, UK.

Clark Electromedical

Clark Electromedical, Reading, Berks, UK.

Clontech

c/o Cambridge Bioscience Ltd., 25 Signet Court, Stourbridge Common Business Centre, Swans Road, Cambridge CB5 8LA, UK.

Clontech Laboratories Inc., 4055 Fabian Way, Palo Alto, CA 94303, USA.

Dagan

Dagan, Minneapolis, MN, USA.

Difco

Difco Laboratories Ltd., Central Avenue, East Molesley, Surrey, KT8 0SE, UK.

Difco Laboratories, PO Box 331058, Detroit, Michigan 48232–7058, USA.

Du Pont-NEN

Du Pont UK Ltd, Wedgwood Way, Stevenage, Herts SG1 4QN, UK.

Du Pont Co. (Biotechnology Systems Division), PO Box 80024, Wilmington, DE 19880–0024, USA.

FMC Marine Colloids

c/o Flowgen Instruments Ltd., Broad Oak Enterprise Village, Broad Road, Sittingbourne, Kent ME9 8BR, UK.

FMC BioProducts Inc., 5 Maple Street, Rockland, ME 04841, USA.

FMC BioProducts (Europe), Risingevej 1, DK-2665 Vallensbaek Strand, Denmark.

Gibco BRL

Gibco BRL (Life Technologies Ltd.), Trident House, Renfrew Road, Paisley PA3 4EF, UK.

Gibco BRL (Life Technologies Inc.), 3175 Staler Road, Grand Island, NY 14072, USA.

Goodfellow Metals

Goodfellow Metals, London, UK.

Heraeus

Heraeus, Hanau, Germany.

Infors

Infors, Basel, Switzerland.

Instrutech

Instrutech, Elmont, NY, USA.

Intelligenetics

Intelligenetics, Palo Alto, CA, USA.

Invitrogen

c/o British Biotechnology Products Ltd, 4–10 The Quadrant, Barton Lane, Abingdon, Oxon, OX14 3YS, UK.

Invitrogen Corporation, 3985 B Sorrento Valley Building, San Diego, CA 92121, USA.

Kobra

Kobra, West Springfield, MA, USA.

Kopf

Kopf, Tujunga, CA, USA.

Laboratory Supplies

Laboratory Supplies, Hickesville, NY, USA.

Laser Science

Laser Science, Cambridge, MA, USA.

LeCroy

LeCroy, Abingdon, Oxon, UK.

List Electronic

List Electronic, Darmstadt, Germany.

Merck

Merck Ltd., Merck House, Poole, Dorset BH15 1TD, UK.

Merck AG, PO Box 4119, Frankfurterstrasse 250, D-6100 Darmstadt, Germany.

Molecular Probes

Molecular Probes, Eugene, OR, USA.

Narashige

Narashige, Tokyo, Japan.

New England Biolabs

c/o CP Labs Ltd., PO Box 22, Bishops Stortford, Herts CM23 3DH, UK.

New England Biolabs (NBL), 32 Tozer Road, Beverley, MA 01915–5510, USA.

New England Nuclear (*see* **Dupont-NEN**)

Pharmacia Biotechnology

Pharmacia LKB Biotechnology Ltd., 23 Grosvenor Road, St. Albans, Herts AL1 3AW, UK.

Pharmacia LKB Biotechnology Inc., 800 Centennial Avenue, PO Box 1327, Piscataway, NJ 08855–1327, USA.

Pharmacia LKB Biotechnology AB, Pharmacia Biotech Europe, Rue de la Fusée 66. B-1130 Brussels, Belgium.

PharMingen

PharMingen, San Diego, CA, USA.

Promega

Promega Ltd., Delta House, Enterprise Road, Chilworth Research Centre, Southampton SO1 7NS, UK.

Promega, 2800 Woods Hollow Road, Madison, WI 53711–5399, USA.

Racal

Racal, Slough, Berks, UK.

RS Components

RS Components, PO Box 99, Corby, Northants, UK.

Schuco International

Schuco International, London, UK.

Seishin Pharmaceuticals

Seishin Pharmaceuticals, Tokyo, Japan.

Sigma Chemical Company

Sigma Chemical Co. Ltd., Fancy Road, Poole, Dorset, BH17 7NH, UK.

Sigma Inc., PO Box 14508, St. Louis, MO 63178, USA.

Spindler and Hoyer

Spindler and Hoyer, Milton Keynes, Bucks, UK.

Sterilin

Sterilin, Feltham, Middlesex, UK.

Stratagene

Stratagene Ltd., Unit 140, Cambridge Innovation Centre, Milton Road, Cambridge CB4 4FG, UK.

Stratagene Inc., 11011 North Torrey Pines Road, La Jolla, CA 92037, USA.

Sutter Instruments

Sutter Instruments, Novato, CA, USA.

United States Biochemicals (see **Amersham International**)

Vetter

Vetter, Rebersburg, PA, USA.

Warner Instruments

Warner Instruments, Hamden, CT, USA.

Worthington

c/o Cambridge Bioscience Ltd., 25 Signet Court, Stourbridge Common Business Centre, Swans Road, Cambridge CB5 8LA, UK.

Worthington Biochemical Corporation, Halls Mill Road, Freehold, NJ 07728, USA.

Index

A23187 (calcium ionophore) 227
abscissic acid 220
acetoxymethyl (AM) esters 225, 226
acetylcholine receptors
  muscarinic 152, 234
  nicotinic 8, 14, 146, 159, 192, 200, 202, 203, 205, 206, 207, 208, 209, 210, 211, 212, 213, 244
access resistance 16
activity coefficient (and ionic activities) 78, 79
adenosine 152
adenovirus 156
ADP-ribosylation 232
adenylate cyclase 234, 235
aequorin 224
affinity chromatography 127–9, 172, 182–5
affinity labelling 140
agar bridges 45, 62, 65–6
alcian blue 34
α-bungarotoxin 203, 207, 211, 213
α-helix 244
aluminium trifluoride 232
amino acid sequence 133, 135, 136, 243
ammonium molybdate 282
AMPA receptor 146
amphiphiles 272
amphoteracin B 12
amplifier
  patch-clamp and bilayer 2, 10, 20–2, 62–4
  signal 23
amplitude histogram 89, 90
analysis software
  for electrophysiological recordings 25, 87
  for image analysis 285
anchored PCR 139
annealing temperature for PCR 135
anomolous mole fraction behaviour 86
antigens
  bacterial fusion proteins as 172–4
  synthetic peptides as 172, 174–6
antipeptide antibodies 172, 177–9, 243
antisense oligonucleotides 212, 233, 234
anti-vibration support 20, 25–6, 64
Arabidopsis thaliana 223
arachidonic acid 234, 238
ATP analogues
  AMP-PNP 235
  ATPγS 235
attenuated total reflection 248
aurothioglucose 286
autofluorescence 229
Autographa californica nuclear polyhedrosis virus (AcNPV) 161, 162; *see also* baculovirus-driven expression
auxin 220
Avena sativa 223

bacteriorhodopsin 244
baculovirus-driven expression 144–5, 161–7
bandwidth 2, 9, 22, 23, 24
BAPTA 11, 236
Barth's medium 148, 153
β-adrenoceptor 243
beta function 90
β-helix 86
β-lactamase 243
β-sheet 244
bilayer fluidity 274
bilayers
  folded 48–52
  in 2D-crystal formation 274
  on patch-pipettes 52–4
  painted 44–8, 70, 71, 72, 101, 126–8, 130
bimolecular diffusion limit 83
binary mixtures of detergents 272
binomial probability distribution 90, 91, 94
biotinylation 180, 182
biphenyl detergents 275
blind-patching 38
bradykinin receptor 237
brain slice
  preparation 29, 34–7
  recording from 8, 37–8
broad-line methods in NMR 261
bursts and burst analysis 93, 94, 106, 107

Ca^{2+}
  as second messenger 219, 236–7
  buffers 11
  measurement 224–30
Ca^{2+} channels
  ligand-gated 193; *see also* inositol 1,4,5-trisphosphate receptor, ryanodine receptor
  voltage activated 78, 79, 152, 193, 233, 234, 235
Ca^{2+}-sensitive microelectrodes 224
cADP-ribose 239
caged cAMP 235
caged IP3, 237

calcium phosphate-mediated transfection 157–8, 163
cAMP 234, 235 236, 237
cAMP, membrane-permeable analogues 234
cAMP phosphodiesterase inhibitors 236
cAMP-dependent protein kinase 235
cAMP-dependent protein kinase inhibitor 235
capacitance
  cancellation and measurement 7, 8–9, 10, 13, 14, 18–19, 47–8
  cell 8–9, 10, 11, 13, 21, 22, 40
  effect on recording resolution 70–2
  electrode and pipette 4, 14, 18, 220; *see also* Sylgard coating
  planar bilayer 22, 47–8, 63
capacitor feedback headstage 21–2
capture radius 72
carbon films, *see* support films for electron microscopy
carbonyl groups 73
cardiac myocytes 235
carrier protein (for haptenization) 175
catalytic efficiency 83, 84
catecholamines 152
CCD camera 37, 38, 229, 286, 287
cell-attached recording 3, 5, 7, 13, 15, 231
cell lysis 205–6, 209
cell membrane potential 3
cellulase 222
cellulysin 221, 222
cGMP 234, 235, 236
cGMP, membrane-permeable analogue 236
cGMP phosphodiesterase inhibitors,
  MB22948, 236
  methylxanthines 236
chambers
  for bilayer recording 45, 50
  for patch-clamping 26–7
CHAPS 61, 126–8, 275, 276
chemical labelling 243, 245
chimeric cDNA 143
chimeric channels 141, 212
CHO cells 157
cholate salts 60, 272, 275
chromaffin cells 10, 30
circular dichroism (CD) spectroscopy 203, 244, 246, 262–3
Cl^--channels
  Ca^{2+}-dependent 152
  Torpedo electroplax 90
  see also GABA-A and GABA-B receptors
cloning vectors 134, 137, 139, 140, 142, 143, 149, 156–7, 159; *see also* plasmid and phagemid vectors
collagen (as a substrate) 30, 31, 32, 39
collagenase 30, 31, 32, 39, 148, 149, 150, 152
collodion films, *see* support films for electron microscopy
column chromatography 127–9; *see also* affinity chromatography
complementary DNA (cDNA)
  chimeric 143–4
  first strand 135
  full-length 137, 138–9, 154
  injection 149, 150–2
  in recombinant virus 165–6
  libraries 134, 135, 138, 139, 155
  probes 134–7, 138
  sequence analysis 137
  subcloning 135–7, 139, 141
  use in hybrid-depletion 154–6
  use in transfection 145, 159–61
computer interface 24–5
concanavalin A
  affinity column 128, 130
  overlay 186
conditional probability 96, 97, 99
conductance 75, 80, 194; *see also* slope conductance
conductance–concentration dependence 88
conductance ratio 75, 80
confocal laser scanning microscopy (CLSM) 229
constitutive promotors 159
continuum models 74, 75, 76, 77–80, 84, 85; *see also* Goldman–Hodgkin–Katz equations
COS cells 157
co-transfection 158, 162, 163, 165, 167, 168
coupling to carrier protein 175, 188
critical micellar concentration (c.m.c.) 272, 276, 277, 282
cryo-transmission electron microscopy (cryo-TEM) 270, 285–8
Cs^+ as K^+ channel blocker 11
CsCl-gradient centrifugation 162
cumulative probability distribution function 92
current convention, patch-clamp 14–15
curvature energy 274
CV1 cells 157
cyclic nucleotide-gated channels 193
cytisine 8
cytomegalovirus (CMV) promotor 157
cytoplasmic messengers, *see* second messengers
cytoskeleton 220

2D-crystals 270, 272, 273, 278–9, 285, 286, 288; *see also* electron crystallography
2D-NMR 260
dehydration–rehydration 60–1
dendrites, recording from 16
desensitization 14, 194
detergents, non-ionic *vs* zwitterionic 272
dextran-conjugated Ca^{2+} indicators 226, 227
diacylglycerol (DAG) 234, 236, 237

dielectric constant 75
diffractogram 277
digital recorders 24
dimyristoyl phosphatidylcholine 253, 254
discrete-state models 74, 75, 76, 80–6, 103
dissecting microscope 148, 150, 151
disulphide bridge formation, *see* post-translational modification
dodecyl-β-D-maltopyranoside (12-malt) 272
dodecyl-octaoxyethylene ($C_{12}E_8$) 272
dopamine (D2) receptors 233
dorsal root ganglion neurones 233, 234
dwell-time 69, 72, 95, 100, 101, 103, 104, 106, 107
dwell-time distribution 101, 103

ECL (enhanced chemiluminescence) detection system 177
edge energy 274
eicosenoids 238
eigenvalue 100, 102, 103, 106, 107, 108
eigenvector 102, 103, 108
electric organ 122, 125
electrodes
  Ag/AgCl 4, 5, 15, 62, 66
  double-barrelled 220
  intracellular 220
electrochemical potential 75
electron crystallography 269, 270–80, 288
electron diffraction 286; *see also* electron crystallography
electron microscope 280, 287–8
electron spin resonance (ESR) spectroscopy 245, 246, 250–9
electronics
  bilayer 62–4
  patch-clamp 21–2
electroporation 156
ELISA (enzyme-linked immunosorbent assay) 176
embryonic chick neurones 233
endoglycosidases 200, 202
endoplasmic reticulum (ER) 191, 203, 212, 213
energy barrier 70, 75, 81; *see also* discrete-state models
enzymatic digestion
  of animal tissues 28–33
  of plant tissues 221–3
eosin 264
exocytosis, *see* secretion
exponential probability distribution 92, 93, 94
expression
  of fusion proteins 172–3
  of ion channels 133, 144–52, 154–68, 176, 192, 193, 194, 202, 204–5, 212, 213
expression cloning 133, 146, 154–6
expression screening 153, 155
expression vectors, *see* plasmid and phagemid vectors
Eyring reaction rate modelling, *see* discrete-state models

Fab fragments 185, 186
Faraday cage 20, 28, 64
fatty acids, direct action on ion channels 338
fatty acylation, *see* post-translational modification
feedback capacitor 21, 22
feedback resistor 2, 21, 22, 62
fetal calf serum 30
Fick's law 79
filter
  digital 23, 70
  electronic 20, 23–4, 64
filtering, effect on analysis 70, 71, 90
fire-polishing (of patch-pipettes) 4, 5, 16, 17, 19–20
flow-dialysis 275–6, 278
flow-dialysis chamber 275
fluctuation analysis, *see* noise, analysis
fluorescein 158, 264
fluorescein di-β-D-galactopyranoside 158
fluorescence energy transfer 263–4
fluorescence microscope 29, 158, 228–9
fluorescence spectroscopy 263–4
fluorescence quenching 263–4
fluorescent indicators (dyes)
  calibration 227–8
  use for measuring intracellular Ca^{2+} 220, 224–30
  see also Fura–2, Indo–1, Quin–2
fluorography 208
forskolin 234, 235
Förster equation 264
Fourier transform infrared (FTIR) spectroscopy 244, 246, 247–51
freeze–thaw 59–60
Fucus serratus 220, 224
functional assay for ion channels 126–7; *see also* expression of ion channels and expression screening
Fura–2, 225, 226, 227, 228
fusion of membrane vesicles with planar bilayers 55–8, 127

GABA-A receptors 140, 146, 159, 167, 168, 171, 180–5, 186, 187, 188, 192, 194, 197, 200, 202, 213
GABA-B receptor 234
gadolinium 152
gap junctions 193
Gaussian (normal) probability distribution 89–90, 91, 93
GDP 232

GDPβS 232
gel filtration, *see* column chromatography
GenBank database 137
gene family 134, 136, 137
geometric probability distribution 93–4
gigaohm seal 2, 3, 6, 13, 15, 40, 220, 221, 222, 223, 230
glass, patch-pipette 15–17
gluconate as impermeant ion in Cl^--channels 11
glutamate receptors 146, 192, 194; *see also* NMDA receptors, AMPA receptor
glycosylation, *see* post-translational modification
Goldman–Hodgkin–Katz (GHK) current equation 79, 82; *see also* continuum models
Goldman–Hodgkin–Katz (GHK) voltage equation 77–9
gonadotrophins 152
Gpp(NH)p 232
G protein (guanine nucleotide binding protein) 152, 231–4, 237, 238
  antibodies to 233
  receptors coupled to 146, 152, 242, 245
Grace's medium 162, 164
gramicidin 86
grounding, of recording setup 27–8
GST gene fusion system 172
GTP 232
GTPγS 232
guanylate cyclase 234, 235

H5 segment 242
headstage (head amplifier)
  bilayer amplifier 62, 63
  patch-clamp amplifier 4, 20, 21, 22
HEK 293 cells 197
heparin 236
Herpes simplex 156
Hill coefficient 88
hippocampal neurones, acutely-dissociated 3
Homarus americanus 270
homologous recombination 161
5-HT (serotonin) receptors 146, 154, 231
hybridization screening 137–8
hydropathy plots 242
hydrophilicity 172, 212, 242
hydrophobicity 242
hydroxyeicosatetraenoates (HETE) 238

IBMX 235
image analysis and processing 283–5
immunization 176
immunoaffinity purification, *see* affinity chromatography
immunoblotting (Western blotting) 176–9, 182, 186–7, 202, 208
immunocytochemistry 188
immunofluorescence 213–15
immuno-immobilization 211
immunoprecipitation 180–2, 193, 195, 202, 203, 205, 206–10, 213
Immunoprecipitin 181
immunostaining 168
independence (in ion permeation) 80
Indo-1, 224, 226, 228, 229
indomethacin 238
inducible promotors 159, 160
infrared absorption by proteins 247
inner-filter effect 263
inositol 1,4,5-trisphosphate (IP_3) 234, 236, 237
inositol 1,4,5-trisphosphate receptor, assay for 118–21
inositol 1,3,4,5-tetrakisphosphate 236
inositol 4,5 bisphosphate 236
input resistance
  of *Fucus* eggs 220
inside–out patch 5, 7, 15, 233
intracellular signalling 219, 234–9; *see also* Ca^{2+}, G proteins, inositol 1,4,5-trisphosphate, second messengers
in vitro crystallization 271
in vitro RNA synthesis 155
ion binding site 73, 74, 75, 76
ion–ion interactions 74, 76, 80, 85, 86–8; *see also* single-ion channel and multi-ion channel
ionomycin 227
isosbestic point 226

junction potential, *see* liquid junction potential

K^+ channels
  voltage-gated 78, 146, 152, 193, 212, 231, 233, 264
  Ca^{2+}-activated 69, 70, 71, 90, 100, 101
  sarcoplamic reticulum 85, 90
Krebs solution 34, 36

L cells 157
lag circuit 10
laminar flow hood 33
laminin 34
Langmuir–Blodgett films 248
laser diffractometer 279, 284
leaching of ions from pipette glass 17
[^{3}H]leucine 204

ligand binding
  assay theory 113–18
  mapping binding sites 185–8; *see also* inositol 1,4,5-trisphosphate receptors, Na^+ channels
ligand-gated ion channels, *see* specific channels
lipid extraction 253
lipid-to-protein ratio (LPR) 273, 274–6, 280
liposome-mediated transfection (lipofection) 156, 163
lipooxygenase pathway 238
liposomes
  formation 59–61
  in ESR 253
  patch-clamping 61–2
liquid junction potential 11–12, 66
lock-in amplifier 10
lucifer yellow 13

magic-angle spectroscopy in NMR 261
manipulator 13, 20, 27, 28; *see also* micromanipulator
MAPS (multiple antigen presentation system) 174–6
Markham rotation 284
mass spectrometry 243
matrix algebra 109–10
m-current 231, 237
mechnical disruption of leaf tissue 221
Mediago sativa 224
membrane-delineated pathways for intracellular signalling 231–4
metabolic labelling 202, 204–5, 213
metallothionein promotor 159
[^{35}S]methionine 204, 206, 212
methylamine tungstate 282
methylation, *see* post-translational modification
mica, use in preparing support films for electron microsocopy 282–3
Michaelis–Menten kinetics 82, 85, 86
microforge 19–20
microinjection
  of anti-G protein antibodies 233
  of fluorescent dyes 227
micromanipulator 7, 13, 20, 27
microscope (light) 20, 26; *see also* dissecting microscope
microscopic reversibility 83
mitochondria 224
mixed micelles 270, 274, 277, 278, 279
modelling of channel protein structure 242
molecular dynamics simulation of ion permeation 75
monoclonal antibodies 203
monolayers, phospholipid 48–9
mouse mammary tumour virus (MMTV) promotor 159, 160
multi-ion channel 74, 75, 85, 86–8
myristic acid 238

Na^+ channels (voltage-dependent) 73, 78, 152, 157, 186, 193, 194, 203
  assay for 119–26
Na^+K^+ATPase 272
n-decane 46
negative co-operativity 87, 88
negative staining 276, 277, 278, 280–
neomycin-resistance 159
Nernst–Planck electrodiffusion model, *see* continuum models
Normarski differential interference contrast (DIC) optics 26, 37
neuroblastoma x glioma hybrid (NG108–15) cells 231, 233, 235, 237
[^{3}H]nicotine 211
nitric oxide 235
NMDA receptors 3, 7, 134, 135, 136, 154, 171, 178–9, 192
noise
  analysis 91
  digitally-generated 106
  electrical 2, 4, 13, 14, 16, 17, 21, 22, 27–8, 64, 70, 71
  gating current 95
  mechanical, *see* anti-vibration support
  open-channel 71, 90
  thermal 2, 21
non-radioactive detection of DNA 137
nordihydroguaiaretic acid (NDGA) 238
n-pentane 49
nuclear magnetic resonance (NMR) spectroscopy 203, 244, 246, 247, 259–62
nystatin 3, 12, 13, 16, 57

octyl-β-D-glucopyranoside (8-Gluc) 272, 276
octyl-pentaoxyethylene (C_8E_5) 276
octyl-(polydisperse) oligooxyethylene (8-POE) 276
Ohm's law 79
oligomeric organization, *see* post-translational modification
oligonucleotide-directed mutagenesis (SDM) 140–1
oligonucleotides, antisense 212, 233, 234
OmpF porin 276, 277
orientation
  of channels in bilayer 58–9
  of transmembrane segments 245
oscillation, electrical 10
oscilloscope 20, 22–3
outside-out patch 5, 11, 13–14, 15, 40, 231

patch-clamp amplifier 20–2, 63
papain 32
paramagnetic ions 258
patch-clamping
  mammalian cells,
  plant cells 221–4
  proteoliposomes 61–2
pea 222
pectolase 221, 222
peritoneal mast cells 10
perforated-patch recording 12–13, 16
permeability, absolute 70, 84, 85
permeability ratio 74, 77, 78, 79, 80, 85, 86
pertussis toxin (Ptx) 232, 233, 234
phase-transition, lipid 46, 272, 273
Phaseolus vulgaris 223
PhoA 243
phorbol esters 237, 238
phosphatidylcholine 59, 60, 271
phosphatidylethanolamine 46, 271
phosphatidylserine 46, 271
phosphoamino acid analysis 197, 199–201
phosphodiesterase inhibitor, *see* IBMX
phospholipase A2, 273
phospholipase C (phosphoinositidase C) 236, 237
phosphotungstic acid 282
photoaffinity-labelling 186–7
photosystem II complex 270
pipette holder 4, 6
pipette pulling 16, 17–18
plaque assay 163–4, 166, 167
plasmid and phagemid vectors
  bacterial expression vector
    pGEX 172–3
  baculovirus transfer vectors
    pVL1392, 162
    pVL1393, 162
  β-galactosidase reporter gene vector
    pSV-β-galactosidase 158
  eukaryotic expression vectors
    pcDNAIAmp 140, 157
    pcDNAIneo 159
    pMAMneo 159
    pMSGneo 159
    pRo/CMV 159
    pXT1, 159
    ZEMRVSP6 159
  see also cloning vectors
Poisson probability distribution 91, 92, 93
polarised luminescence spectroscopy 264
poly-L-lysine 34, 158
polymerase chain reaction (PCR) 135, 136, 137, 139, 141, 142, 143, 144, 165, 166
polymerase chain reaction mutagenesis 141–4
polymerase chain reaction primers 135, 136, 139, 141, 165
poly-ornithine 34
post-translational modification 133, 141, 152, 191, 193–4, 215
  disulphide bond formation 202–3
  fatty acylation 202
  glycosylation 153, 176, 186, 200–2, 242, 243
  methylation 243
  phosphorylation 71, 153, 194–200, 231, 243
  proteolytic cleavage 198
  subunit assembly 144, 145, 154, 203, 211–12, 245
power spectrum 261
probability density function (pdf)
  of the exponential distribution 92, 100, 101, 103, 104, 105, 107
  of the Gaussian distribution 89
  see also Q-matrix analysis
probenacid 226
proline isomerization 193
prostaglandins 152, 238
protein folding 195, 203, 246
protein kinase C 236, 237, 238
protein kinase C activators, *see* phorbol esters
proteoliposomes 59–62, 127
proteolytic cleavage, *see* post-translational modification
proton channel 244
protonmotive ATPase, vacuolar 244, 277
protoplasts 220, 221, 222, 223
pulse-chase (pulse-labelling) 204
  see also metabolic labelling
purification
  of GABA-A receptor 183–5
  of sarcoplasmic reticulum channels 126–30

Q-matrix analysis 94–110 (including a summary of Matrix Algebra)
Quin-2, 226

radiation sesitivity 287
rapid amplification of cDNA ends (RACE) 139
rate constants 75, 76, 80, 82, 83, 84, 89, 91, 92, 96, 97, 103, 104, 106, 107
ratio fluorimetry 225, 230
ratio imaging 225, 229
reconstitution of ion channels 43, 55–62, 126
redundancy of PCR primers 135, 136
reporter enzymes 158, 243
reporter genes 158
resistor-feedback headstage 21
restriction endonucleases and use 135, 139, 140, 141, 142, 143, 144, 161
reversal potential 7, 74, 78, 79, 85
rhodamine 264

rhodopsin 243, 272
rise time 62
RNA
antisense, *see* oligonucleotides, antisense
in situ hybridization compared to immunoreactivity 188, 189
messenger RNA (mRNA or Poly(A)$^+$ RNA) 135, 144, 146, 149, 150, 151, 152, 153, 154, 155
rotational correlation time 257
ryanodine receptor/Ca^{2+}-release channel 236

sarcoplasmic reticulum (SR)
purification of ion channels from 127–9
vesicles 59
saturation transfer ESR (STESR) 251, 255–7, 258, 264
saxitoxin 122–6; *see also* Na^+ channel assay
Scatchard plot 116
scrape-loading 233
secondary structure 133
second messengers 12, 152, 219, 230–2, 234–8
secretion, monitoring by cell-capacitance measurement 10–11
selection markers 159
selectivity, ionic 73, 73, 74, 75, 78, 83; *see also* permeability, absolute, permeability ratio
self-block 88
series resistance and series resistance compensation 9, 10, 13, 18, 20, 21, 62
Sf9 cells 162, 163, 164, 166, 167, 168
single-electrode discontinuous voltage-clamp (switch-clamp) 220
single-filing region 72, 73, 74
single-ion channel 76, 80–5, 86
single-particle averaging 284
site-directed antibodies 133; *see also* antipeptide antibodies
site-directed mutagenesis (SDM) 73, 133, 139, 140–4, 157, 196, 202, 212
size-fractionation of mRNA 154, 155
slope conductance 3, 7
solubilization of ion channels 60, 61, 126–7, 181, 205–6
somatostatin receptors 234
source resistance 2
specificity constant 85
Spinacia oleracea 270
spin-labelled lipids 253–5, 259
spin-labelled proteins 255, 256
Spodoptera frugiperda 162; *see also* Sf9 cells
spot-scan imaging 287
stably transfected cells 133, 144, 145, 158–61
stirrer, bilayer 63
stoichiometry (of multi-subunit ion channels) 146, 192, 193, 212
stomatal guard cell 220, 221–2
stretch-activated channels 40, 152, 153
structure-function studies 133; *see also* site-directed mutagenesis, chimeric channels
subunit assembly, *see* post-translational modification
sucrose-gradient centrifugation 154, 203, 208–11
sulphydryl reagents 243, 245
superior cervical ganglion neurones 233
support films for electron microsocopy 280–3, 285
surface charge 72, 73
surface potential 73
suspension culture 162
Sylgard coating 18–9

T4 DNA polymerase 140
T7 DNA polymerase 140
T25-ras oncogene protein 233
tape recorder 20, 24–5
tempering 273, 280
test-pulse (t-pulse) 6, 7, 8, 13
tetrodotoxin, use in Na^+ channel assay 121–6
thermolysin 32
thin-layer chromatography in peptide mapping 198–201
thromboxanes 238
time constant
of cell membrane capacitance transient 220
of electrode capacitance transient 220
of patch-clamp response 9, 21
of single-channel dwell-time 93, 103, 104
tissue dissociation 28–33
topography of channel proteins 133, 242, 243
transfection, *see* stably transfected cells and transiently transfected cells
transiently transfected cells 133, 144–5, 156–8; *see also* baculovirus-driven expression
transfer vector, baculovirus 161, 162, 163
transition rate 96–9
transmembrane segments 242, 243
trypsin 30, 31, 32
trypsin inhibitor 31
tunicamycin 200, 202
two-microelectrode voltage clamp 9, 153

unitary (single-channel) current 69
uranyl acetate 282, 286
uranyl formate 282

vaccinia 156
vacuole, plant 220
vasoactive intestinal peptide 152
vectors, *see* plasmid and phagemid vectors
vibration 25–6, 64
Vicia faba leaves 221, 222
vesicles, membrane 43, 49, 58, 59, 70, 127
vestibule 72, 73, 76
vitelline membrane 39–40, 153
vitreous water 285
voltage-dependent block 77, 88
voltage-gated channels, *see* specific channels
washout of cell contents 12

whole-cell recording 5, 8–11, 21, 22, 233, 235, *see also* perforated-patch recording
Woodhull model 77

Xenopus oocyte 14, 39–40, 144, 145, 146–156, 192, 212